ATOMIC SPECTRA AND COLLISIONS IN EXTERNAL FIELDS

PHYSICS OF ATOMS AND MOLECULES

ATOMIC SPECTRA AND COLLISIONS IN EXTERNAL FIELDS

Edited by
K. T. Taylor
University of London
Royal Holloway and Bedford New College
Egham, Surrey, United Kingdom

M. H. Nayfeh
University of Illinois at Urbana–Champaign
Urbana, Illinois

and
C. W. Clark
National Institute of Standards and Technology
Gaithersburg, Maryland

PLENUM PRESS • NEW YORK AND LONDON

Library of Congress Cataloging in Publication Data

Atomic spectra and collisions in external fields / edited by K. T. Taylor, M. H. Nayfeh, and C. W. Clark.
    p.   cm. — (Physics of atoms and molecules)
  "Proceedings based on a satellite meeting of the Fifteenth International Conference on the Physics of Electronic and Atomic Collisions, entitled: Atomic Spectra and Collisions in External Fields 2, held July 30–31, 1987, at the University of London, Royal Holloway and Bedford New College, in Egham, Surrey, United Kingdom" — T.p. verso.
  Includes bibliographies and index.

  ISBN-13: 978-1-4612-8315-7   e-ISBN-13: 978-1-4613-1061-7
  DOI: 10.1007/978-1-4613-1061-7

  1. Nuclear excitation — Congresses. 2. Electric fields — Congresses. 3. Magnetic fields — Congresses. 4. Photoionization — Congresses. 5. Multiphoton processes — Congresses. 6. Lasers — Congresses. I. Taylor, K. T. II. Nayfeh, Munir H. (Munir Hasan) III. Clark, C. W. (Charles W.), 1952- . IV. International Conference on the Physics of Electronic and Atomic Collisions (15th: 1987: Brighton, England) V. Series.
QC794.6.E9A863   1989   89-3522
539.7 — dc19   CIP

Proceedings based on a satellite meeting of the Fifteenth International Conference on the Physics of Electronic and Atomic Collisions, entitled: Atomic Spectra and Collisions in External Fields 2, held July 30–31, 1987, at the University of London, Royal Holloway and Bedford New College, in Egham, Surrey, United Kingdom

© 1988 Plenum Press, New York
Softcover reprint of the hardcover 1st edition 1988
A Division of Plenum Publishing Corporation
233 Spring Street, New York, N.Y. 10013

In December 1986, Coulter McDowell retired prematurely from his position as Professor of Applied Mathematics and Head of Department at Royal Holloway and Bedford New College in the University of London. Coulter made important contributions to several areas represented in this volume, particularly the study of atoms in very strong magnetic fields. He was well known for his open, inquisitive approach to difficult questions and his enthusiastic encouragement of junior workers in the field. We are gratified to be able to dedicate this volume to Coulter on the behalf of his many friends in atomic physics, as a small token of their affection and appreciation.

K. T. Taylor
M. H. Nayfeh
C. W. Clark

PREFACE

This volume contains papers associated with the conference "Atomic Spectra and Collisions in External Fields II", that took place July 30–31 1987 at Royal Holloway and Beford New College. The first meeting of this name was held at the National Bureau of Standards in Gaithersburg, Maryland in 1984, and, if any tradition can yet be said to have been established in the series, it is that the proceedings be written after the conference. We hope thereby to preserve some impression of the discussions that took place, which in both cases were vigorous and unihibited.

Both meetings happen to have convened in proximity to major developments in the field. At the time of the first conference, results of experimental measurements of dielectronic recombination in electron-ion beams were beginning to appear. These showed large discrepancies with theoretical calculations, which were attributed to the effects of rather weak electric fields on the highly-excited states that mediate the recombination process. This conjecture gave rise to widespread concern in the plasma physics community that the representation of dielectronic recombination in existing plasma models, in which it plays an important role in energy and ionization balance, might be seriously in error due to neglect of the effects of electric and magnetic fields. The subject of field effects on recombination processes was thus a major focus of the 1984 meeting. During the next three years experimental and theoretical work moved toward reasonable agreement, and although there still remain difficulties of quantitative comparison, the major effects of external fields on recombination processes appear to be well understood.

The year preceeding the second meeting saw extremely rapid progress on one of the major unsolved problems in the subject, the spectrum of a simple atom in a uniform magnetic field. The main stimuli of these developments have been the high-resolution experiments on atomic hydrogen carried out by the Bielefeld group. Theoretical response to their work has produced spectacular agreement of experimental and computed specta over a wide range of energy, has elucidated the nature of the quantum mechanical counterpart of classical chaos in this system, and has articulated a new approach to spectroscopy based upon isolated, periodic classical orbits. First steps have also been taken towards a practical quantum defect theory treatment of the diamagnetic spectra of complex atoms, and the threshold behavior of photoelectrons in magnetic fields has been greatly clarified. Contributions from most of the groups active in these areas are contained in this volume.

Another area of rapid development, represented here but not at the first meeting, is the interaction of atoms and molecules with strong laser fields. The phenomenon of above-threshold ionization (ATI), first observed in 1979, has received enormous attention since the reports of extensive continuum structure in 1983; just at the time of this meeting the first results were appearing on ATI with sub-picosecond pulses, in which a radical change of the photoelectron spectrum was observed. The exploitation of molecular dissociation as a femtosecond clock for timing strong-field ionization dynamics is another notable technique of which some early results were reported at this meeting.

We do not venture to predict which subjects will appear most prominently in the next meeting of this type. It seems certain, however, that many beautiful phenomena still await discovery in "classical" areas that have hardly been explored: e.g. the spectra of simple molecules in d.c. fields, and atoms in crossed electric and magnetic fields.

Atomic Spectra and Collisions in External Fields II was held as a satellite meeting of the XV International Conference on the Physics of Electronic and Atomic Collisions, and received financial support from: The Royal Society of London; the European Research Office of the U.S. Army Research, Development and Standardization Group; Cray Research Inc.; the U.S. National Bureau of Standards; Lumonics Ltd.; and Spectra Physics, Inc. We thank these organizations for their assistance.

We appreciate the collaborative efforts of Jeannine Adomaitis, Debra Oberg and Sue Lynn White of the University of Illinois for their technical expertise in word processing.

K. T. Taylor

M. H. Nayfeh

C. W. Clark

CONTENTS

I. SPECTROSCOPY AND COLLISION PHYSICS OF ATOMS IN D.C. FIELDS

A. MAGNETIC FIELDS

B. ELECTRIC FIELDS

C. COMBINED ELECTRIC AND MAGNETIC FIELDS

LASER SPECTROSCOPY OF THE DIAMAGNETIC HYDROGEN ATOM IN THE CHAOTIC REGIME

A. Holle, J. Main, G. Wiebusch, H. Rottke, and K. H. Welge

Fakultät für Physik
Universität Bielefeld
D-4800 Bielefeld, FRG

Atomic diamagnetism is still an essentially open problem of fundamental importance. Since the discovery of atommic quasi-Landau resonances by Garton and Tomkins[1] it has been a subject of intense experimental and theoretical research.[2] With its purely Coulombic field, the hydrogen atom has served as basis for virtually all theoretical work.[3] The general interest arises from the non-separability of the Schrödinger equation, even with the simplest Hamiltonian, containing the Coulomb and diamagnetic term only ($\gamma = B/(2.35 \times 10^5 \mathrm{T})$):

$$H = \frac{1}{2} p^2 - 1/r + \frac{1}{8} \gamma^2 \rho^2 \qquad (1)$$

Experiments with the diamagnetic H-atom have recently been performed for the first time. The results have been compared with the exact quantum mechanical calculations in the energy range below the ionization threshold.[4] Measurements in the energy range around threshold have revealed the existence of a multitude of further quasi-Landau resonance types.[5,6] Briefly, hydrogen atoms are excited to even parity final states with m = 0 and m = -1 by resonant two-photon absorption through Paschen-Back resolved 2p state:

$$H(1s, m = 0) + h\nu(\lambda = 121.6 \text{ nm}) \rightarrow H(2p, m = 0 \text{ or } m = -1) \qquad (2)$$
$$H(2p, m = 0 \text{ or } m = -1) + h\nu(\lambda \simeq 365 \text{ nm}) \rightarrow H^*(m = 0 \text{ or } m = -1)$$

The pulsed vacuum-ultraviolet laser light for the first excitation step was linearly polarized parallel or perpendicular to the magnetic field to excite 2p states with m = 0 or m = -1 and had a bandwidth of 3GHz. While taking a spectrum, the vuv wavelength was fixed to the respective Lyman-α transition. The uv laserlight for the second excitation step had a bandwidth of 1 GHz and was always linearly polarized parallel to the magnetic field, to excite final m = 0 and m = −1 states. During measurement it was scanned through an energy range around threshold. Electrons produced by photon- or field ionization were detected with a surface barrier diode (Fig. 1a and Fig. 1b).

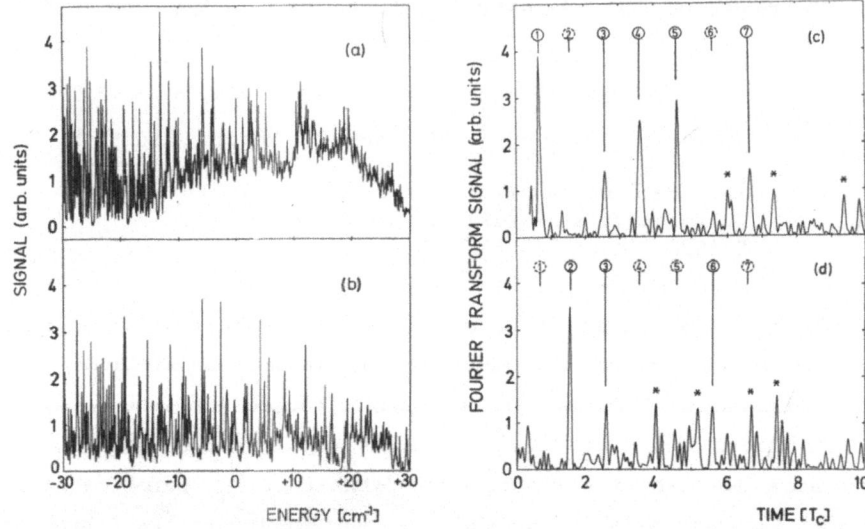

Fig. 1. (a,b) Excitation ionization spectrum of H-atom Balmer series in a state homogeneous magnetic field of 5.96 T, with m = 0 and m = −1. (c,d) Fourier transformation of (a,b), plotted is the absolute value squared. T_c is the cyclotron period. (From Ref. 6.)

The discovered new resonances, i. e. modulations of the energy spectrum, were found by Fourier transforming the experimental data into a time spectrum (Fig. 1c and Fig. 1d). The resonance types 1 to 7 were explained by classical trajectory calculation (6), figure 2 shows their shape at E = 0.

A detailed discussion of energy dependence and quantization of the corresponding trajectories is given in Ref. 7. As an example, Fig. 3a and Fig. 3b each show 21 trajectories at a magnetic field of 6 Tesla at different energies in their α−z projection. These trajectory types in Fig. 3a and Fig 3b explain quantitatively the position of peaks 2 and 4 at 1.57 T_c and 3.60 T_c, respectively. Each trajectory type (besides trajectory 1) degenerates at a certain negative energy E_{cut} to a linear motion along the z-axis. This "cut-off" behavior is extensively discussed in Ref. 7 and can also be seen in Fig. 4, which gives the energy dependence of recurrence time T/T_c, energy spacing $\Delta \varepsilon$ and starting angle θ with respect to the z-axis for the trajectories 1 to 7.

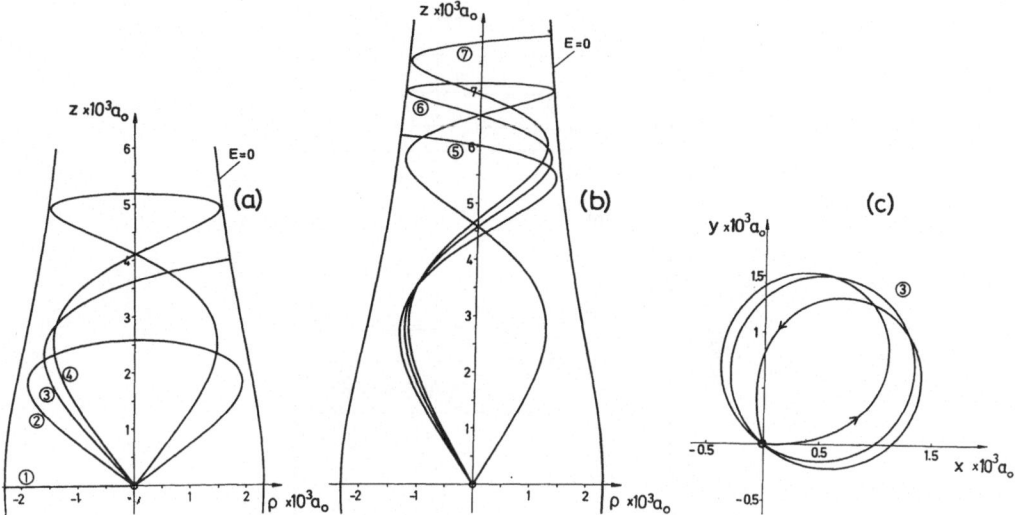

Fig. 2. (a,b) Calculated closed trajectories of electron motion at energy E = 0, corresponding to the first seven resonances 1 to 7. Projection onto (ρ,z) coordinates. E = 0: Potential of field-free ionization threshold. (c) Closed trajectory 3 at E = 0 in projection onto the z = 0 plane. Final state |m = 0⟩. (From Ref. 6.)

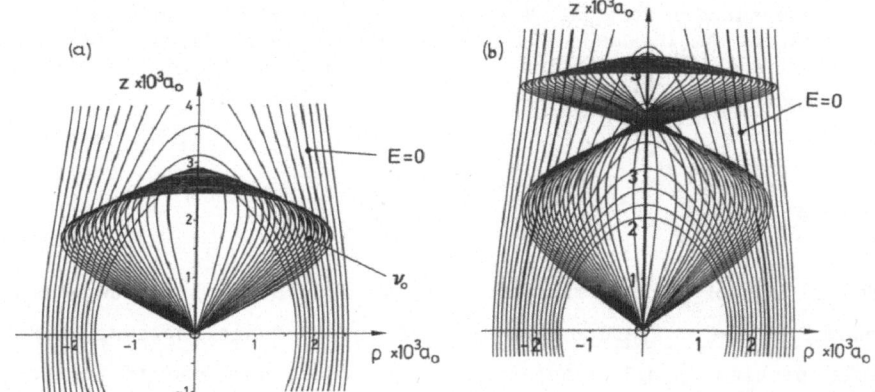

Fig. 3. Calculated closed classical trajectories of electron motion at energies drawn in increments of 5 cm^{-1}. Projection on (ρ,z) coordinates. Equidistant (10 cm^{-1}) potential lines, E = 0: field-free ionization potential. a) trajectory 2, b) trajectory 4.

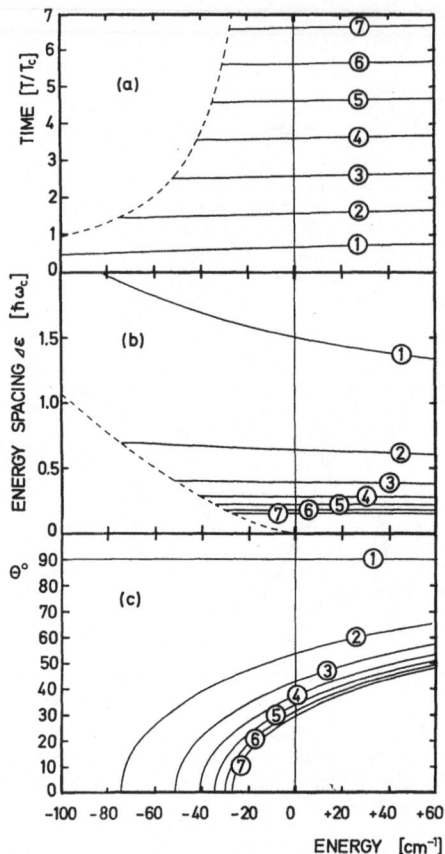

Fig. 4. (a) Energy dependence of recurrence times at B = 6T. Time scale in units of cyclotron period $T_c = 2\pi/\omega_c$. Dashed line indicates cut-off limits. (b) Energy dependence of energy spacing of resonances 1 to 7 at B = 6T. Energy scale in units of Landau spacing $\hbar\omega_c$. Dashed line indicates cut-off limits. (c) Starting angle as function of energy at B = 6T for resonances 1 to 7. (From Ref. 7.)

The method of Fourier transforming an absorption spectrum measured at <u>constant magnetic field</u> over a defined energy range, however, bears a principle problem: all quasi-Landau resonances are energy dependent. This leads to a broadening of the Fourier peaks, as the Fourier integral averages the dependence. Strongly energy dependent resonance types vanish or interfere in an undefined way, if the energy range is too large. But a shortening of the measured range again leads to a broadening of the Fourier peaks, because of the smaller width of the finite Fourier integral. Thus, this method allows to discover only those resonance types which depend only little on energy. Indeed, the resonance types discussed abover are characterized by being nearly independent of energy over the measured range from -3 cm^{-1} to $+30$ cm^{-1}, as seen in Fig. 4a.

This principle problem, arising from the energy dependence of the quasi-Landau resonances, can be solved by taking the scaling properties of the classical Hamiltonian

$$H = (1/2)(p_\rho^2 + p_z^2) + (1/8)\gamma^2\rho^2 - (\rho^2 + z^2)^{-1/2} \qquad (3)$$

with zero z-component of the angular momentum ($m = 0$) into account. Employing the appropriate scaling transformations[8] $\tilde{\vec{r}} = \gamma^{2/3} \vec{r}$; $\tilde{\vec{p}} = \gamma^{-1/3} \vec{p}$ one obtains

$$H = \gamma^{2/3} \tilde{H} \qquad (4)$$

with $\tilde{H} = (1/2)\tilde{p}_\rho^2 + (1/8)\tilde{\rho}^2 - (\tilde{\rho}^2 + \tilde{z}^2)^{-1/2}$ the scaled Hamiltonian. The EBK quantization condition[9] for multidimensional non-separable systems $\oint \sum_i p_i dq_i = 2\pi\hbar(n + \alpha/4)$, which in the diamagnetic case is written as $\oint(p_\rho d\rho + p_z dz) = 2\pi\hbar n$ results in

$$\oint(\tilde{p}_\rho d\tilde{\rho} + \tilde{p}_z d\tilde{z}) = 2\pi\hbar n \cdot \gamma^{1/3} \qquad (5)$$

after scaling. The ring integral has to be taken over a closed classical trajectory, n is the EBK-quantum number and the Maslov index α is set to 0. As the left side of Eq. (5) depends on the scaled energy $E_{scal} = E\gamma^{-2/3}$ alone, the results of EBK-calculations can simply be represented as functions

$$n\gamma^{1/3} = f(E\gamma^{-2/3}) \qquad (6)$$

for each type of closed orbits. Thus equidistant resonances are obtained, when varying the magnetic field in steps proportional to $\gamma^{-1/3}$ at fixed scaled energy. Fourier transformation of experimental data measured in this way leads to sharp peaks which can be identified by scaled classical trajectories and their EBK-quantization. (For a detailed derivation, see Ref. 7). Thus, for a proper detection of energy dependent resonance types (e.g. the ones denoted by asterix in Fig. 1c and Fig. 1d), it is necessary to measure at <u>constant scaled energy.</u> Fig. 5a shows a scaled spectrum at $E_{scal} = -0.45$. During measurement, the magnetic field was scanned with a rate proportional to $\gamma^{-1/3}$ and the

5

laser was stepped synchronously to satisfy the scaling relation. The Fourier transformation of the scaled spectrum is given in Fig. 5b. It shows clear structures and each peak can be explained by a classical trajectory. For the four most prominent resonance types the shapes of the corresponding trajectories are given in the figure. The $n\gamma^{-1/3}$-values for peaks and trajectories agree to better than 0.5%.

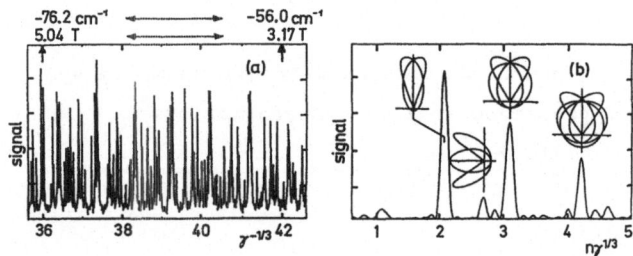

Fig. 5. (a) Scaled excitation spectrum at E_{scal} = -0.45. (b) Fourier transformation of (a), plotted is the absolute value squared, for the peaks at $n\gamma^{1/3}$ = 2.10, 2.70, 3.13, 4.13 the corresponding trajectories are shown in ρ-z projection.

In future measurements, where the scaled energy shall be increased in small steps, we will study the energy development of these peaks to energies above threshold and hope to get a deeper understanding of the correspondence between classical and quantum mechanics and classical and quantum chaos for the diamagnetic H-atom.

References

1. W.R.S. Garton and F. S. Tomkins, Astrophys. J. 158, 839 (1969)

2. J. C. Gay, in "Progress in Atomic Spectroscopy, Part C," edited by H. J. Beyer, and H. Kleinpoppen, (Plenum, New York, 1984), pp. 177-246; C. W. Clark, K. T. Lu, and A. F. Starace, in "Progress in Atomic Spectroscopy, Part C," edited by H. J. Beyer, and H. Lein-poppen, (Plenum, New York, 1984), pp. 247-320.

3. G. Wunner, U. Woelk, I. Zech, G. Zeller, T. Ertl, F. Geyer, W. Schweitzer, and H. Ruder, Phys. Rev. Lett. 57, 3261 (1986); D. Wingten, and H. Friedrich, Phys. Rev. Lett. 57, 571 (1986); M. L. Du, J. B. Delos, Phys. Rev. Lett. 58, 1731 (1987); D. Delande, and J. C. Gay, Phys. Rev. Lett. 57, 2006 (1986); D. Wintgen, Phys. Rev. Lett. 58, 1589 (1987); A. R. Edmonds and R. A. Pullen, Preprint Imperial College, London (1980) unpublished; M. A. Al-Laithy,

P. F. O'Mahony, and K. T. Taylor, J. Phys. B**19**, L773 (1986); D. Wintgen and H. Friedrich, Phys. Rev. A**36**, 131 (1987).

4. A. Holle, G. Wiebusch, J. Main, K. H. Welge, G. Zeller, G. Wunner, T. Ertl, and H. Ruder, Z. Phys. D**5**, 279 (1987); D. Wintgen, A. Holle, G. Wiebusch, J. Main, H. Friedrich, and K. H. Welge, J. Phys. B**19**, L557 (1986).

5. A. Holle, G. Wiebusch, J. Main, B. Hager, H. Rottke, and K. H. Welge, Phys. Rev. Lett. **56**, 2594 (1986).

6. J. Main, G. Wiebusch, A. Holle, and K. H. Welge, Phys. Rev. Lett. **57**, 2789 (1986).

7. J. Main, A. Holle, G. Wiebusch, and K. H. Welge, Z. Phys. D**6**, 205 (1987).

8. H. Hasegawa, S. Adachi, and H. Harada, J. Phys. A**16**, L503 (1983).

9. J. B. Keller, Ann. Phys. (NY)**4**, 180 (1958).

STRONGLY MAGNETIZED ATOMS – IN THE LABORATORY AND IN THE COSMOS

G. Wunner, G. Zeller, U. Woelk, W. Schweizer, R. Niemeier, F. Geyer, H. Friedrich, and H. Ruder

Lehrstuhl für Theoretische Astrophysik
Universität Tübingen
D-7400 Tübingen, Federal Republic of Germany

1. INTRODUCTION

Recent years have seen tremendous progress in studies of the properties of atoms in strong magnetic fields. Decisive stimulus came from the discovery of huge magnetic fields in astrophysical "laboratories," viz. field strengths of order $\sim 10^7 - 10^9$ T in neutron stars and of order $\sim 10^2 - 10^4$ T in white dwarf stars. At these field strengths the magnetic forces acting on an atomic electron outweigh the Coulomb binding forces even in low-lying states, and thus atomic structure is completely changed. On the other hand, the rapid advancement of high-resolution laser spectroscopy has made it possible to produce atoms in highly excited states, with principal quantum numbers ranging, presently,[1] up to n ≅ 520, and therefore Rydberg atoms can be used to investigate the effects of magnetic dominance on atomic structure also in terrestrial laboratories with magnetic fields of a few Tesla, or less.

A detailed description of the astrophysical scenarios which finally led to the conclusive determination of magnetic field strengths up to $5 \cdot 10^8$ T in neutron stars, and an account of theoretical results for atomic level schemes, transition rates, and wavefunctions at these field strengths have been presented by us elsewhere.[2] Here we will concentrate on new results. We just briefly review, in Section 2, the basics of the theory, and then present, in Section 3, results for the wavelength spectrum of the hydrogen atom in magnetic fields of arbitrary strength. We describe in which way the unravelling of this spectrum has laid the foundation for a new branchof stellar atomic spectroscopy, viz. the spectroscopy of "stationary lines," which has led to the detection of the largest magnetic field strength ever found in white dwarf stars to date, namely B ~ 35000 T in the object Grw +70°8247, and B ~ 50000 T in the object PG 1031+234. We then turn to atoms in terrestrial laboratories and discuss, in Section 4, the progress that has recently been made in the theoretical description of magnetized Rydberg atoms. By use of complete basis sets and efficient diagonalization techniques it is now possible to compute routinely energies, wavefunctions, and, thus, transition probabilities, for levels way up shortly below the field-free ionization threshold as functions of the magnetic field strength. We compare spectroscopic predictions of theory with results of actual experiments and demonstrate, in Section 5, that phenomena which have turned out characteristic of the onset of "quantum stochasticity" in investigations of model Hamiltonian systems are recovered in the quantal properties of magnetic Rydberg atoms. This emphasizes the point that atoms in strong magnetic fields represent ideal systems in which to study – not only in theory but also in experiment – phenomena that are expected to be typical of the quantum behavior of nonintegrable systems in general. These studies nicely complement the investigations of chaos for hydrogen Rydberg atoms in intense microwave fields.[3-5]

2. HYDROGENIC ATOMS IN MAGNETIC FIELDS OF ARBITRARY STRENGTH

2.1 The Hydrogen Atom in Three Characteristic Domains of the Magnetic Field Strength

The Hamiltonian which describes the motion of an electron under the combined influence of a fixed Coulomb potential and a uniform magnetic field $B = B \cdot e_z$ reads (in atomic units, $\beta = B/B_0$ with $B_0 = 2(\alpha m_e c)^2/(e\hbar) \cong 4.70 \cdot 10^5$ T),

$$H = -\Delta - \frac{2}{r} + 2\beta l_z + \beta^2(x^2 + y^2). \tag{1}$$

The reference magnetic field B_0 is chosen in such a way that at B_0 the cyclotron energy of the electron becomes equal to four times the Rydberg energy. Energies and wavefunctions calculated from the infinite-nuclear-mass Hamiltonian (1) can be related to those belonging to finite core mass by use of the appropriate mass scaling laws which hold in the presence of amagnetic field (see, e.g., Pavelov-Verevkin and Zhilinskii,[6] Wunner et al.[7]). The basic difficulty in solving the Schrödinger equation belonging to the Hamiltonian (1) for <u>arbitrary</u> field strength

10

lies in the fact that the spherical symmetry of the Coulomb potential, on the one hand, and the cylindrical symmetry of the magnetic field on the other, prevent a separation of variables so that closed-form analytical solutions are not possible in general. To put it differently, the Hamiltonian (1) belongs to the class of nonintegrable Hamiltonians. On account of the different symmetries one can distinguish, in a natural way, three domains of the magnetic field strengths.

i) The weak-field regime, where the Coulomb potential predominates and the magnetic field can be considered as a small perturbation. This range, the regime of the linear and quadratic Zeeman effect, where the magnetic field mixes angular momentum states of a given n manifold, was clarified already in the early days of quantum mechanics. It should be pointed out, however, that even in this regime progress has been made in recent years in understanding a) the separability of the Zeeman wavefunctions in momentum space, b) the existence of an additional, approximate, constant of motion for given principal quantum number, and, related to this, c) the symmetry classification of the states of a diamagnetic band in terms of a rotational-vibrational scheme (see, e.g., Herrick,[8] Gay et al.,[9] Wintgen and Friedrich,[10] and Wunner[11]).

ii) The intense-field regime, where the magnetic field predominates, and the Coulomb potential can be treated as a perturbation which influences the (slow) motion of the electron parallel to the magnetic field, but does not affect the (rapid) gyrations - quantum mechanically described by Landau states - in the plane perpendicular to the field. This situation is adequately accounted for by the well-known "adiabatic approximation" introduced by Schiff and Snyder[12] as early as 1939.

iii) The intermediate- (often called "strong-") field regime, where the electron experiences electric and magnetic forces of comparable strength. It is obvious that the elaboration of energies and wavefunctions in this regime causes the greatest difficulties, and is incomplete even today. Moreover, the absolute sizes of the field strengths at which one lies in this regime evidently depend on the state of excitation of the electron. By considering the equality of Coulomb and Lorentz forces for an electron in a circular Bohr orbit with principal quantum number n one obtains as a rough measure $B_n \cong B_0/(2n^3) \cong (8.7 \text{ T})(30/n)^3$. Thus for white dwarf and neutron star magnetic fields low-lying states are found to be subjected to a strong-field, or even intense-field, situation, while at laboratory field strengths studies of the strong-field regime are restricted to Rydberg states.

2.2 Numerical Methods

The nonintegrability of the Hamiltonian (1) implies that one is forced, in the complete quantum theoretical treatment of the poblkem, to resort to numerical methods. To obtain "exact" (as opposed to variational) solutions it is suggestive to invoke basis function expansions which are inspired by the symmetries of the limiting cases $B \to 0$ and $B \to \infty$. These are, in the first case, expansions in terms of oscillator functions in semiparabolic coordinates (adapted to the $SO(2,2) = SO(2,1) \oplus SO(2,1)$ dynamical symmetry of the Coulomb problem, cf. Englefield[13]), or expansions in terms of spherical harmonics and a set of complete radial functions (ordinary $SO(3)$ symmetry), while the case $B \to \infty$ suggest expansions in terms of Landau functions with complete longitudinal basis. The different expansions are employed depending on whether one approaches the intermediate–field regime from the side of low or intense fields, and overlapping results are expected in the transitional region.

Except at very high fields $(\beta > 0.1)$[14] calculations using the Landau function expansion have confined themselves so far to low–lying states. The longitudinal parts of the wavefunctions were determined by direct integration of the system of coupled differential equations that follow from Schrödinger's equation.[15] By contrast, calculations using expansions based on the field–free group properties have been performed for both low–lying and highly excited states, and in this case energies and wavefunctions were obtained by either direct integration, or diagonalization of the Hamiltonian matrix in large basis sets. Inour own computations in this regime we expanded the wavefunctions in terms of spherical harmonics, $\psi_m = \sum h_\ell(r) Y_{\ell,m}(\theta,\phi)$ (m is the magnetic quantum number), and the radial functions in the complete, orthonormal set of functions

$$G_{n\ell}^{(\zeta)}(r) = \zeta^{3/2} \left[\frac{n!}{(n + 2\ell + 2)!}\right]^{1/2} \exp[-\zeta r/2](\zeta r)^\ell L_n^{(2\ell+2)}(\zeta r), \qquad (2)$$

where ζ denotes an inverse–length parameter, and the $L_n^{(2\ell+2)}$ are generalized Laguerre polynomials. Matrix elements with respect to this basis can be expressed in closed anlytical form and give rise to a banded Hamiltonian matrix which can be diagonalized by efficient standard algorithms. Our choise of basis bears some resemblance with the Sturmian basis used previously by Edmonds[16] and Clark and Taylor[17] but avoids the difficulties associated with the nonorthogonality of the latter. As compared to the oscillator basis in semiparabolic coordinates (used e.g.

by Delande and Gay,[18] and, extensively, by Wintgen and Friedrich[19]), where the diagonalization produces eigenstates each of which belongs to a different value of the magnetic field strength (solutions are obtained in the E − B plane along straight lines E/B = const.), in our basis one diagonalization procedure yields the spectrum at <u>fixed</u> B, in accordance with most experimental situations. In our calculations we used basis sizes of up to 6400 for determining both eigenvalues and eigenvectors of the first ~ 700 states in given m−parity subspaces in the intermediate-field regime. Convergence was established by varying the size of the basis and the scale parameter ζ.

As a result of our efforts we are able to continuously trace the energy values and oscillator strengths of low−lying states (n < 5) from zero field up to neutron star magnetic fields (for accurate numerical values we refer the reader to Rösner et al.,[15] Forster et al.,[20] and Ruder et al.[21]; for a graphical interpolation of the energy values of the 14 lowest states in the m = 1 subspaces see Wintgen and Friedrich[19]), while we can calculate the energies and oscillator strengths of Rydberg states in laboratory fields of a few Tesla up to shortly below the fiel-free ionization threshold. The following sections present examples of the results obtained so far.

3. LOW−LYING STATES IN MAGNETIC FIELDS OF ARBITRARY STRENGTH

3.1 Wavelength Spectrum of the Hydrogen Atom

The behavior of the energy values and wavefunctions over the range from laboratory fields up to neutron star magnetic fields has been discussed in great detail elsewhere (Ruder et al.[2]). Here we concentrate on the behavior of <u>wavelengths</u> and present, in Fig. 1, the spectrum of the hydrogen atom between 20 nm and 10000 nm as a function of β in the range 10^{-3} < β < 10^3 which we have computed from our energy values for a total of 364 transitions between states with n < 5. This is the most comprehensive compilation of the hydrogen spectrum to date.

The Lyman lines exhibit a rather smooth behavior, and are continuosly shifted to shorter wavelengths, with the exception of the transition originating in $2p_{-1} \rightarrow 1S_0$, which first runs through a broad maximum of λ = 134.1 nm at β ≈ 0.12 before starting the monotonous descent essentially caused by the strong energetic lowering of the $1S_0$/000 final state. Turning to Balmer and Paschen transitions, we clearly recognize, in the region of small magnetic field strengths

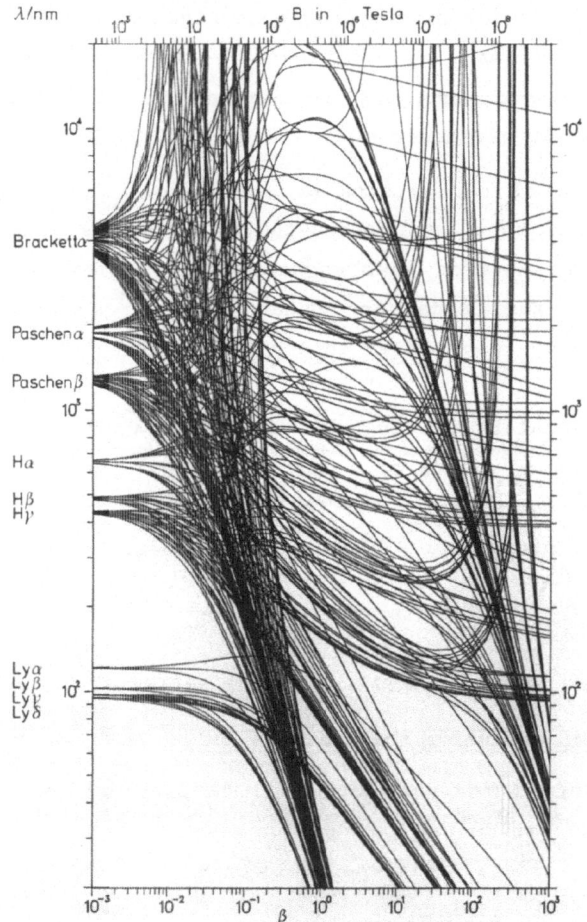

Fig. 1. The wavelength spectrum of the hydrogen atom from the soft X-ray range (20 nm) up to the far infrared (10000 nm) as a function of the magnetic field strength in the interval 470 T to $4.7 \cdot 10^8$ T. Effects of the finiteness of the proton mass are taken into account. The two rapidly declining bunches of lines correspond to cyclotron-like transitions of electrons (left-hand bunch) and protons (right-hand bunch) respectively. Note the occurrence of "stationary" lines in the intermediate region.

($\beta \approx 10^{-3}$), the splittings of the unperturbed lines into three equidistant Zeeman components. For larger β these components continue splitting by the quadratic Zeeman effect. The onset of the quadratic Zeeman effect is shifted to smaller β values with increasing wavelengths. Beyond this region ($\beta \approx 10^{-2}$), where the perturbation theory treatment breaks down, the lines are completely torn apart by the magnetic field within one β decade and the spectrum becomes totally distorted. Since the energy levels of states with different m and different z parity are allowed to cross, the wavelengths of corresponding transitions go to infinity at certain values of β.

Order reappears only in intense fields, indicative of the fact that in the limit $B \to \infty$ the level scheme approaches that of the one-dimensional Coulomb problem, which consists of tightly bound levels and levels whose energies equal those of the field-free H atom. As a consequence numerous lines tend to the wavelengths of the unperturbed hydrogen series at the right-hand side of Fig. 1. Further ordering is evident from the clustering of many lines into two conspicuous bunches, which decline proportional to $1/\beta$. The left-hand bunch comprises all electron cyclotron transitions with $\lambda_{cycl}^{(e)} \cong 22.8$ nm/β. The shortest wavelengths then correspond to cyclotron transitions of the electron from the first-excited to the ground-state Landau level. For neutron star magnetic fields ($\beta = 10^3$), the appropriate photon energies lie at 54 keV, and thus in the X-ray region. For photon energies like these, which amount to more than 10% of the rest energy of the electron, of course relativistic effects of the same order are to be expected. This is indeed confirmed by the rigorous relativistic treatment (c.f. Daugherty and Ventura[22]; Herold et al.[23]) for cyclotron transitions in which the transverse state of excitation of the electron changes; for the Coulomb binding energies, however, in which the longitudinal motion plays the essential role, relativistic effects remain well below 0.1% even at magnetic fields of 10^9 T (Lindgren and Virtamo[24]), which is understandable on account of the smallness of the ratio between the Coulomb binding energy and the rest energy of the electron. The bunch on the right-hand side is due to the finiteness of the proton mass by which levels with adjacent azimuthal quantum numbers are shifted with respect to each other by the proton cyclotron energy, giving rise to transitions with wavelengths $\lambda_{cycl}^{(p)} = (m_p/m_e)\lambda_{cycl}^{(e)} \cong 4.18 \cdot 10^4$ nm/β. All lines in the proton cyclotron bunch would either tend to wavelengths of the unperturbed H atom or even go to infinity (in the case of energy levels coinciding in the $B \to \infty$ limit of the one-dimensional Coulomb problem), if the finiteness of the proton mass were neglected. The lines below ~ 100 nm between the two cyclotron bunches correspond to transitions to tightly bound states. The behaviour of the wavelengths in the intermediate regime appears fully disordered.

Clearly, any attempt to observe, and resolve, a line spectrum of hydrogen at a given magnetic field strength in the intermediate regime is doomed to failure. An element of order is, however, brought in even in this domain by several transitions whose wavelengths go through minima and maxima in certain intervals of the magnetic field strength, that is, they are less sensitive to variations of the magnetic field than the many fast running components. An inhomogeneous field with a variation of, say, a factor of two (as is the case for a dipolar field) around extrema of wavelengths will therefore filter out exactly these stationary components, thus opening the possibility of observing, in this instance, a clearly arranged spectrum with few well resolved features. Speculative as this may sound, nature has indeed provided us with a cosmic laboratory to test this hypothesis, and in fact this has opened a totally new era of stellar atomic spectroscopy, namely the "spectroscopy of stationary lines."

3.2 Spectroscopy of Stationary Lines

Some 42 light years away in the stellar configuration Draco lies a 13 magnitude star long suspected of having a strong magnetic field. Its spectrum had consistently defied interpretation ever since the first shallow absorption features were discovered in the spectrum of this white dwarf star, known as Grw+70°8247, more than four decades ago. The circular polarization of its optical continuum[25] had given a clue to the existence of a strong magnetic field in the vicinity of this object, and Angel[26] first proposed a tentative identification of a few of the features in terms of stationary lines using variational energy values of Praddaude[27] available at that time. On the basis of our very accurate computations of the energy values and transition rates of the hydrogen in magnetic fields of arbitrary strength it became possible to positively identify all observed features as hydrogen in a magnetic field whose value was pinned down to between 17000 and 35000 T.[28-30] No other previously known white dwarf had a magnetic field even one tenth this value. Figure 2 and Fig. 3 demonstrate the excellent agreement between the wavelength positions of the extrema of stationary components of H_α, H_β, H_γ and absorption features in the red and blue part of the spectrum of Grw+70°8247. In particular, the sharp blue edges and "red-shaded"

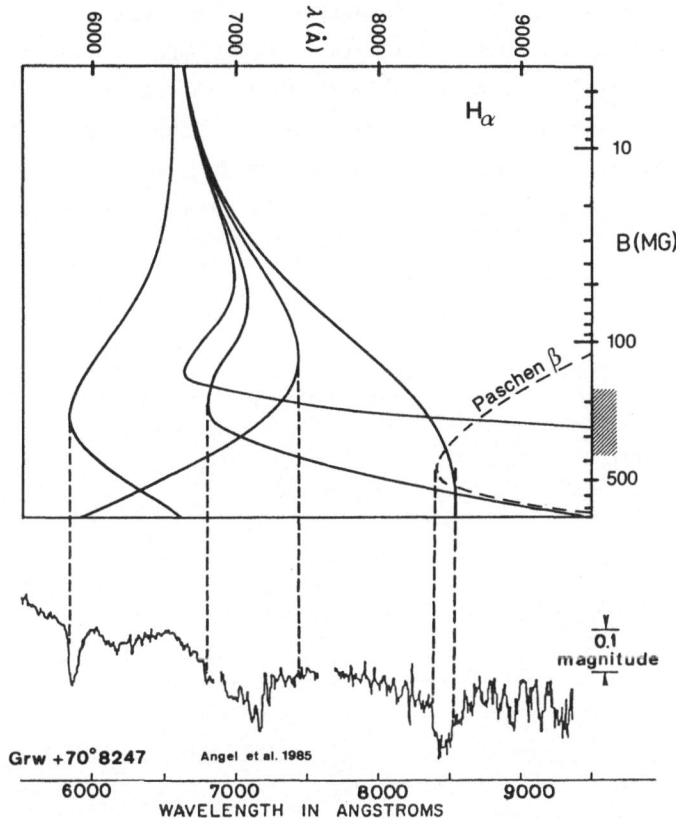

Fig. 2. Stationary H_α transitions of the hydrogen atom in magnetic fields from 4 to 700 million Gauss in comparison with the red part of the optical spectrum of the white dwarf star Grw+70°8247 taken by Angel et al.[28] The sharp blue edge of the feature at 5800 Å evidently coincides with the position of the wavelength minimum of a single stationary H_α transition, as indicated by the dashed line, and the red extension of the feature can be explained, in a natural way, by the variation of the wavelength around the minimum in an extended magnetic field whose strength varies from ~ 170 to ~ 350 million Gauss (see the corresponding hatching along the B ordinate). Such a variation is present, e.g., in a dipolar magnetic dipole field. The broader feature around 7000 Å is accounted for by a blend of two stationary H_α components in the same range of field, while to the feature around 8400 Å even a strongly "blue-shifted" stationary Paschen β transition contributes.

extensions of the features at 3650 Å, 4135 Å, and 5800 Å are well accounted for by the minimum character of the corresponding stationary components. More complicated structures of other features are produced by blends of stationary components, such as the broad feature around

8500 Å to which a stationary transition of H_α and one of Paschen β contribute. Thus for the first time <u>low-lying states exposed to a strong-field situation</u> were actually observed in nature. In the meantime a second object has been found (PG 1031+234, Schmidt et al.[32]), in the spectrum of which stationary components of hydrogen lines, in a field ranging up to ~ 50000 Tesla, have positively been identified.

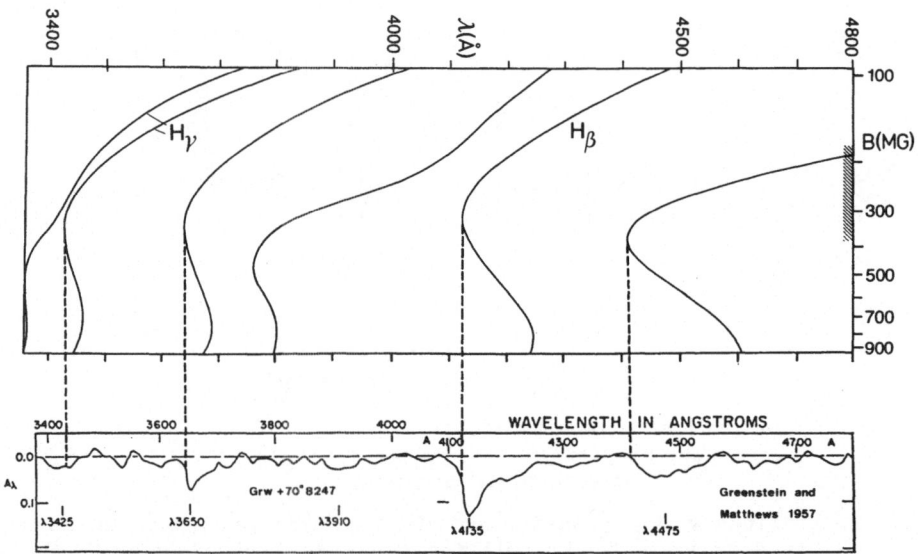

Fig. 3. Stationary H_β and H_γ transitions of the hydrogen atom in magnetic fields from 100 to 900 million Gauss in comparison with the short-wavelength part of the spectrum of Grw+70°8247 taken by Greenstein and Matthews.[31] Again all the features are explained in a consistent way in terms of stationary Balmer transitions of hydrogen in the range of magnetic field found in Fig. 2 (cf. the hatching along the B ordinate).

Here the observed phase modulations of the spectrum yield additional information from which the morphology of the magnetic field of this star can be extracted.

Evidently quantitative calculations of synthetic spectra will provide a sensitive tool to explore the exact emmission conditions in these strongly magnetized objects. In addition to wavelenths, these computations require, as in input from atomic physics, the strengths of the different transitions as functions of the magnetic field (for tables see Wunner et al.[33]), and the rates of bound-free transitions. The latter have not yet been determined in the interesting region

$\beta \sim 10^{-2} - 10^{-1}$, but evidently theoretical work in this direction is urgent.

4. HIGHLY EXCITED STATES IN STRONG LABORATORY FIELDS – COMPARISON BETWEEN EXPERIMENTAL AND THEORETICAL SPECTRA

For many years research of magnetized Rydberg atoms was characterized by the situation that experimental work concentrated on nonhydrogenic atoms (e.g. Ba, Na, Li) while most of the theoretical papers were devoted to hydrogenic systems. Semiclassical and quantal calculations were carried out to account for the nature of the quasi-Landau-resonances observed experimentally around and above the field-free ionization threshold, but so far no detailed quantitative comparisons were possible for the rich and complicated line structure seen in experimental spectra of magnetized Rydberg atoms. Quite recently this situation has changed considerably in that, on the one hand, actual experiments have been performed with highly excited <u>hydrogen</u> and <u>deuterium</u> atoms in 4-6 T fields by the Bielefeld group (Holle et. al.[34]), and, on the other, the computer codes developed over the last years allow an almost routine calculation of highly accurate energy values and wavefunctions of one-electron systems in uniform laboratory magnetic fields up to shortly below the field-free ionization threshold. To illustrate the state of the art, Fig. 4 provides a comparison, in different energy intervals, between the experimental spectrum of $\Delta m = 0$ Balmer transitions to $m = 0$, even-parity Rydberg states of deuterium taken by the Bielfeld group at $B = 5.96$ T, and the corresponding theoretical spectrum computed by us. The interval considered in Fig. 4 covers a range of energy where one lies deeply in the in the n-mixing regime. The agreement between theory and experiment can be considered excellent; moreover, theory reveals where neighboring lines were no longer resolved in the experiment. We note that the comparison of relative intensities is slightly marred by the fact that saturation effects occurred in the experiment in strong lines. As one moves further up in energy in Fig. 4, the line density clearly grows and the structure of the spectrum becomes increasingly complicated. Nevertheless it is also here that we find practically complete agreement between theory and experiment, with the experiment being limited, as regards the number of detectable lines, by the finite resolution. This is particularly evident from Fig. 5, which shows a direct comparison between the experimental and

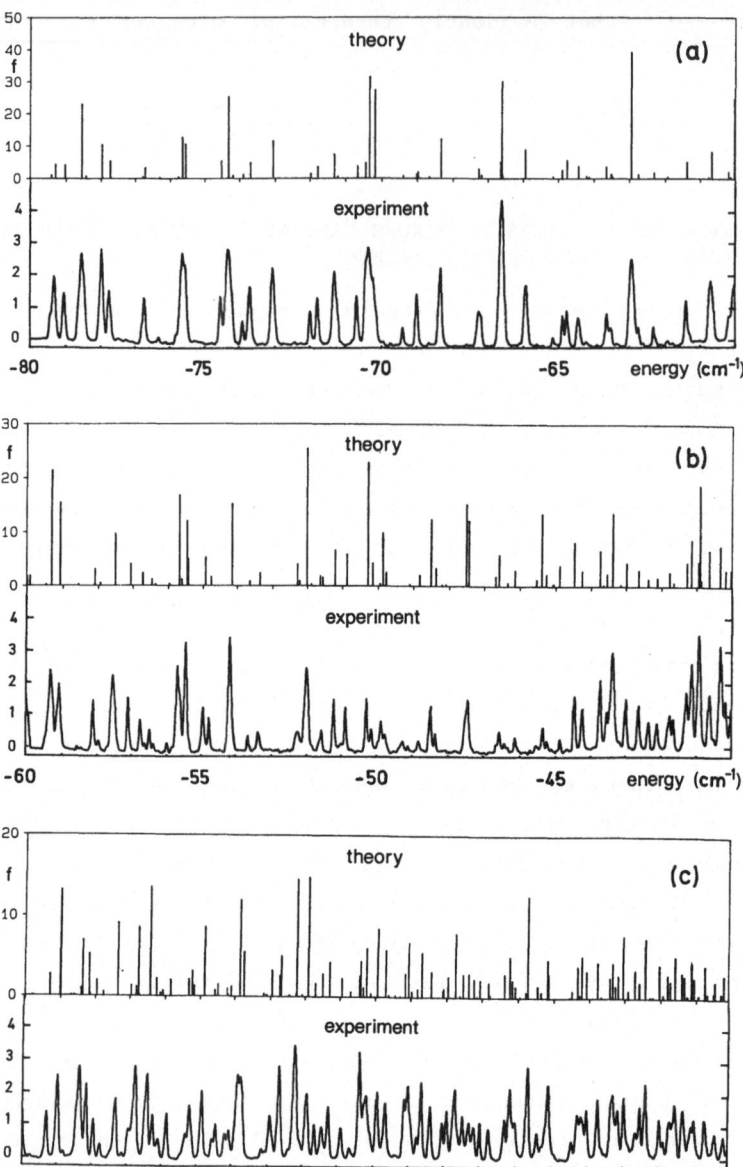

Fig. 4. Deuterium Rydberg atoms in a magnetic field of 5.96 T: comparison between the theoretical oscillator spectrum and the experimental photoabsorption spectrum (Holle et al.[34]) for $\Delta m = 0$ Balmer transitions to $m = 0$, even-parity final states, over the range of energy from -80 cm^{-1} to -20 cm^{-1}. The oscillator strengths are given in units of 10^{-6}, the experimental intensity scale is in arbitrary units. A total of 267 theoretical lines contributes to the spectrum shown. Note that at the end of the energy range classical motion becomes completely irregular (cf. Fig. 6). (The exact value of β for which the theoretical calculation was performed is $\beta = 1.2675 \cdot 10^{-5}$ and the mass scaling for deuterium[6,7] is taken into account.

theoretical spectra in the range -24 cm^{-1} to -12 cm^{-1}, at 6 T. The broad experimental feature around -13 cm^{-1}, e.g., is found to be composed of ~ 10 lines of different intensities. Obviously this calls for increased experimental efforts to refine resolution and actually check the theoretical spectra also in these ranges of energy.

All in all we have compared successfully more than 1000 lines so far in the π and σ spectra in the strong-field regime, and this certainly can be considered a hallmark of modern quantitative spectroscopy in magnetic fields.

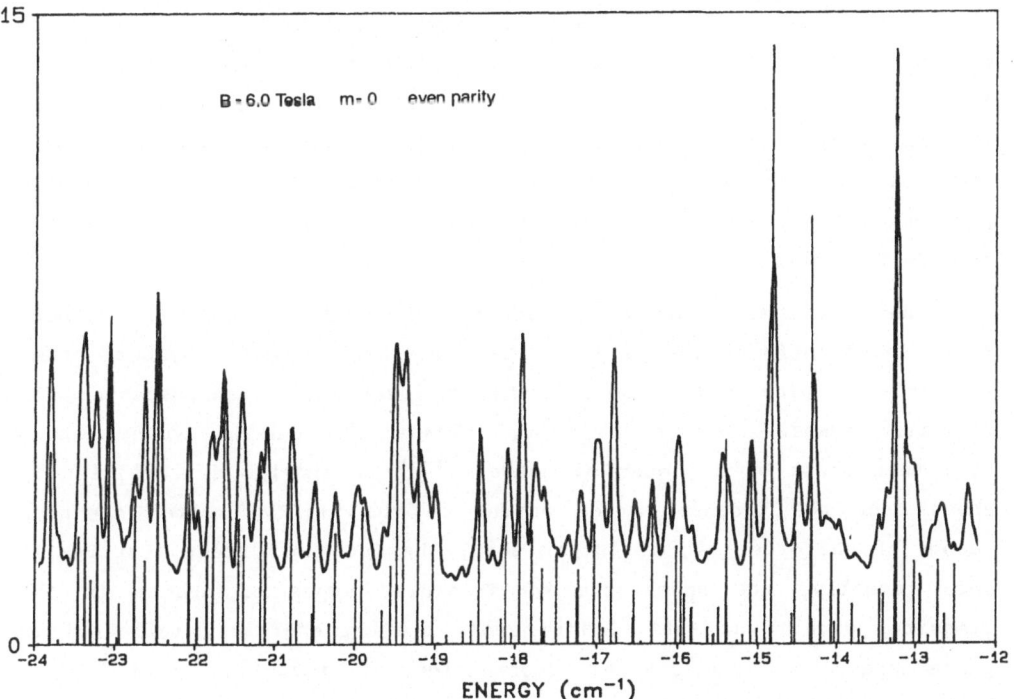

Fig. 5. Comparison between the experimental spectrum and the theoretical photoabsorption spectrum of $\Delta m = 0$ transitions from $2p_0$ to even parity final Rydberg states with energies between -12 cm^{-1} and -24 cm^{-1} in a magnetic field of 6 T. For the theoretical results (straight lines) the ordinate represents the oscillator strength in units of 10^{-6}. The experimental results have kindly been provided by the Bielefeld group.[34] Note that the wealth of theoretical spectral structure can no longer be resolved by the experiment. The range of energy shown lies in the domain of chaotic motion of the corresponding classical system.

5. MAGNETIC RYDBERG ATOMS - AN OBJECT LESSON IN "QUANTUM CHAOLOGY"

5.1 Quantum Physics in Classically Chaotic Regions

It has been realized, in recent years, to an increasing extent that in classical Hamiltonian systems the occurrence of "chaotic" behavior – in the sense of exponential separation of initially infinitesimally neighboring trajectories in phase space – is the rule rather than the exeption. Furthermore, theoretical and experimental techniques have progressed to a point where it is now possible to study in detail quantal systems in the limit of large quantum numbers, in which the laws of quantum theory should reduce to the laws of classical physics. All this has combined in creating a new field of research, for which the term "quantum chaology" has been coined by Berry,[35] who defines it as "the study of semiclassical – but still <u>nonclassical</u> – behavior characteristic of systems whose classical motion exhibits chaos." It should be emphasized that the term "quantum chaos," which has become rampant in literature, still lacks a rigorous definition. Although finally aiming at such a definition quantum chaology is taking, at present, a more empirical point of view: what one is looking for is archetypal quantum phenomena in a range of parameters where the classical analogs of the systems turn chaotic.

One of the most intriguing features of the spectroscopy of Rydberg atoms in magnetic fields, lies in the fact that magnetic Rydberg atoms are able to play a leading role in the search for manifestations of "quantum" chaos. On the one hand, "classical" magnetic Rydberg atoms have been shown by a number of authors[36-38] to undergo a transition to chaos once the Lorentz forces acting on the highly excited electron become of the order of, or larger than, the Coulomb forces. Putting it more formally, the appearance of irregular types of orbits in the classical problem at sufficiently small binding energies, reflected in phase space by the gradual destruction of invariant tori, is a consequence of the nonintegrable nature of the Hamiltonian (1). On the other hand, high-resolution spectroscopic experiments on "real," quantal, magnetic Rydberg atoms are performed by a number of groups, and, as demonstated in the foregoing section, large-scale computations have reached a point where it is possible to determine theoretically the energetic positions of, and oscillator strengths of transitions to, Rydberg states in Tesla fields up to almost the field-free ionization limit. Therefore magnetic Rydberg atoms indeed offer themselves as a paradigm for studying, both theoretically and experimentally, the basic

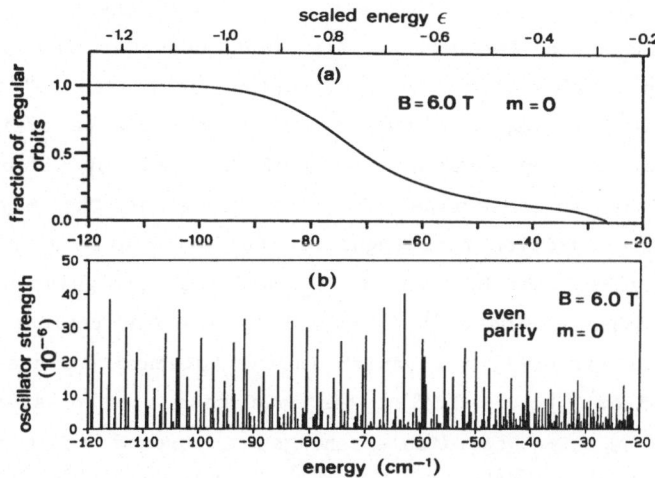

Fig. 6. The area fraction of the Poincaré surfaces of section of classical motion at z = 0 which are filled by regular orbits as a function of orbital energy (a) compared with the oscillator strengths of Δm = 0 transitions from $2p_0$ to hydrogen Rydberg states (b). The breakdown of regularity in classical motion is reflected by an increasing complexity of the quantal spectrum.

question of how quantum systems behave in a range of parameters where their classical counterparts exhibit phenomena of chaos.

A very convenient way of visualizing the appearance of chaotic orbits in a classical problem is to look at the Poincaré surfaces of section of the trajectories in phase space. To find the range of energy where, for the parameters used in the Bielefel experiments, the transition from regularity to chaos occurs in the classical magnetic Kepler problem we computed trajectories as a function of energy for a sufficiently large number of random initial conditions, and determined the area fraction of regular orbits in the Poincaré surfaces of section at z = 0. In Fig. 6a the breakdown of regularity in the magnetic Kepler problem is quantified by plotting the fraction of regular orbits as a function of energy for $\beta = 1.275 \cdot 10^{-5}$ (B = 6.0 T) and m = 0. Figure 6a is in fact universal, and not restricted to this set of parameters. The reason for this is a scaling property[37] of the (classical) Hamiltonian (1): by the replacements $r \to \beta^{2/3} r$, $p \to \beta^{-1/3} p$ the Hamiltonian is brought into a form where it no longer depends on <u>two</u> parameters, viz. magnetic field strength and energy, but solely on one parameter, the "scaled" energy $\varepsilon = E/(2\beta)^{2/3}$ (E in Rydbergs). The intuitive meaning of this transformation is that the relative sizes of the Coulomb attraction and the diamagnetic interaction are kept constant for given scaled energy, which implies that classical motion is "frozen in" for fixed ε. Converting, in Fig. 6a, absolute to scaled energies we arrive at the ε scale shown at the top horizontal axis of Fig. 6a. By aid of this scale it is possible to determine, for any value of the magnetic field strength, the transition from regularit to irregularit as a function of aboslute energies. For $B = 6$ T it is found that irregular orbits become noticeable around ~ -104 cm^{-1} ($\varepsilon \approx -1.10$), and the regions filled with regular orbits virtually all vanish in the Poincaré surfaces of section at $E_c \approx -24.5$ cm^{-1} ($\varepsilon_c \approx -0.25$). We note that the transformation laws imply that the time coordinate changes according to $t \to \beta \cdot t$, and angular momentum according to $1 \to \beta^{1/3} 1$. From the latter relation it follows that in laboratory fields ($\beta \sim 10^{-5}$) even for nonvanishing z-component of the angular momentum the <u>scaled</u> angular momentum still can be very small, which makes Fig. 6a also applicable to these cases.

Figure 6b shows the quantal oscillator strength spectrum of $\Delta m = 0$ transitions from $2p_0$ to even-parity Rydberg states for the same magnetic field strength as in Fig. 6a. The comparison between classical chaos and the oscillator strength spectrum conveys a <u>qualitative</u> impression of the increasing complexity of the quantum spectrum as one penetrates into ranges of energy where classical motion becomes more and more chaotic.

5.2 Statistical Analysis of Energy Level Sequences

To put the notion of increasing complexity of the quantum spectrum in the classically chaotic region on a more <u>quantitative</u> footing it is useful to undertake a statistical analysis, in the spirit of Bohigas et al.,[39] of the level fluctuations of the sequence shown in Fig. 6. Our results for the distribution of the spacings of adjacent energy levels, and the results of the Dyson-Mehta Δ_3 statistics for the nearest-neighbor spacings, in three successive energy intervals have been presented elsewhere (Wunner et al.[40]). The histograms of the nearest-neighbor distributions exhibit a clear transition from a Poisson-like to a Wigner-like distribution as one proceeds to intervals of energies where classical motion becomes increasingly irregular. In an analogous manner the Δ_3 statistics display a transition from the Poisson case to that characteristic of the predictions of random-matrix-theory (Gaussian orthogonal ensemble, GOE). A Poisson form of the nearest-neighbor histogram implies maximum probability, and a Wigner form zero probability, for finding two levels at the same energy value. Therefore level clustering (~ uncorrelated energy levels) proves characteristic of quantum systems in the classically regular regime, while level repulsion (~ correlated levels) proves typical of quantum systems in the classically chaotic regime. The universality of the characterization of the onset of "quantum" stochasticity by these level fluctuation rules, which had before only been found in model systems such as the stadium problem or the Sinai billiard, is thus reinforced also in magnetic Rydberg atoms. This finding is confirmed by independent work of Wintgen and Friedrich[41] and Delande and Gay.[42]

5.3 Fit Formulae for Nearest-Neighbor Spacing Histograms

The question of what functional form best fits the nearest-neighbor spacing histograms in the transition region from regularity to irregularity has been of topical interest in the wake of a paper by Wintgen and Friedrich,[43] who analyzed large ensembles of quantum levels of the hydrogen atom in strong magnetic fields at fixed scaled energy (cf. also their contribution to this book). They pointed out that the family of functions, which Berry and Robnik[44] had derived to describe the level spacing distributions inthe semiclassical limit of any system with two degrees of freedom when one chaotic region predominates, does not properly fit the numerical histograms in the near-regular regime. The reason for this lies in the fact that for any deviation, however small, from complete integrability residual interactions between the levels will cause them to repel at very small energetic distances, and thus make the probability for exact degeneracy vanish – a feature not contained in the original Berry-Robnik formula. Robnik[45] has recently presented an improved formula, and another formula has been proposed by Hasegawa[46] et al. who adopted a stochastic differential equations approach. The "stochastic" formula for nearest-neighbor distributions of quantum levels of nonintegrable systems reads

$$P_\lambda(S) = \frac{N}{\sqrt{\dfrac{1}{e^{-2pS}} + \lambda^2 \dfrac{1}{\rho^2 S^2 e^{-\alpha^2 \rho^2 S^2}}}} . \tag{3}$$

The underlying idea is that there exist two different types of noises in the system, one responsible for level clustering, and the other responsible for level repulsion. The average level densities associated with the two types of noises are given by ρ and $\alpha \cdot \rho$, respectively. The parameter λ measures the relative strength of the two noises. The normalization constant, N, and ρ can be eliminated from (3) by the conditions $\langle 1 \rangle = 1$, $\langle S \rangle = 1$ (S is the level distance). Limiting forms of the stochastic formula are the Poisson distribution ($\lambda = 0$) and the Wigner distribution ($\lambda \to \infty$). The formula does contain the desired nonanalytic behavior at $S = 0$: $P_\lambda(0)$ is finite for $\lambda = 0$, while $P_\lambda(0) = 0$ for any $\lambda \neq 0$. The parameter α can be interpreted as representative of the nonuniversal behavior of the specific system under consideration in the transition region between complete regularity and irregularity.

We have tested the usefulness of the stochastic formula by performing least-squares fits to the histograms for the nearest neighbor-distributions of the energy levels of the hydrogen atom in a magnetic field at fixed scale energies $\epsilon = -0.8$, -0.6, -0.4, and -0.2 as given by Wintgen and Friedrich.[43] Figure 7 shows the results obtained choosing $\alpha = 1$. It is seen that the stochastic formula is indeed able to produce reasonable fits to the histograms also in the near-regular regime. Also shown in Fig. 7 are the results which we obtained using the Berry-Robnik[44] formula, and the empirical Brody[47] formula. The "overshooting" effect of the stochastic formula near maximum is mitigated if in addition to λ we also vary α in the fitting process. The results obtained in this way are shown in Fig. 8, and reveal the potential of the stochastic formula.

5.4 Parameter Sensitivity of the Spectra

A manifestation of irregularity in quantum spectra which is more directly perceptible to the eye than is the statistical analysis of level sequence fluctuations, an in addition to energy values includes the probing of wavefunctions, is found when one asks for the variation of photoabsorption spectra with the magnetic field strength. To answer this question we have computed the oscillator strength spectra of Balmer transitions from $2p_0$ to $m = 0$, even-parity final states in the range -35 cm^{-1} to -11 cm^{-1} for 21 equally spaced magnetic field strengths in the interval 5.95 T to 6.06 T. The range of energy extends from the end of the transition region, where the classical system retains only a small fraction of regular orbits, into the classically completely chaotic regime. The detailed graphical representations of the results, which are shown in the contribution of G. Zeller et al.[48] to this book (see also Fig. 9), clearly demonstrate that on account of numerous avoided crossings of the levels small variations of the field strength can cause extreme fluctuations of the line strengths and, thus, lead to drastic changes in the outward appearances of the observable spectra.

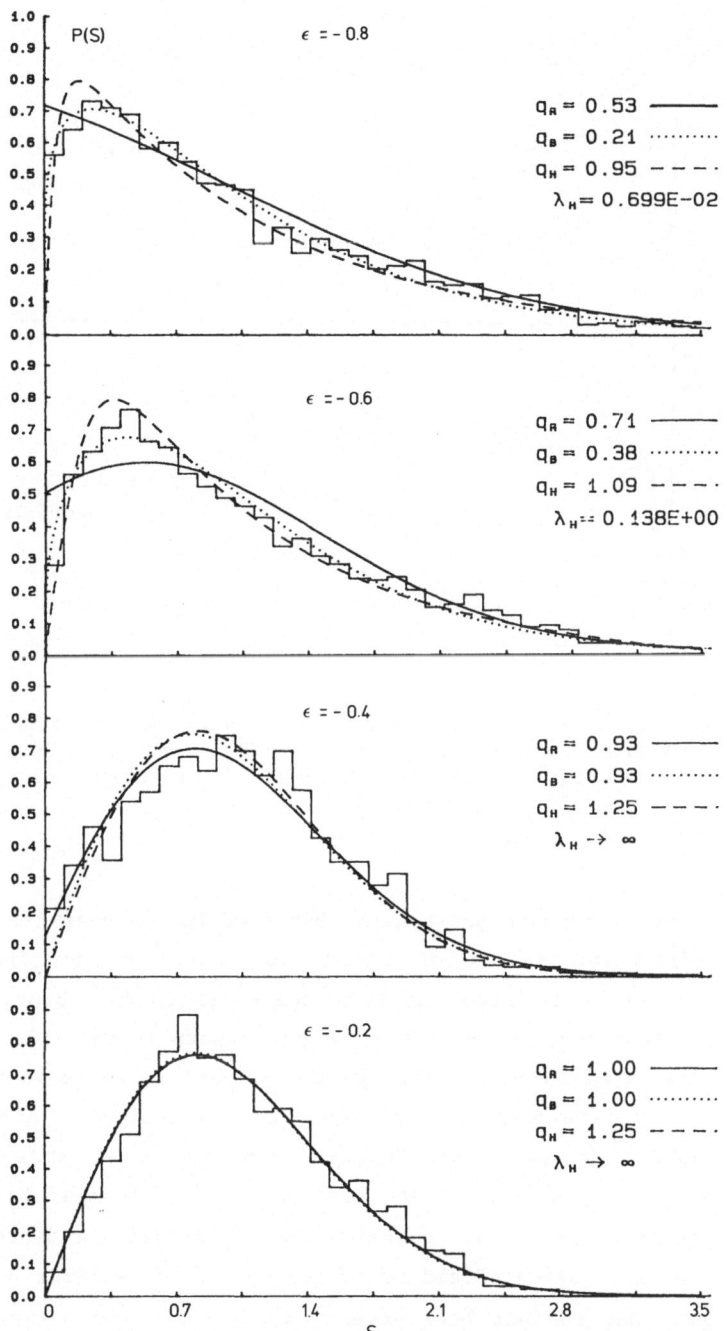

Fig. 7. Nearest neighbor energy spacing histograms for hydrogen Rydberg levels calculated at different values of the scaled energy ε. The smooth curves show the results of least-squares fits to the histograms using different fit formulae (solid line: Berry and Robnik formula; dotted line: Brody formula; dashed line: Hasegawa's stochastic formula, evaluated for α = 1). The q and λ values denote the corresponding best fit parameters.

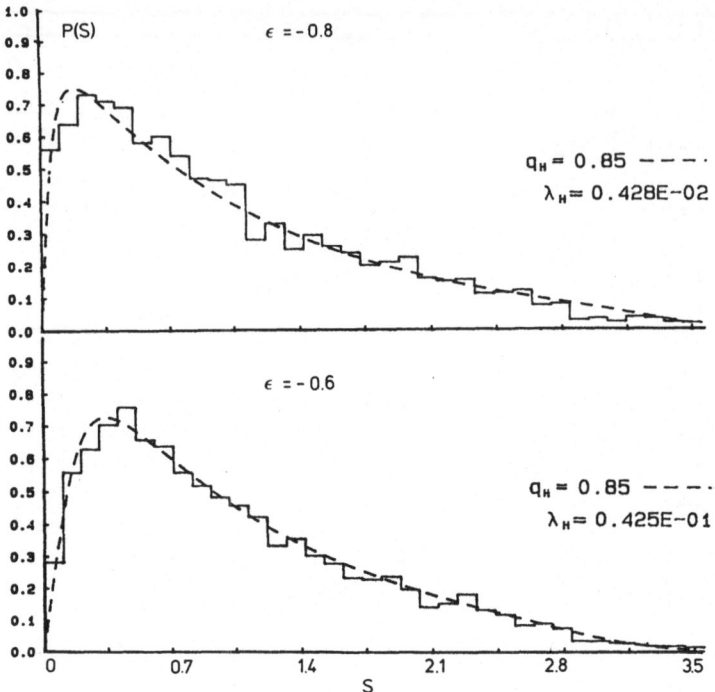

Fig. 8.   Same as Fig. 7, but with the results of the stochastic fit
           formula obtained by varying both λ and α. Note that the
           overshooting near maximum is slightly mitigated by
           additionally varying α.

The qualitative difference between the behavior in the near-irregular and
completely irregular ranges of energy is that, in the first, line
strengths can still be more or less conserved as one passes through
single or multiple avoided crossings, which renders it possible to trace
individual states diabatically through the crossings, while in the latter
line strengths behave much more irregularly, in that e.g. strong lines
"die" in avoided crossings and "revive" somewhat later at a displaced
position of the spectrum.   In particular it becomes increasingly
difficult, if not impossible, to follow the individual states through the
crossings.   At much higher field strengths ($\beta > 10^{-3}$) similar behavior of
line strengths had in fact been noted[19] in investigations exploring the
limits of quasi-separability of the Hamiltonian (in the sense of the
existence of approximate quantum numbers) in this range of field
strength.   Our investigations show that this behavior persists at
laboratory field strengths, and, moreover, that the breakdown of quasi-
separability in the quantal problem is obviously closely related to the

Fig. 9. Fourier spectra of the theoretical photoabsorption spectra with $m = 0$ in the range -35 cm^{-1} to -13 cm^{-1} for four slightly different magnetic field strengths. Peaks are indicative of corresponding quasiperiodic classical orbits with different frequencies (given in units of the cyclotron frequency in the top frame). Note the structural change in the transforms at small variations of the magnetic field strength.

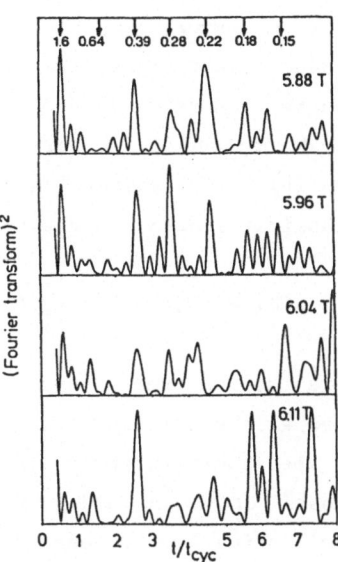

onset of chaos in the classical problem. We note that similar sensitivity of quantum spectra to small variations of an external parameter (here the magnetic field strength) in the classically near-irregular and chaotic regime, associated with the appearance of "robust" and multiple avoided crossings of the energy levels, had before been found only in nonintegrable model Hamiltonian systems, for example in the stadium problem (cf. Berry[49]).

5.5 Importance of Periodic Classical Orbits to the Quantal Spectrum

A key role in the interpretation of the experimental and theoretical oscillator strength spectra can be assigned to unstable classical periodic orbits present even in the otherwise completely irregular regime: these orbits are found to give rise to modulations in the quantal spectra with frequencies equal to their classical orbital frequencies[50] ("quasi-Landau resonances"). The existence of the modulations can be seen from corresponding peaks in the Fourier transforms of the oscillator strength spectra, shown in Fig. 9 for 4 slightly different magnetic field strengths. Orbital frequencies of the "fundamental" series[50] are indicated by arrows in the top frame. The rapid changes in the modulation patterns at small variations of the field again reflect the strong parameter sensitivity of the spectra.

A rigorous theory highlighting the importance of periodic orbits to quantal spectra in the semiclassical, nonclassical regime is provided by work of Gutzwiller,[51] Balian and Bloch,[52] and Berry,[49] who showed that closed classical orbits give riese to long-range modulations of the level density. More explicitly, the level density as a function of energy can be decomposed in a part representing the variation of the mean level density with energy, and an oscillatory part to which all closed classical orbits contribute, viz.

$$n(E) = \overline{n}(E) + \sum_j \sum_m a_{jm}(E) \cos[m(S_j(E) - \alpha_j)/\hbar].$$ (4)

Here the sum over j runs over all closed classical orbits, the sum over m over all repetitions, S_j is the action and α_j an appropriate phase shift of the primitive orbit j. The most important feature of (4) is that each closed orbit contributes an oscillation to the level density. The oscillation has an energy "wavelength" ΔE given by $1/\hbar \; d(mS_j)/dE = 2\pi$. Noting that the derivative of action with respect to energy yields the inverse of the orbital frequency, ν_{orb}, we have $\Delta E = h\nu_{orb}/j$, so that modulations of the quantal level denstiy can appear at energy spacings that are given by the classical orbital frequencies and their subharmonics. For more details we refer the reader to the review paper of Berry.[49] An application of this theory to the hydrogen atom in magnetic fields has recently been given by Wintgen[53] and Du and Delos.[54]

As a consequence, the energy dependence of quasi-Landau resonances can quite easily be determined by computing the change of the frequencies of the corresponding classical orbits as a function of energy. A nice example of this conclusion, which essentially is already implied in the pioneering work of Edmonds[55] on the Garton-Tomkins-type[56] quasi-Landau resonance, has recently been presented by Rinneberg et al.[1] who measured, for a magnetic field strength of B = 0.0736 T, the spacings of ~ 180 Garton-Tomkins-type quasi-Landau resonances as a function of energy around the zero-field threshold. The variation of the spacings by almost a factor of 4 over the range of energy measured in the experiment is quantitatively reproduced by classical trajectory calculations.[1] This is an excellent demonstration of how prominent aspects of the quantum mechanical system in this semiclassical regime can indeed be accounted for by simply solving Newton's equations of motion.

5.6 Liapunov Exponents Associated with Closed Classical Orbits, and "Scarring" of Wavefunctions

A measure of the instability of a classical trajectory in the chaotic region is provided by the Liapunov exponent, which characterizes the velocity at which trajectories in the vicinity of a given trajectory __exponentially__ separate from that trajectory in phase space. Rigorous computations of Liapunov exponents can be performed by following the stability matrix of a trajectory as a function of time, and etailed results for the orbits associated with the "fundamental" series of the hydrogen atom in magnetic fields obtained using this method are presented by Schweizer et al.[57] in this book. (Special cases had before been considered by Wintgen.[58]) Interest in the size of Liapunov exponents of periodic orbits stems from the fact that, on the one hand, they determine the behavior of the expansion coefficients α_{jm} in the modulation formula (4), and thus decide on the actual strength with which the frequencies of the individual primitive orbits become visible in the quantum spectrum. On the other, studies by Heller[59] in model systems indicate that periodic orbits "scar" the wavefunctions of the quantal system according to the smallness of λ/ν_{orb}, where λ is the Liapunov exponent. In Table 1 we present sample results for the size of Liapunov exponents of the first 6 periodic orbits of the fundamental series (characterized by their orbital frequency) at E = 0. Surprisingly the orbit running in the plane perpendicular to the direction of the field is found to be most unstable at this energy, while the degree of instability decreases as one goes up the series. On the other hand, the "discrepancy per period," λ/ν_{orb}, is seen to be the smallest for the planar orbit and to increase for the longer-periodic orbits. Thus if scarring occurs in these states around E = 0, one would expect that the orbit running in the plane perpendicular to the direction of the magnetic field (which had already previously[53] been associated with the quasi-Landau resonance of $\sim 1.5\ \nu_{cyc}$ around E = 0) should most prominently scar the wavefunctions, while the scarring should become less pronounced for the longer-periodic orbits.

Table 1

Liapunov exponents of periodic orbits at energy E = 0

ν_{orb}/ν_{cyc}	1.50	0.64	0.39	0.28	0.22	0.18
$\lambda/(\pi\nu_{cyc})$	0.63	0.59	0.46	0.38	0.32	0.27

Wavefunctions around E = 0 in this problem have not yet been fully analyzed with respect to the presence or absence of scars, but we can present results of our analyses of wavefunctions below this limit. Figure 10 shows the wavefunctions of the state which can diabatically be traced back to the state n = 36, K = 0 in the low-B limit at a magnetic field strength of 5.96 T (the energy of the state at this field strength is $E \sim -32\ cm^{-1}$, corresponding to a scaled energy of $\varepsilon \sim -0.3$, and thus lies in the near-irregular regime, cf. Fig. 6). This state can be associated with the periodic orbit in the plane perpendicular to the direction of the magnetic field, although the neighborhood of the classical orbit also contributes to the wavefunction of course. Even though most of the planar orbit is still found to vanish (stable orbit), and hence the very pronounced concentration of the wavefunctions along the classical periodic orbit is the expected correspondence familiar from situations where WKB (or EBK) quantization is still applicable. In the

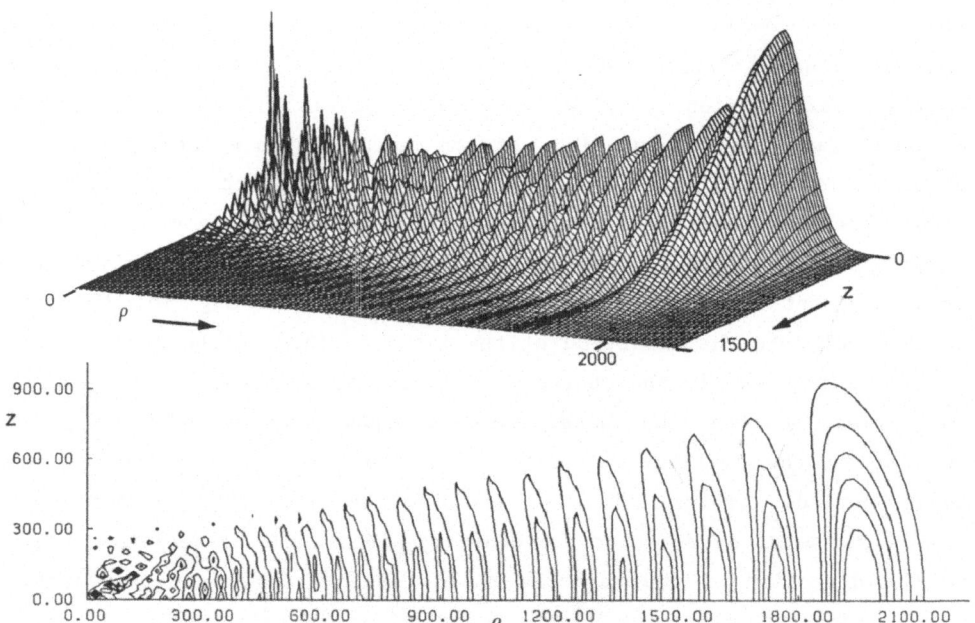

Fig. 10.   3d-plot and contour plot of $\rho\psi^{*}\psi$ of the state that can diabatically be traced back to the state n = 36, K = 0 in the low-B limit at a magnetic field strength of B = 5.96 T ($E \sim -32$ cm^{-1}). Length scales are in Bohr radii. Evidently the wavefunction of this state is strongly concentrated along the periodic orbit that runs perpendicular to the direction of the magnetic field, which is still stable at this energy.

light of scar theory this correspondence can be considered the limiting ($\lambda \to 0$) case of scarring. Figure 11 gives an example of a wavefunction in the completely irregular regime ($E \sim -15$ cm^{-1}). The contour plot shows vague indications of enhanced probability of presence along both the planar orbit and the first nonplanar orbit, which are both found to be unstable at this energy. Obviously the instability has grown so large as to wash out the precise tracks of the periodic orbits over an extended fraction of space in the wavefunction.

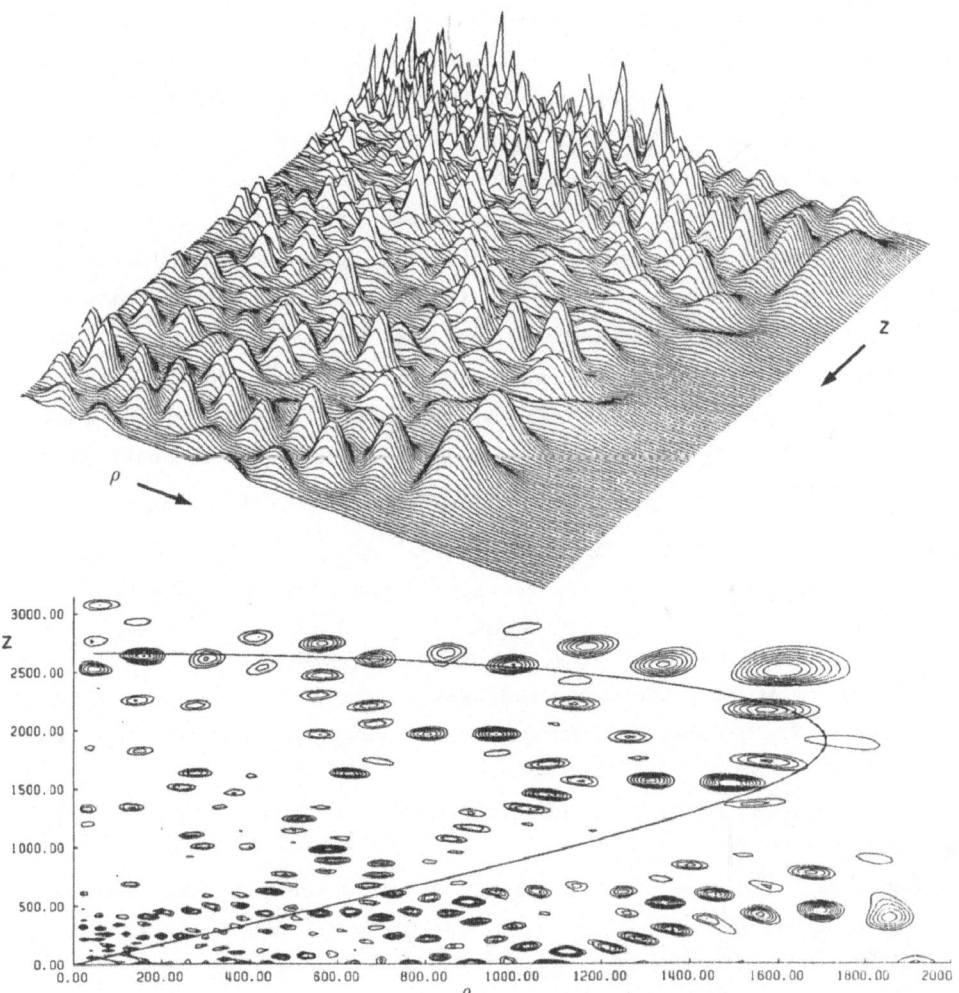

Fig. 11. Same as Fig. 10 for a state in the completely irregular regime (B = 5.96 T, E ~ -15 cm^{-1}). The contour plot reveals vague scars caused by both the planar and the first nonplanar orbit (solid curve), which are unstable at this energy.

6. OUTLOOK

In spite of the immense progress which investigations of strongly magnetized atoms have made over the last years, a number of serious problems still remain. We only mention the description of continuum states in magnetic fields of arbitrary strength. These are of importance in both astrophysical fields, where the determination of synthetic spectra from radiative transfer calculations for the atmospheres of white dwarfs require bound-free opacities, and in laboratory field strengths, where experiments with Rydberg atoms easily pass the ionization threshold, and only there reveal the whole wealth of quasi-Landau structures in the spectra. Calculations along these lines are urgent. Moreover, the example discussed in Section 5 have given ample evidence of the fact that magnetic Rydberg atoms are indeed ideal systems in which to perform case studies in quantum chaology, and further progress in this direction can be expected in the near future. It goes without saying that these investigations derive much of their attraction from the fact that they are able to illustrate in a very beautiful way the continuity of thought that exists between the world of classical physics and the world of quantum theory.

This work was supported by the Deutsche Forschungsgemeinschaft.

REFERENCES

1. H. Rinneberg et al., these proceedings.
2. H. Ruder, H. Herold, W. Rösner and G. Wunner, Physica 127B, 11 (1984).
3. J. E. Bayfield and L. A. Pinnaduwage, Phys. Rev. Lett. 54, 313 (1985).
4. K.A.H. van Leeuwen, G. V. Oppen, S. J. Renwick, B. Bowlin, P. M. Koch, R. V. Jensen, O. Rath, D. Richards and J. G. Leopold, Phys. Rev. Lett. 55, 2231 (1985).
5. G. Casati, B. V. Chirikov, D. L. Shepelyanski and I. Guarneri, Phys. Rev. Lett. 57, 823 (1986).
6. V. B. Pavlov-Verevkin and B. I. Zhilinskii, Phys. Lett. 78A, 244 (1980).
7. G. Wunner, H. Ruder and H. Herold, Astrophys. J. 247, 374 (1981).
8. D. R. Herrick, Phys. Rev. A 26, 323 (1982).
9. J. C. Gay, D. Delande, F. Biraben and F. Penent, J. Phys. B: At. Mol. Phys. 16, L693 (1983).
10. D. Wintgen and H. Friedrich, J. Phys. B: At. Mol. Phys. 19, 1261 (1986).

11. G. Wunner, J. Phys. B: At. Mol. Phys. 19, 1623 (1986).

12. L. I. Schiff and H. Snyder, Phys. Rev. 55, 59 (1939).

13. M. J. Englefield, Group Theory and the Coulomb Problem (Wiley, New York, 1971).

14. H. Friedrich and M. Chu, Phys. Rev. A 28, 1423 (1983).

15. W. Rösner, G. Wunner, H. Herold and H. Ruder, J. Phys. B: At. Mol. Phys. 17, 29 (1984).

16. A. R. Edmonds, J. Phys. B: At. Mol. Phys. 6, 1603 (1973).

17. C. W. Clark and K. T. Taylor, J. Phys. B: At. Mol. Phys. 15, 1175 (1982).

18. D. Delande and J. C. Gay, J. Phys. B: At. Mol. Phys. 16, L335 (1984); 19, L173 (1986).

19. D. Wintgen and H. Friedrich, J. Phys. B: At. Mol. Phys. 19, L99 (1986); 19, 991 (1986); Phys. Rev. lett. 57, 571 (1986).

20. H. Forster, W. Strupat, W. Rösner, G. Wunner and H. Herold, J. Phys. B: At. Mol. Phys. 17, 1301 (1984).

21. H. Ruder, F. Geyer, H. Herold and G. Wunner, Physics Reports, in press (1987).

22. J. K. Daugherty and J. Ventura, Astron. Astrophys. 61, 723 (1977).

23. H. Herold, H. Ruder and G. Wunner, Astron. Astrophys. 115, 90 (1982).

24. K.A.U. Lindgren and J. T. Virtamo, J. Phys. B: At. Mol. Phys. 12, 3465 (1979).

25. J. C. Kemp, J. Swedlund, J. D. Landstreet and J.R.P. Angel, Astrophys. J. 171, L77 (1970).

26. J.R.P. Angel, Ann. Rev. Astron. Astrophys. 16, 487 (1978).

27. H. C. Praddaude, Phys. Rev. A 6, 1321 (1972).

28. J.R.P. Angel, J. Liebert and H. S. Stockman, Astrophys. J. 292, 260 (1985).

29. J. L. Greenstein, R.J.W. Henry and R. F. O'Connell, Astrophys. J. 289, L47 (1985).

30. G. Wunner, G. Rösner, H. Herold and H. Ruder, Astron. Astrophys. 149, 102 (1985).

31. J. L. Greenstein and M. S. Matthews, Astrophys. J. 126, 14 (1957).

32. G. D. Schmidt, S. C. West, J. Liebert, R. F. Green and H. S. Stockman, Astrophys. J. 309, 218 (1986).

33. G. Wunner, F. Geyer and H. Ruder, Astrophys. Space Sci. 131, 595 (1987).

34. A. Holle, G. Wiebusch, J. Main, B. Hager, H. Rottke and K. H. Welge, Phys. Rev. Lett. 56, 2594 (1986); K. H. Welge, these proceedings.

35. M. V. Berry, "The Bakerian Lecture 1987," in Proceedings of the Royal Society (1987).

36. M. Robnik, J. Phys. A: Math. Gen. 14, 3195 (1981).

37. A. Harada and H. Hasegawa, J. Phys. A: Math. Gen. 16, L259 (1983).

38. J. B. Delos, S. K. Knudson and D. W. Noid, Phys. Rev. A 30, 1208 (1984).

39. O. Bohigas, M. J. Giannono and C. Schmit, Phys. Rev. Lett. 52, 1 (1984).

40. G. Wunner, U. Woelk, I. Zech, G. Zeller, T. ERtl, F. Geyer, W. Schweizer and H. Ruder, Phys. Rev. Lett. 57, 3261 (1986).

41. D. Wintgen and H. Friedrich, Phys. Rev. Lett. 57, 571 (1986).

42. D. Delande and J. C. Gay, Phys. Rev. Lett. 57, 2006 (1986).

43. D. Wintgen and H. Friedrich, Phys. Rev. A 35, 1464 (1987).

44. M. V. Berry and M. Robnik, J. Phys. A: Math. Gen. 17, 2413 (1984).

45. M. Robnik, J. Phys. A: Math. Gen. 20, L495 (1987).

46. H. Hasegawa, H. J. Mikeska and H. Frahm, preprint (1987); H. Hasegawa, these proceedings.

47. T. A. Brody, Lett. Nuovo Cim. 7, 482 (1973).

48. G. Zeller, U. Woelk, G. Wunner and H. Ruder, these proceedings.

49. M. V. Berry, in Chaotic Behavior of Deterministic Systems, eds. G. Iooss et al. (North-Holland, Amsterdam, 1983) p. 171.

50. J. Main, G. Wiebusch, A. Holle, K. H. Welge, Phys. Rev. Lett. 57, 2789 (1986).

51. M. C. Gutzwiller, J. Math. Phys. 12, 343 (1971); Phys. Rev. Lett. 45, 150 (1980).

52. R. Balian and C. Bloch, Ann. Phys. (NY) 59, 76 (1972).

53. D. Wintgen, Phys. Rev. Lett. 58, 1589 (1987).

54. M. L. Du and J. B. Delos, Phys. Rev. lett. 58, 1731 (1987).

55. A. R. Edmonds, J. Physique, 31, C4-71 (1970).

56. W.R.S. Garton and F. S. Tomkins, Astrophys. J. 158, 839 (1969).

57. W. Schweizer, G. Wunner, R. Niemeier and H. Ruder, these proceedings.

58. D. Wintgen, J. Phys. B: At. Mol. Phys. 20, L511 (1987).

59. E. J. Heller, Phys. Rev. Lett. 53, 1515 (1984).

IMPLICATIONS OF NEW EVIDENCE ON QUASI-LANDAU SPECTRA

FOR A UNIFIED TREATMENT OF COLLISIONS AND SPECTRA

U. Fano

Dept. of Physics & James Franck Institute
University of Chicago
Chicago, IL 60637

ABSTRACT

Novel evidence, sparked by Bielefeld measurements of two-photon Quasi-Landau excitations, is interpreted in the context of its relevance to more general phenomena. The interpretation implies that resonant states localized in regions of high potential energy decay to lower potential regions at large radial distances, where the period of classical orbits exceeds the half-life of localization. The localized states are now seen to be stabilized mainly by repulsion from states localized at low potentials rather than by short-range mechanisms, an unanticipated conclusion. These findings require an adaptation of earlier theory, which is outlined together with initial steps that are reported elsewhere.

1. INTERPRETATION OF EVIDENCE

A. Background

Photoionization of atoms in a magnetic field $\vec{B}$ ejects electrons along the axis $\hat{z} = \vec{B}$ in alternative "Landau" channels, whose thresholds are separated by the cyclotron frequency ω_c. These channels correspond to different levels of oscillation in directions ρ orthogonal to $\hat{z}$. Photoexcitation levels converging to the Landau thresholds are, however, overlaid by series of "Quasi-Landau" resonant states that propagate mainly in ρ directions, with oscillations along z astride the z = 0 plane. The intense Quasi-Landau resonances have even parity under reflection through this plane.

For lower Rydberg levels, the magnetic field amounts only to a weak perturbation. Numerical diagonalization of the degenerate perturbation matrix for a fixed value of the quantum number n has yielded two sets of eigenstates with the characteristics of Landau and Quasi-Landau channels, as well as intermediate states that are not clearly identified in optical spectra owing to approximate selection rules.[1] Further calculations have shown that the Landau and Quasi-Landau states maintain their identity as n increases far beyond the range where B amounts to a weak perturbation. The Schrödinger equation is thus seen to separate approximately both astride the z axis and astride the z = 0 plane, which are the loci of Landau and Quasi-Landau states, respectively.[2]

B. Bielefeld Observations and Accompanying Theories

High resolution, two-photon absorption spectra of H and D in fields of ~6T have revealed that the orderly arrangement of Quasi-Landau resonances breaks down within ~30cm^{-1} of the ionization thresholds, demonstrating a progressive deterioration of quasi-separability.[3] (Vestiges of the Quasi-Landau order are nevertheless apparent at low resolution up to 50-100 cm^{-1} beyond the threshold.) The spectrum of intervals between adjacent levels evolves concomitantly from an exponential to a Wigner distribution demonstrating directly the disappearance of any approximate constant of the motion other than the energy.[4] Extension of the calculations of Ref. 2 to energies closer to the ionization threshold also showed an incipient departure from approximate separability.[5]

Novel evidence on the evolution of the Quasi-Landau spectrum also emerged from its Fourier transform,[3] which displays striking periodicities in the time response of atoms excited by pulses of < 1ps. A main set of pulses includes the period of oscillation of superposed Quasi-Landau resonances followed by multiple oscillations with the cyclotron ("Landau") period; this is the pattern to be expected if Quasi-Landau states decay into coherent superpositions of Landau channels.

This further discovery has spawned manifold successful searches for semiclassical trajectories that would diplay the observed periodicities, with results reported extensively at this Conference. The correspondence of such trajectories to the final state of photoabsorption transitions remains unclear at this time. Bundles of classical trajectories, typically those that are tangent to a caustic, are known to identify a local separability of eigenfunctions in coordinates parallel and orthogonal to the caustic.[6] Single trajectories, on the other hand, are not consistent with uncertainty relations. Related studies at this Conference deal with the correspondence between the breakdown of the approximate separability of Quasi-Landau states and the onset of classical chaos. A discussion of the semiclassical aspects of Rydberg states is deferred to a separate paper.[7]

C. Analysis of Experimental Findings

The positions and oscillator strengths of very numerous levels observed at Bielefeld have been reproduced in extraordinary detail by numerical solution of the Schrödinger equation in Refs. 3 and 5, and particularly in extensive calculations led by Wunner.[8] These results verify the consistency of experimental data with the Schrödinger equation. Here I attempt a quantum mechanical qualitative analysis of the same data with the goal of extracting elements relevant to general mechanisms of isolation, production and decay of resonant states localized in regions of high potential.

The occurrence of Quasi-Landau and Landau states in the spectra of atoms in magnetic fields implies the existence of projection operators Q and L that select those states and that commute approximately with the Hamiltonian. To evaluate the approximate separability of the Schrödinger equation $H\Psi = E\Psi$ we evaluate here the influence of the Hamiltonian's commutator with a projection operator $P \rightarrow \{Q, L, \ldots\}$ by applying P to the equation, and casting it into the form

$$PH\Psi = HP\Psi - [H,P]\Psi = EP\Psi. \tag{1}$$

Viewing (1) as a Schrödinger equation for $P\Psi$ we note that its operator $[H,P]$ may or may not be treated as a small perturbation depending on the ratio of its expectation value to the separation of the eigenvalues of H

$$\langle\Psi|[H,P]|\Psi\rangle/(E_{n+1}-E_n). \tag{2}$$

The existence of an approximate joint eigenfunction Ψ of the operators H and P with eigenvalue E_n hinges thus on the ratio (2) being much smaller than unity in magnitude. (Recall that $[H,P]$ is antihermitian, with imaginary mean values.) In time-domain language, the reciprocal $1/(E_{n+1}-E_n)$ represents the period of a classical orbit and $1/\langle\Psi|[H,P]|\Psi\rangle$ the half-life of $P\Psi$'s localization, the ratio (2) is then small when the half-life exceeds the orbital period. Accordingly I attribute the observed breakdown of the Quasi-Landau classification to the sharp decrease of level spacing (i.e., increase of orbital period) associated with the divergence of atomic level density at the ionization threshold. The observed level density of $\sim 5/cm^{-1}$ at $E \sim 30cm^{-1}$ implies that

$$\langle \Psi | i[H,Q] | \Psi \rangle = O(0.1 cm^{-1}). \tag{3}$$

Note that the divergence of the level density at threshold stems from the Landau channels localized about the z axis where the potential is lowest, in contrast to the Quasi-Landau channels whose level density would remain nearly constant, of $O(1/\omega_c)$. The breakdown of the Quasi-Landau classification thus implies that the admixture of Quasi-Landau and Landau channels becomes appreciable in most eigenfunctions.

Once it becomes impossible to construct an eigenfunction Ψ with a single dominant term $Q\Psi$, the requirement of ensuring the continuity of a superposition $a_Q Q\Psi + a_L L\Psi + \ldots$ may well cause its coefficients to vary erratically from one level to the next. Whether such variations are actually erratic or instead systematic to some degree is a matter of conjecture at this time. Whether erratic behavior should be properly labeled "quantum chaos" is, of course, a matter of semantics.

The breakdown of the classification of single levels as Quasi-Landau is very rapid along the spectrum in accordance with the very rapid rise of the level density ($\propto n^3$). On the other hand this breakdown does not imply that the concept of Quasi-Landau states becomes irrelevant. Non-stationary Quasi-Landau states can indeed be constructed by superposing eigenfunctions of a number of closely spaced levels. The modulation of absorption spectra with Quasi-Landau spacings observed at low resolution across and beyond the ionization threshold reflects indeed the relevance of such states. The decreasing amplitude of the modulations with increasing energy suggest that the half-life of Quasi-Landau states near the threshold amounts to an appreciable fraction of ω_c^{-1}.

From this analysis I draw two main qualitative conclusions:

1) The decay of Quasi-Landau (or analogous) resonant states, localized in high-potential regions, takes place at <u>large</u> radial distances where the level density of Landau (i.e., low-potential) channels diverges.

(2) The decay is mediated by the admixture of states with different localization in the eigenstates whose energy approaches the ionization threshold.

A third, most important, conclusion could have been drawn from earlier calculations, e.g., from Ref. 1, independently of the Bielefeld experiments: The localization of approximate eigenstates in regions or high- and low-potential energies stems from a single process of matrix diagonalization rather than from local circumstances prevailing in their respective separate regions. Evidence for this conclusion, available for several years, seems to have come into focus only through a development to be reported in the final Sec. IIC.

II. REFORMULATION OF GENERAL THEORY

A. Background

Collision processes may be viewed as involving two reciprocal states: The formation of a "complex" by initially separate reactants (one of which may be a single electron) and "fragmentation" of the complex into separate products. A photoexcited state involves much the same stages, with the framentation aborted by insufficient energy to attain full separation. A remarkable finding has been that structural parameters -- e.g., quantum numbers -- remain largely invariant during each states, except at critical "level crossings" where alternative configurations of a complex have equal energy. Structural change, involving energy exchanges among different degrees of freedom, occur mainly at such critical loci. Analytical descriptions have been developed that utilize this structural stability interrupted by localized structural changes; they are loosely dubbed "adiabatic procedures."

An adiabatic procedure has been outlined which encompasses the representations of collision and of photoexcitation processes.[9] It parametrizes the course of fragmentation and contraction of an N-particle complex through a single "hyperradius"

$$R = [\sum_i m_i r_i^2 / \sum_i m_i]^{1/2} \tag{4}$$

where r_i is the distance from the center of mass to the ith particle with mass m_i. Each of the 3N-4 internal coordinates $\{\omega\}$ orthogonal to R extends only over a finite range. Motion along these coordinates have accordingly a discrete spectrum. The procedure of Ref. 9 relied on approximate separability of the coordinate R from the others using approximate eigenfunctions of the form $F(R)\Psi(R;\omega)$ where Ψ is an eigenfunction of the Hamiltonian with a fixed value of R. Efforts have been in progress to identify a more effective separation procedure.[10]

Ref. 9 anticipated accumulations of level crossings to occur in regions of high potential ("potential ridges") giving rise to localized resonant states through novel, still unidentified nonadiabatic processes. The developments outlined in Sec. I -- and in the following Sec. IIC -- suggest now that localization on ridges (and in "valleys") arises from pre-diagonalization of the Hamiltonian for the discrete spectra of motions approximately orthogonal to R. Nonadiabatic effects of the ridge should then reduce the more standard level crossing phenomena. Optimization of approximate separation of R should also result.

B. Current Goals

Remarks in Sec. IC and IIA suggest the opportunity of developing the program of Ref. 9 in semianalytical form. I mean by this to represent, e.g., wavefunctions by analytical expressions with parameters determined numerically as was done for one-dimensional problems in a recent paper[11] and in the following Sec. IIC.

A first step would recast earlier results in the following light. Reference 2 verified that eigenstates of the degenerate perturbation of H by the magnetic field remain valid for higher excitations than expected. This circumstance suggests that the eigenstates' structure remains approximately invariant as the electron propagates to larger radial distances (until its energy approaches the ionization threshold closely). The results of Ref. 2 might thus actually rest on an adiabatic approximation. Recasting of the degenerate perturbation treatment into an adiabatic procedure is now under investigation.

More remote problems will be met in parametrizing the mechanism and rate of decay of Quasi-Landau into Landau channels and in determining the distribution of photoelectrons among Landau channels. (This distribution hasn't yet been investigated experimentally.) The decay of Quasi-Landau states proceeds presumably through a chain of avoided crossings. A prototype example of such a chain has been studied theoretically long ago[12] but its eventual relevance to our context can hardly be conjectured. An open question is whether and how the aggregate action of many level crossings can be formulated as a macroscopic process.

C. Preliminary Results

A preliminary to the program indicated above has recast into semianalytic form the degenerate perturbation treatment of the magnetic field action on Rydberg states with a fixed quantum number n. This step was stimulated by striking features of the plot of the eigenvalue spectrum of this treament shown in Fig. 1 and obtained by numerical diagonalization of the magnetic energy matrix. The spectrum consists of two analogous branches with opposite curvatures and with different scales, features that could presumably be devied analytically.

The matrix of magnetic energy has indeed a simple analytic tridiagonal form derivable from the algebra of a Lie group. Its diagonalization can be obtained by solving a second order difference equation with a remarkable conjugation property. The lower branch of the eigenvalue plot in Fig. 1 represents the eigenvalues of an oscillator-like equation with the potential shown in the lower branch of Fig. 2. The upper branch of the eigenvalue plot pertains instead to an "inverted oscillator" problem, with the potential in the upper portion of Fig. 2, obtained by a simple analytical transformation of the equation's eigenvectors.[13]

The eigenvalues and eigenvectors of the energy matrix are represented approximately by standard analytical expressions with parameters calculated and adjusted according to Ref. 11. The eigenfunctions are represented, in terms of the matrix eigenvectors $a_{\ell 1}$ and of hydrogenic wave functions $f_{n\ell}(r)P_\ell(\cos\theta)$, by

$$\Psi(r,\theta) = \sum_\ell a_\ell f_{n\ell}(r)P_\ell(\cos\theta). \tag{5}$$

A very limited mapping of these wave functions has been obtained showing that those with lowest and with highest eigenvalues have the characteristics of Landau and Quasi-Landau functions respectively. A more systematic study of the wave functions (5) is in progress.

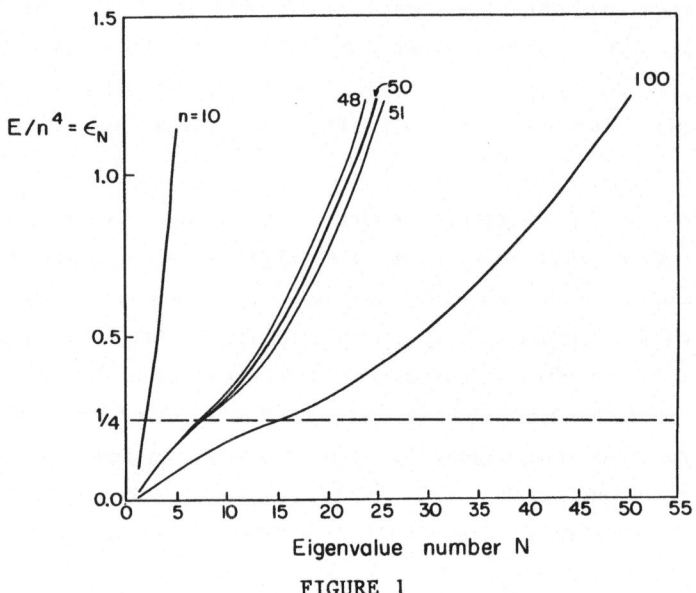

FIGURE 1

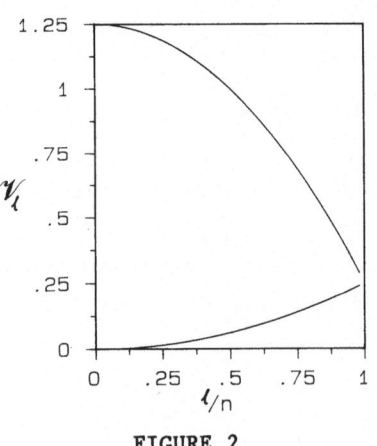

FIGURE 2

REFERENCES

1. C. W. Clark, K. T. Lu and A. F. Starace, in "Progress in Atomic Spectroscopy," Part C, ed. by H. J. Beyer and H. Kleinpoppen (Plenum, New York, 1984) p. 247.

2. D. Wintgen and H. Friedrich, J. Phys. B **19**, 1261 (1986).

3. A. Holle et al., Phys. Rev. Lett. **56**, 2594 (1986) and Z. Phys. D **5**, 279 (1987); J. Main et al., Phys. Rev. Lett. **57**, 2789 (1986) and Z. Phys. D, in press.

4. D. Wintgen and H. Friedrich, Phys. Rev. Lett. **57**, 571 (1986); D. Delande and J. C. Gay, ibid, p. 2006.

5. D. Wintgen and H. Friedrich, J. Phys. B **19**, L557 (1986).

6. N. DeLeon and E. J. Heller, Phys. Rev. A **30**, 5 (1984); S. Watanabe and P. F. O'Mahony, J. Phys. B **20**, 223 (1987).

7. U. Fano in "Festschrift in Honour of Val Telegdi," ed. K. Winter (North Holland, Amsterdam, 1987).

8. G. Wunner, this volume and Ref. 3.

9. U. Fano, Phys. Rev. A **24**, 2402 (1981).

10. U. Fano in "Semiclassical Descriptions of Atomic and Nuclear Collisions," ed. J. Bang and J. de Boer (Elsevier, Amsterdam, 1985) p. 367.

11. F. Robicheaux et al., Phys. Rev. A **35**, 3619 (1987).

12. Yu. N. Demkov and I. V. Komarov, Zh. Eksp. Teor. Fiz. **50**, 286 (1966) [Sov. Phys. JETP **23**, 189 (1966)].

13. U. Fano, F. Robicheaux and A.R.P. Rau, submitted to Phys. Rev. A, July 1987; also Abstracts, XV ICPEAC, Brighton, July 1987.

ACKNOWLEDGEMENTS

The preparation of this paper was supported by the NSF Grant PHY 86-10129 and by travel support by the XV ICPEAC and by the II Conf. on Atomic Spectra and Collisions in External Fields. I am indebted to P. F. O'Mahony and F. Robicheaux for useful remarks.

MAGNETIC ROTATION SPECTROSCOPY WITH SYNCHROTRONS AND WITH LASERS

J. P. Connerade,[a,b] W.R.S. Garton,[a,b]
J. Hormes,[b] Ma-Hui,[a] Shen-Ning[a] and T. A. Starakas[a]

[a]Blackett Laboratory, Imperial College, London SW7, U.K.
[b]Physikalisches Institut Universität Bonn
Nussallee 12, 53 Bonn, West Germany

The present paper constitutes an interim report on experiments being conducted at Imperial College and at the Physikalisches Institut of the University of Bonn. These experiments involve high magnetic fields but, in contrast with most experimental investigations which have been concerned with structure, aim to reveal modifications in atomic f-values due to an externally applied magnetic field. It is perhaps fitting to recall here that, when our project was initiated, only one contemporaneous study of the change of oscillator strength as a function of applied field strength was known to us, namely the theoretical work of Professor M.R.C. McDowell and his collaborators.[1]

We have set up two classes of experiments to measure atomic f-values accurately in the presence of a magnetic field. The first of these is based on synchrotron radiation and the second on an ultraviolet laser. Since the experiments are closely similar, they provide us with an interesting practical comparison between these two light sources, which have complementary properties.

The first experiment was set up at the 500 MeV synchrotron of the Physikalisches Institut, at the University of Bonn. Since it has already been reported in the literature,[2] we give only a brief description here, and refer to earlier papers[3,4] for a full description of the theory on which the method is based. The general layout of the experiment is shown in Fig. 1. The optical arrangement is the classical one with crossed polarizer and analyzer for the observation of Faraday rotation dispersed as a function of wavelength. In the present instance, the polarizer is polarized in the plane of the orbit,[5] while the analyzer is a high

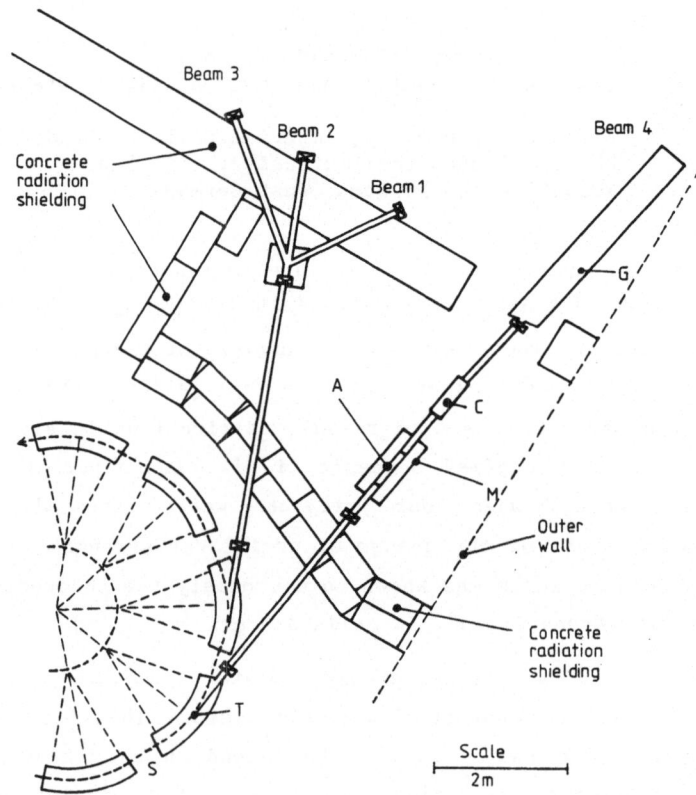

Fig. 1 Experimental setup for magneto-optical studies at the 500 MeV electron synchrotron in Bonn.

dispersion holographic grating which we have found to possess a marked sensitivity to the polarization of the incident light.

Near an absorption line, the rotation angle ϕ is a strong function of the detuning. Since the transmission through the crossed polarizer and analyzer depends on $\sin^2\phi$, one observes modulations of the intensity on both sides of the Zeeman pattern which provide a direct, self calibrated measure of the rotation angle as the field-free line center is approached: the maxima and minima occur at $\frac{\pi}{2}$, π, $\frac{3\pi}{2}$ ... (see Fig. 2). Notice that no actual measurement of the intensity in the far wing is necessary to locate these extrema and that, since the rotation angle accumulates along the atomic column, its value is independent of the optical depth.

While a rotation measurement in the far wing is attractive for its simplicity (see the formula of Mitchell and Zemansky), it is not a sufficient basis for an accurate method of determining the f-values of weak lines (i) because the departure from the far-wing approximation is greatest for the oscillations approaching the Zeeman pattern, where can be measured accurately, while the far-wing theory applies well away from the Zeeman pattern, where the measurement of _____ is much less accurate and (ii) because the pattern of intensity oscillations often merges with the Zeeman pattern, overlapping with it completely in energy, especially for the weaker lines, so that the far wing approximation again breaks down.

For both reasons, we therefore set up a complete theory, which applies over the whole extent of the line profile, by combining the effects of magneto-optical rotation (the Faraday effect) at arbitrary detuning. This is readily achieved by calculating the absorption coefficients:

$$a_\pm = \frac{e^2 Nf}{mc} \frac{\Gamma/4\pi}{(\nu_o - \nu \pm \alpha)^2 + (\Gamma/4\pi)^2}$$

in a standard notation, as well as the refractive indices

$$n_\pm - 1 = \frac{e^2 Nf}{4\pi m} \frac{\nu_o - \nu \pm \alpha}{(\nu_o - \nu \pm \alpha)^2 + (\Gamma/4\pi)^2}$$

over the whole profile, and then combining them in the intensity formula:

$$I = \frac{I_o}{4} \left\{ \left(e^{-\frac{a_+ Z}{2}} - e^{-\frac{a_- Z}{2}} \right)^2 + 4e^{-\frac{(a_+ + a_-)Z}{2}} \sin^2\phi \right\}$$

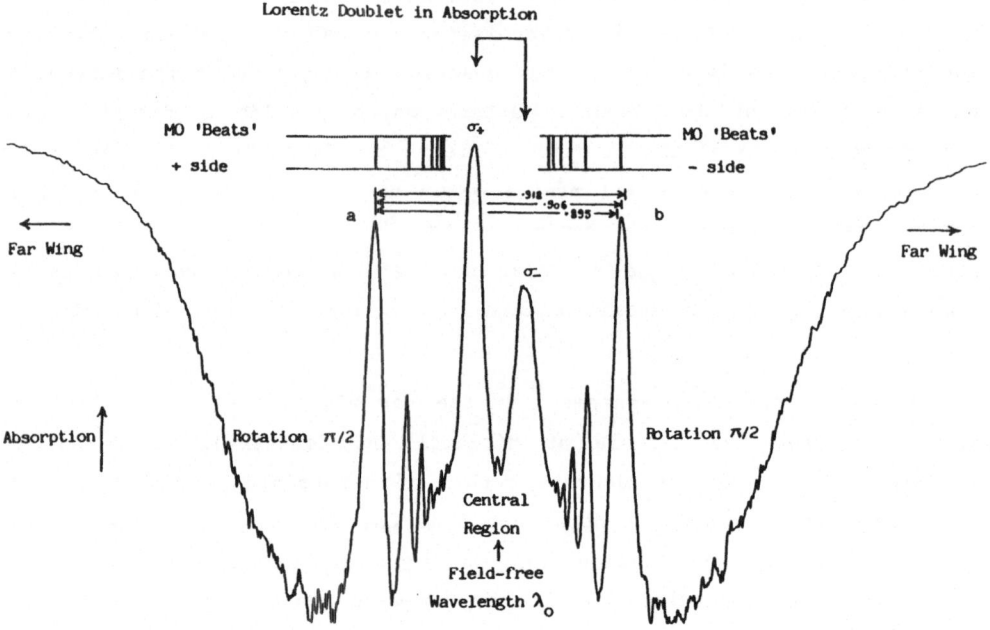

Fig. 2 A typical magneto-optical pattern as recorded in synchrotron radiation experiments in Bonn.

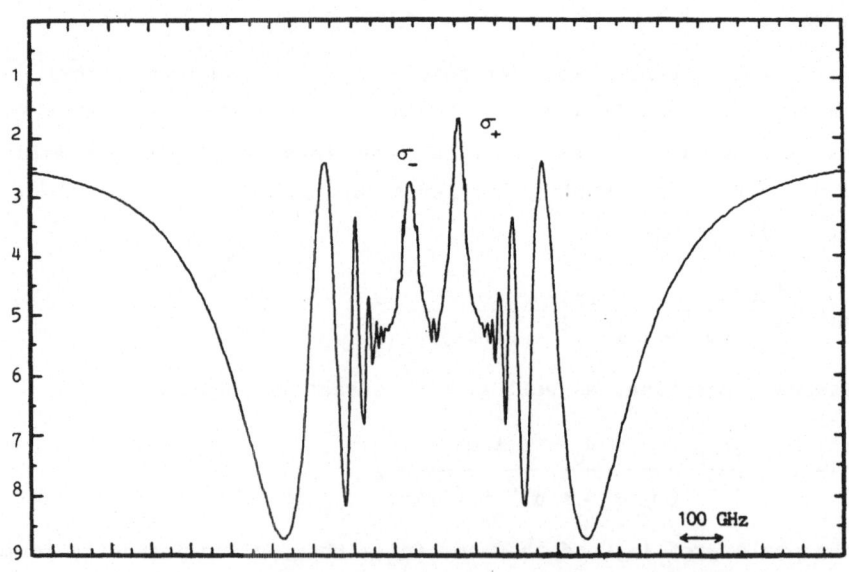

Fig. 3 A simulated spectrum obtained by the computational procedure outlined in the text.

in a conventional notation (see e.g. Ref. 3) to compute simulated spectra (see e.g. Fig. 3). The computation is facilitated by exploiting the Kramers-Kronig relations and computing a complex error function at each point of the profile.[5]

As a by-product, the more sophisticated theory also yields an improved explicit formula for ϕ, applicable over a wide range:

$$\phi(\nu) \approx - \frac{Nf\ell Be^3}{8\pi m^2 c^2} \frac{\nu(\nu - 2\nu_o)}{(\nu_o^2 - \alpha^2)[(\nu - \nu_o)^2 - \alpha^2]}$$

which is superior to the far-wing approximation.

Using the approach just outlined, and the exact form of ϕ one finds that the patterns are extremely sensitive in the region near the line center, while they vary slowly in the wings as a function of oscillator strength (see Fig. 4). This is the basis of the Magneto-Optical Venier (MOV) method, in which a measurement is made in the fast changing cycles at the center of the pattern and the cycle is determined by a measurement further out into the wings, thus providing high accuracy and sensitivity.[5]

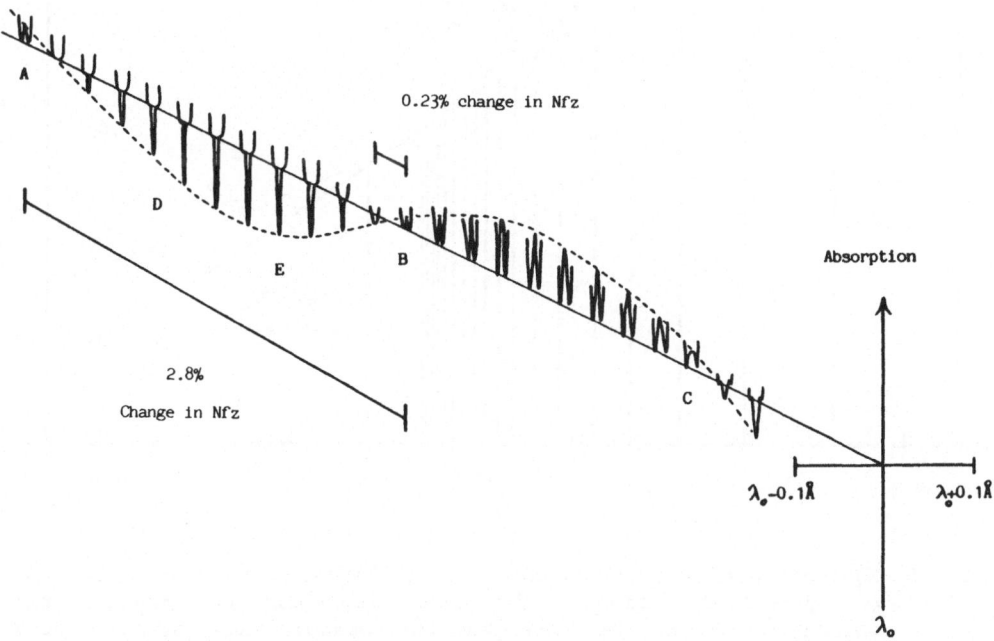

Fig. 4   A plot of the modulations at the center of the magneto-optical pattern, illustrating the principle of the MOV method.

When the data for the principal series of Sr I obtained by the MOV methods are plotted in the style of quantum defect theory (see Fig. 5) one expects a straight line plot, but two departures from linearity are observed. The first, near n = 18, is readily accounted for as additional rotation due to a calcium impurity line (calcium is always present as an impurity in strontium metal, even after extensive purification).

The second departure is more interesting: at high n-values, we find a <u>decrease</u> in f-values as compared to the expected linear law. We had previously speculated that this might be due to the influence of the magnetic field itself on the atomic f-values. This explanation has recently received spectacular confirmation through the theoretical study of the Quadratic Zeeman Effect in Non-hydrogenic Systems by O'Mahony and Taylor,[6] who have found that, around n = 29 there is "a much more drastic redistribution of oscillator strength ... than in the vapour experiments."

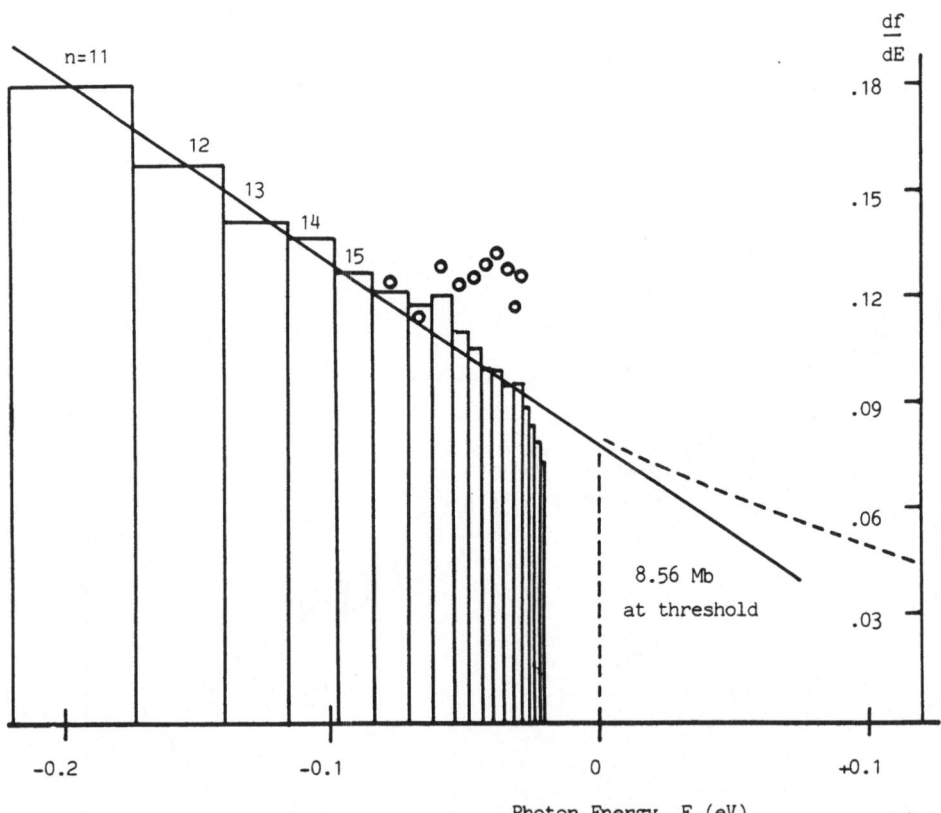

Fig. 5 MQDT-style plot of the course of oscillator strengths in SrI for the principal series. The boxes represent our experimental values; open circles are earlier measurements by the hook method. Note the departure from a straight line around n = 18 and again at high values of n, as discussed in the text.

The latter experiments are of course the photo absorption studies reported by Lu, Tomkins and Garton,[7] who used high vapour pressures to observe the weak satellite structure, and therefore produced data which are distorted by optical thickness effects near the strong parent lines.

Our data confirm the conclusion reached theoretically by O'Mahony and Taylor[6]: around n = 29, oscillator strength is "peeled off" the parent lines (which we observed in our rotation studies) and lost to the satellite structure (which is not included in our simulations of the rotation spectra). In principle, we therefore have set up a méthod of measuring the changes in f-value due to a strong magnetic field.

The second experiment is an attempt to extend our synchrotron method by exploiting a tunable laser as the light source and replacing the superconducting magnet by a pulsed Bitter Coil. The advantages of this approach are (i) that higher resolving power is available from a laser than from the combination of synchrotron radiation with a high dispersion holographic grating and (ii) that a pulsed Bitter coil can readily be used to reach rather higher magnetic fields than a superconducting solenoid. Unfortunately, there are also disadvantages in our second experiment, which are not immediately obvious and to which we return below.

Our experimental setup is illustrated in Fig. 6, which is an overall block diagram of the laser, the control system and pulsed Bitter coil. The period of the capacitor bank and coil is altogether of the order of milliseconds, i.e. much longer than the duration of the laser pulse (several nanoseconds) so that the magnetic field is essentially constant during the time of observation.

Much attention has been devoted to the quality of the pulsed laser system, since most of the advantage of laser operation is connected with a narrow linewidth. With a pulsed laser, under the conditions we are working in, the theoretical optimum line width is 600 MHz. We have in fact achieved about twice this figure. Our setup is illustrated in Fig. 7. The excimer-pumped dye laser oscillator is built inside a pressure vessel, which also contains an intracavity Fabry-Pérot etalon. The cavity is of the Littman design, with coarse tuning by mechanical movement and fine tuning over 3Å by varying the pressure in the cell. Only in the fine tuning range can we preserve an essentially single mode structure giving 1.2 GHz resolution, the optimum our system can achieve. The spectral purity of the laser is checked during the experiment by passing the light down a 3 meter spectrometer crossed with a Fabry-Pérot etalon.

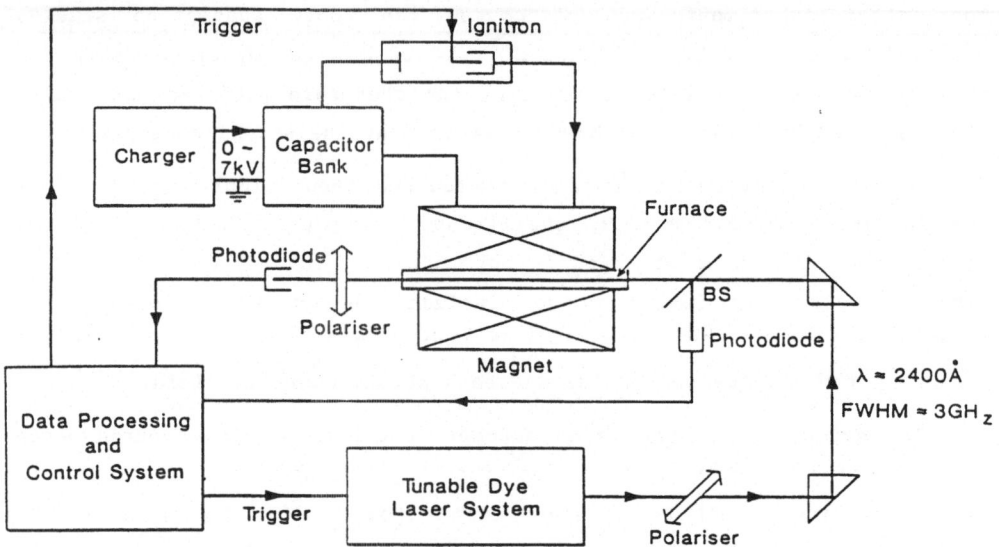

Fig. 6 Overall diagram of the laser and pulsed field system.

The quality of the overall system can be judged by the fact that we were able (see Fig. 8) to observe the spectrum of the principal series of Ba I up to n = 125. For a pulsed laser system, this represents a good experimental resolution, and is certainly superior to the resolution so far achieved by classical spectroscopy using synchrotron radiation.

The laser is then synchronized with the pulsed magnet, and the whole system is computer controlled so that spectra can be accumulated shot by shot. In general, we arrange that, for each step in wavelength of the laser, several measurements are made at different field strengths by varying the delay between the magnet and the laser pulse. This ensures that any long term drift in the furnace conditions does not influence our measurements of field dependence.

At this point it is perhaps worth remarking that the resolution of the whole system does not only depend on the quality of the laser profile in a single shot. The apparatus function also includes shot to shot instabilities of the laser intensity, of the magnetic field circuit and of the detector response. In spite of all these uncertainties, the resulting magneto-optical patterns (see e.g. Fig. 9) demonstrate a significant improvement in resolution and in observed detail as compared with synchrotron data. Also, since the detector is linear, our experimental spectra bear a very close resemblance to the simulated spectra which we have calculated (see e.g. Fig. 10).

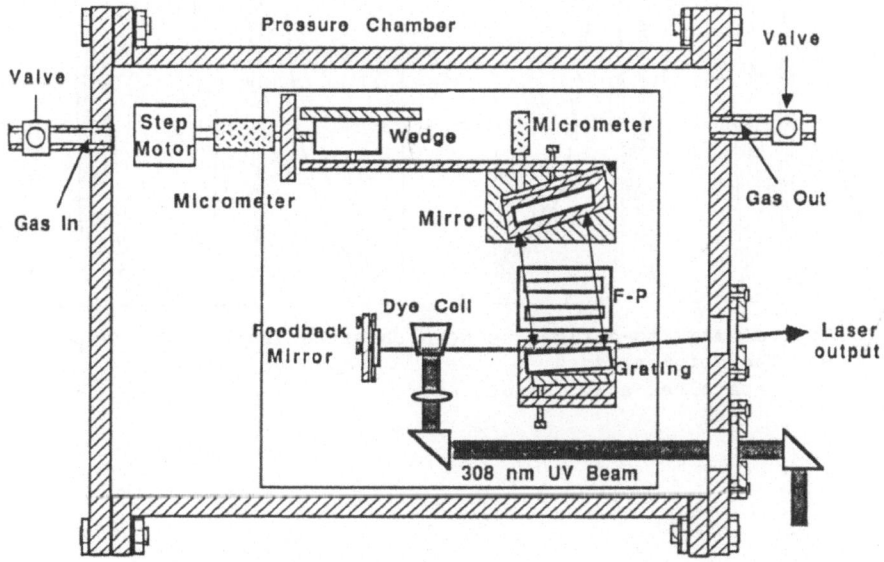

Fig. 7 Construction of the pressure tuned dye laser system.

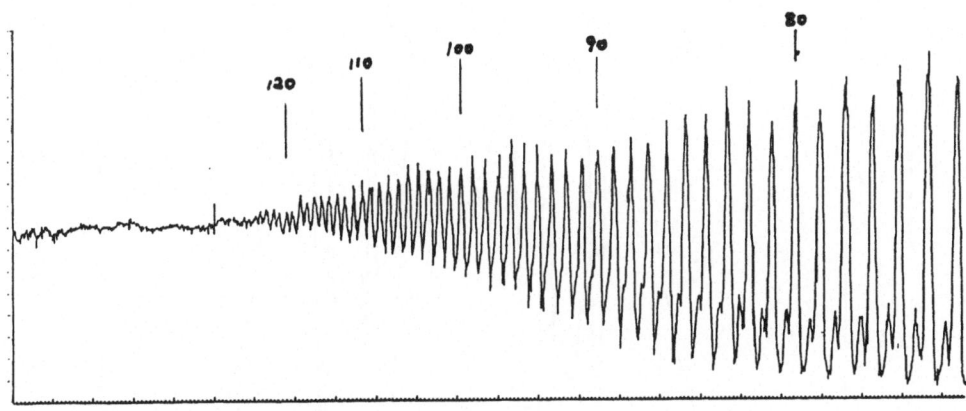

Fig. 8 Showing high series members in the principal series of BaI observed with the pulsed laser.

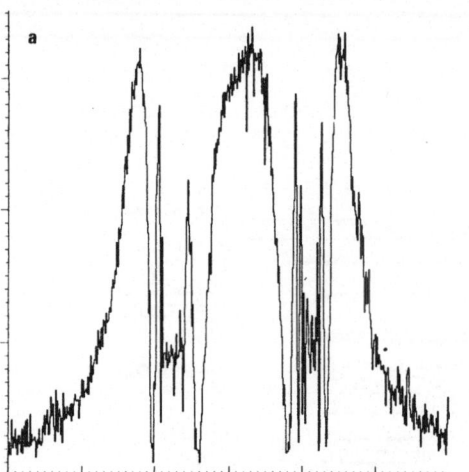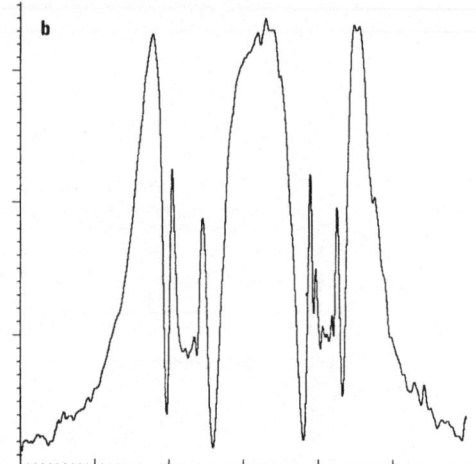

Fig. 9 Typical magneto-optical pattern observed in the laser experiment.
a) The original data
b) Data after a 353-smoothing.

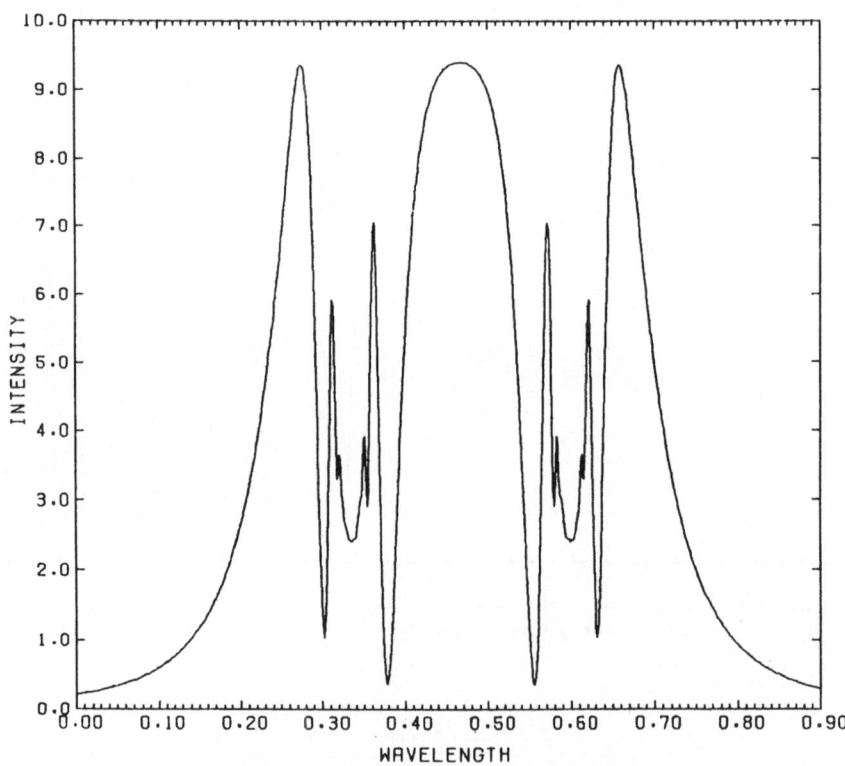

n=17 B=4.8 Tesla   Nfl=2.76*10^{13}

Fig. 10 Simulated spectrum for the experimental conditions in Fig. 9.

As remarked at the outset, this is an interim report on a continuing investigation: we have extended measurements of the f-values in Ba I to much higher n values in the principal series than previously, and we are studying field variations both in the f-values and in the profile of autoionizing lines.

In the meantime, we can already compare the two exponential approaches. It is clear that:

(i) For a given experimental setup, the synchrotron covers a much wider wavelength range than is readily accessible to the laser.

(ii) The synchrotron experiment can be performed very quickly. Under favourable conditions, some twenty transitions can be recorded at a single field strength in about 10 minutes, thus eliminating variations due to instabilities of furnace conditions, particularly as detection is performed in parallel at all wavelengths (multiplex advantage). Against this is the fact that, for optimum resolution, the detector has a nonlinear response. With the laser, on the other hand, many shots are required to cover a single pattern (up to three hours experimental time) but the detection is linear and several field strengths can be covered in one laser sweep under identical furnace conditions.

(iii) The spectral resolution available with the laser is typically up to an order of magnitude better than with the synchrotron + spectrograph.

(iv) Pulsed operation opens up new prospects for the study of atoms at very high field strengths.

ACKNOWLEDGEMENTS

It is a pleasure to acknowledge many useful discussions with Drs. Taylor and O'Mahony during the course of this work. Financial support was received from the SERC (U.K.) and from the B.M.F.T. under special funds for synchrotron radiation research.

REFERENCES

1. S. M. Kara and M.R.C. McDowell, J. Phys. At. Mol. Phys. 11, 1719 (1981).

2. J. P. Connerade, W.R.S. Garton, M. A. Baig, J. Hormes, T. A. Stavrakas and B. Alexa, J. de Physique C2 43, 317 (1982).

3. J. P. Connerade, J. Phys. B. At. Mol. Phys. 16, 399 (1983).

4. J. P. Connerade, T. A. Stavrakas and M. A. Baig, <u>Proceedings Scottish Summer School</u> (1985).

5. G. A. Schott, "Electromagnetic radiation and the mechanical reactions arising from it," <u>Adams Prize Essay</u> (Cambridge University Press, 1912).

6. P. F. O'Mahony and K. T. Taylor, Phys. Rev. Lett. <u>57</u>, 2931 (1986).

7. K. T. Lu, F. S. Tomkins and W.R.S. Garton, Proc. Roy. Soc. <u>A362</u>, 421 (1978).

PHOTOABSORPTION AND PHOTOIONIZATON OF AN ATOM IN A MAGNETIC FIELD

P. F. O'Mahony, M. A. Al-Laithy and K. T. Taylor

Department of Mathematics
Royal Holloway and Bedford New College
(University of London)
Egham, Surrey TW20 0EX, England

1. INTRODUCTION

There have been many new developments in the analysis of the quadratic Zeeman effect, both in theory and experiment since the last meeting (Nayfeh and Clark.[1] New high resolution experiments on hydrogen have revealed a dense series of sharp lines near and above the ionization threshold by Holle et al.,[2] and Main et al.[3] showing that the lower resolution quasi-Landau modulations of Garton and Tomkins[4] are full of structure. A Fourier transform of the absorbtion spectrum also revealed new modulations. Theoretically the large basis set calcultions for hydrogen by Clark and Taylor[5] have been taken higher in energy towards threshold by Wintgen and Freidrich,[6] Delande and Gay,[7] and Wunner et al.[8] There has followed a parallel development in the semi-classical treatment of the problem by Al-Laithy et al.,[9] and Wintgen and Freidrich.[10] The quadratic Zeeman effect for atoms has proved to be a physically realizable quantum system whose classical Hamiltonian exhibits chaotic motion. The consequences of this will be discussed in detail in Section 3, but initially in Section 2 we will describe a new approach which extends the basis set methods to non-hydrogenic systems (O'Mahony and Taylor.[11]

2. NON-HYDROGENIC ATOMS IN THE INTER-ℓ AND INTER-n MIXING REGIMES

The basis set methods for hydrogen are not readily applicable to non-hydrogenic atoms because of the multi-electron core potentials. Recently a method has been developed to extend the hydrogenic calculations to multi-electron atoms.[11] The hydrogenic results are recovered when the multi-electron potentials vanish.

For a typical laboratory strength magnetic field of 47kG we can divide the space into different regions. (The radial distances are given in atomic units.)

(I) a complicated multi-electron region where field effects are negligible, $0 < r < 10$ ($=r_I$)

(II) a Coulomb field only, $10 < r < 500$ ($= r_{II}$)

(III) Coulomb plus external field $500 < r < \infty$

The value of r_{II} could of course be reduced for stronger fields. In nonhydrogenic systems we have to describe the motion of the photoelectron in the complicated multi-electron core (I). However the effect of this region on the electron escaping into region II is embodied in the phase shift δ_ℓ for the ℓth partial wave. Each spherical component of the wavefunction in region II can thus be written as a phase shifted Coulomb function as follows.

$$\Psi_{\epsilon\ell m}(\vec{r}) = f_{\epsilon\ell m}(\vec{r})\cos\delta_t - g_{\epsilon\ell m}(\vec{r})\sin\delta_t \tag{1}$$

where $f_\ell(\vec{r})$ and $g_\ell(\vec{r})$ are the regular and irregular Coulomb solutions respectively.

The phase shift $\delta_\ell = \pi\mu_\ell$ where μ_ℓ is the quantum defect. $\mu_\ell = 0$ for $\ell < 4$ for most atoms.

In region III the electron's motion is described by the "hydrogenic" Hamiltonian for the quadratic Zeeman effect

$$[-\frac{1}{2}\nabla^2 - \frac{1}{r} + \beta L_z + \frac{1}{2}\beta^2 r^2 \sin^2\theta]\Psi = E\Psi \tag{2}$$

as the atomic potentials are negligible in this region. We construct solutions in region III by diagonalizing the full quadratic Zeeman Hamiltonian Eq. (2) over the limited region $r_{II} < r < \infty$. A basis set of Sturmian functions is used over this region. These satisfy the correct boundary conditions at infinity as they decay exponentially. The radial integrals that occur in evaluating the Hamiltonian in the basis of Sturmian functions can be evaluated analytically. Using the eigenvalues and eigenvectors that result from the diagonalization over this outer region we can construct a logarithmic derivative matrix or an R-matrix $R_{\ell\ell'}$ (the orbital angular momentum acting as a channel index) at r_{II}. $\underline{R}$ gives the relationship between a function and its derivative.

$$\vec{F}(r_{II}) = \underline{R}(E) \frac{d\vec{F}}{dr}(r_{II}) \tag{3}$$

We know from region II that the wavefunction $\vec{F}$ is some linear combination of the solutions (1) therefore

$$\vec{F}(r_{II}) = \underline{P}\vec{A} \tag{4}$$

where $\underline{P}$ is a diagonal matrix with diagonal elements equal to (1). $\vec{A}$ are coefficients which are determined by the matching at r_{II}. Substituting (4) in (3) we obtain

$$(\underline{P} - R(E)\underline{P'})\vec{A} = 0 \tag{5}$$

This equation has solutions, or in other words the logarithmic derivatives match, when the determinant of the term in brackets in (5) vanishes. This occurs only at certain discrete energies E_n. A search for the zero's of the determinant gives the bound state energies E_n. An efficient algorithm to search for the zeros of the determinant has been developed by Seaton.[12] $\vec{A}$ is subsequently obtained from (5) and the nth eigenfunction is thus completely determined. The oscillator strength is given by the field free oscillator strength modulated by the square of the amplitude factor A. The μ_ℓ's and field free oscillator strength needed for the calculations with magnetic field can be calculated ab initio for light atoms or taken from experiment for heavier atoms.

The calculated spectra are compared with experiment for strontium and barium in Figs. 1 and 2.

Figures 1(a) and 2(a) are the ground state photoabsorption spectra in zero magnetic field showing the Rydberg series converging to the ionization threshold. Figures 1(b) and 2(b) are the experimental spectra in a field of 4.7 Tesla.[4] Figures 1(c) and 2(c) show the theoretical spectra using the above theory. The agreement between experiment and theory is excellent. There is a big difference between thes pectra of barium and strontium in the inter-ℓ and inter-n mixing regions of the spectrum. This is due to the different behavior of their significant quantum defects. In strontium the $\ell = 1$ and $\ell = 3$ quantum defect, μ_p and μ_f, are very different: $\mu_p \sim 0.72$ and $\mu_f \sim 0.08$.[13] The "p" states therefore stand out from the manifold and hardly mix with the other states in the energy range below 29p. In fact the principal line in

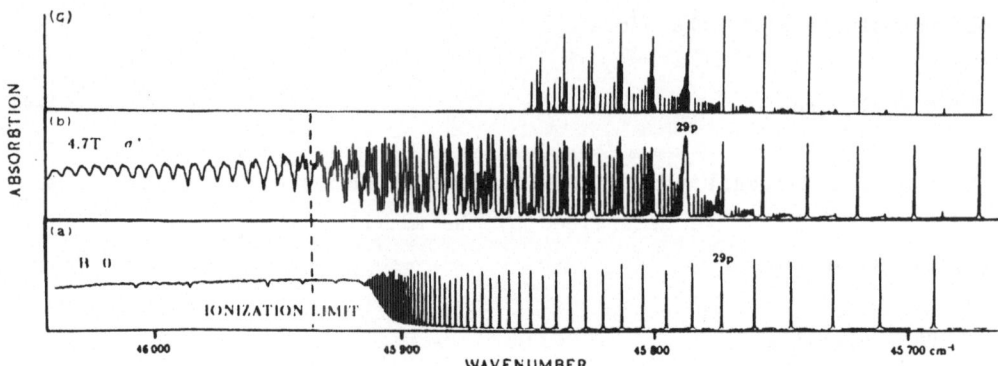

Figure 1   Strontium absoprtion spectrum vs transition wave number from
the ground state: (a) field-free spectrum: (b) experimental
densitometer tracing for strontium in a magnetic field of
4.7 T: (c) theoretical photoabsorption spectrum in a field of
4.7 T. The theoretical results give the absoulte oscillator
strengths, but for n less than 29 the strongest lines have
been reduced in size to facilitate comparison with the
nonabsolute experimental measurements.

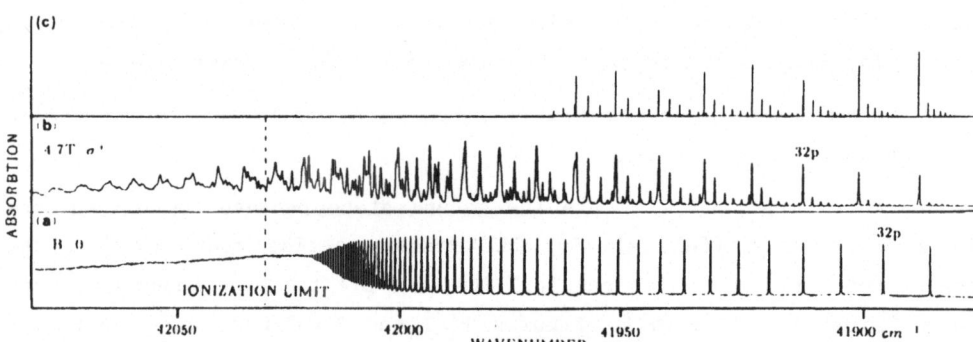

Figure 2   Barium absorption spectrum vs transition wave number from the
ground state: (a) field-free spectrum: (b) experimental
densitometer tracing for barium in a magnetic field of 4.7 T:
(c) theoretical photoabsorption spectrum in a field of 4.7 T.

Fig. 1(c), below 29p, is much larger than indicated and is truncated to facilitate the comparison with experiment. Barium on the other hand, being strongly perturbed by the configuration 5d8p near threshold, has quantum defects $\mu_p \sim 0.92$, $\mu_f \sim 1.00$ for n $\sim$ 32.[14-15] Therefore the quantum defects are almost zero (modulo 1) and this explains the "hydrogenic" pattern seen in barium. Strontium is more typical of the general atom. The field free oscillator strengths used in the above calculations were taken from experiment for strontium[16] and barium.[17] The calculations described could be pushed closer to threshold as was done in hydrogen by Holle et al.[18] but this has not been pursued.

3. NEAR THRESHOLD PHOTOABSORPTION AND PHOTOIONIZATION FOR A HYDROGEN ATOM IN A MAGNETIC FIELD

The spectra in Figs. 1 and 2 become increasingly more complicated as one approaches the ionization threshold yet above threshokld one recovers a simple modulation witha spacing of about 1.5 $\hbar\omega_c$. These so-called quasi-Landau modulations were believed to dominate the spectrum in the threshold region and they were associated with wavefunctions localized orthogonal to the field direction, i.e. about the z = 0 plane. However the recent high resolution experiments on hydrogen[2-3] have shown the above to be an over simplification of the problem. The spectra contain many sharp lines above and below the ionization threshold with no obvious modulation of 1.5 $\hbar\omega_c$. Main et al.[3] analyzed their spectra by taking the real part of the Fourier transform of the cross section. This is shown in Fig. 3.

The peak marked (1) corresponds to a frequency (=2π/T) of about 1.5 $\hbar\omega_c$. This peak is absent in Fig. 3(b) because the corresponding spectrum is $m^\pi = (-1)^1$ and is odd in parity about the z = 0 plane. (Parity, π, here refers to reflection symmetry about the z = 0 plane.) Many other frequencies are present however in Fig. 3. These new frequencies have been found theoretically in an analysis of the classical Hamiltonian.[3,9] The Hamiltonian can be written in a scaled form as follows

$$H = \frac{1}{2}\left[P_r^2 + \frac{P_\theta^2}{r^2} + \frac{L_z^2}{r^2\sin^2\theta}\right] + \frac{1}{2}r^2\sin^2\theta - \frac{1}{2r} = E \qquad (6)$$

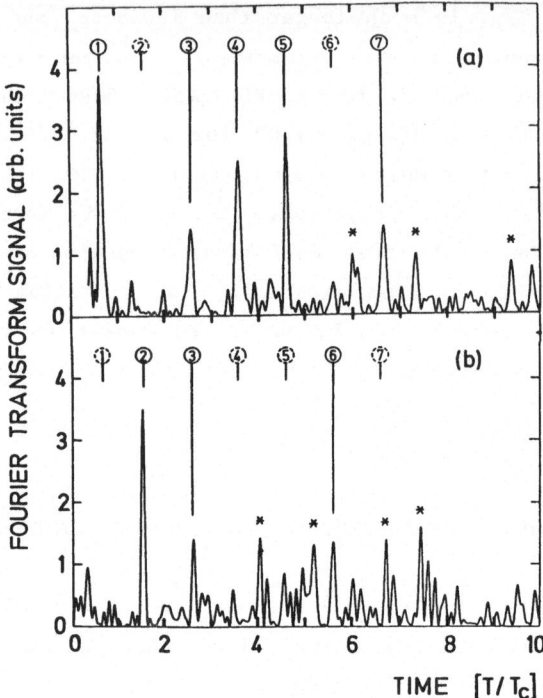

Figure 3 Absolute value squared of the Fourier transform of the absorption spectrum for hydrogen in a field of 5.96 T as a function of time measured in units of the cyclotron period. Two different final states are shown (a) $m^\pi = (0)^0$ and (b) $m^\pi = (-1)^1$.

The scaled energy is $E = CB^{-2/3}\varepsilon$ where C is a constant depending on the units used, ε is the unscaled energy and B is the magnetic field. As E increases towards threshold the classical motion becomes irregular. However initially new torii are created which have periodic orbits at their center. There are many such torii but we concentrated on those trajectories which had periodic orbits that begin at the nucleus as these should be the most relevant to photoabsorption experiments from low lying excited states. These periodic orbits are shown in Fig. 4.

The frequencies associated (in units of $\hbar\omega_c$) with these orbits are 0.64, 0.39, 0.28, 0.22, 0.18, 0.15, etc. These are exactly the frequencies labelled 2, 3, 4, 5, 6, 7 in Fig. 3 and interpreted independently by Main et al.[3] In addition we show in Fig. 5 the

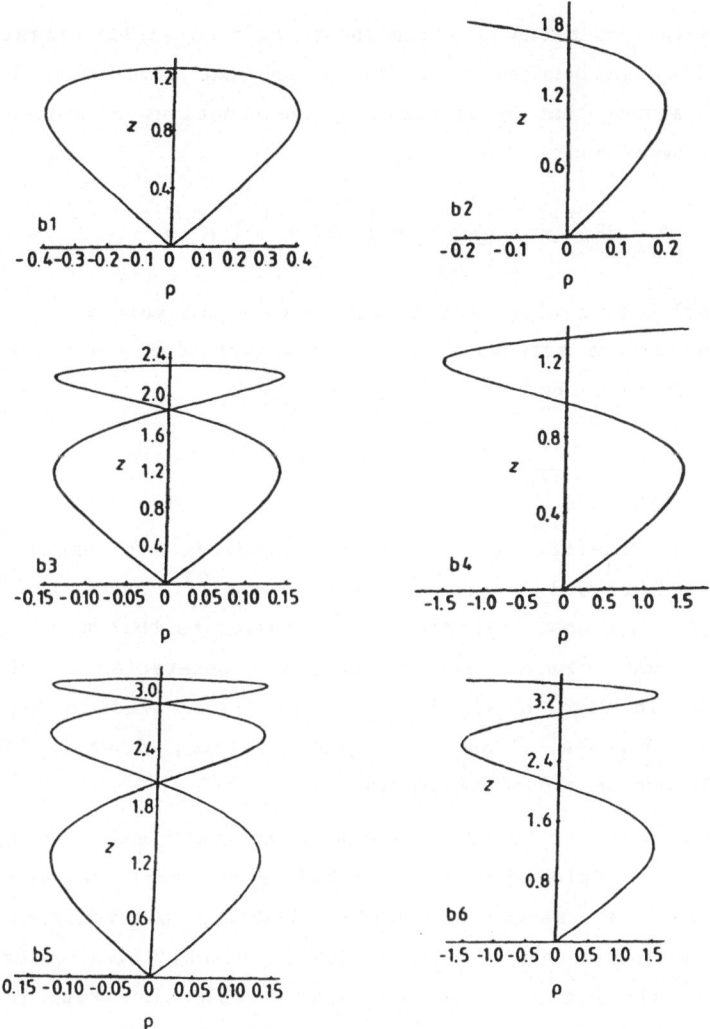

Figure 4 Six classical periodic orbits plotted as a function of ρ and z.

energy dependence of the frequencies of the orbits and at which energy each type of orbit begins.[9]

The orbits have a very weak dependence on energy. This is crucial as modulations which change rapidly as a function of energy would be averaged over in a Fourier transform spectrum.

The energy positions at which these begin coincides classically with an instability in bundles of trajectories that move along the z-axis. Sumetskii[19] showed that by linearizing the equations of motion about the z-axis i.e. by expanding V as follows

$$V(\rho,z) = +\frac{1}{2}\beta^2\rho^2 - \frac{1}{\sqrt{\rho^2 + z^2}} \approx \frac{1}{2}(\beta^2 + \frac{1}{z^3})\rho^2 - \frac{1}{z} \tag{7}$$

the classical motion along the z-axis becomes unstable as E is increased and then oscillates between stability and instability until E = 0. The instability energies he found to be given by

$$E_n = -\frac{1}{2}\frac{1}{2(n-\alpha)^{2/3}} \tag{8}$$

where n is an integer greater than zero. Using some approximations he found[20] $\alpha = 2/3$ and Al-Laithy and Farmer[21] have found $\alpha \approx 0.4$ numerically). The most important point however is that the instabilities in the z motion coincide exactly with the generation of the periodic orbits shown in Figs. 4 and 5. This is illustrated in Fig. 6, which shows the bifurcation of the new periodic orbits, shown in Fig. 4, from the unstable motion along the z-axis.

The new periodic orbits are stable initialy and then they become unstable. They period double again before the orbit becomes unstable. One then gets a cascade of period doubling bifurcations which is difficult to follow numerically. These instabilities occur when the frequency of the motion in ρ, frequencies which are typical of quasi-Landau states, become degenerate with frequencies of the Coulomb motion in z. Then one has coupling between the modes creating the instability and the new orbits. The 2/3 power in Eq. (8) therefore just comes from the energy dependence of the frequency of the Coulomb motion in z, which goes as $\omega_z \sim E^{3/2}$, the frequency in ρ, ω_ρ, being constant and independent of energy. Quantum mechanically the above phenomena can be viewed as follows: as one approaches threshold the quasi conservation of the K quantum number breaks down and the quasi-Landau modes in ρ mix with the states concentrated to zero order around the z-axis.

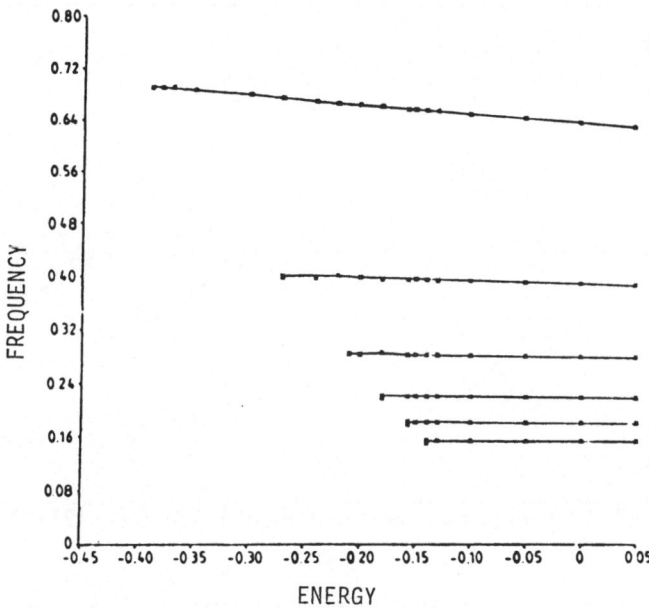

Figure 5. Classical frequency, in units of $\hbar\omega_c$, as a function of scaled energy E for each orbit shown in Fig. 4.

In pursuit of a quantum mechanical solution to the problem we extend the ideas of Section 2. That is we define a third region $r_{II} < r < r_{III}$ and a fourth region $r_{III} < r < \infty$ where the potential can be expanded as in (7). The solution in region IV can be written in cylindrical coordinates as

$$\Psi = \sum_i \Phi_i(\rho,\phi)F_i(z) \tag{9}$$

where Φ_i is the ith Landau state. The solutions in region III, in spherical coordinates, are then matched to those in IV at some $z = z_0$. The problem is complicated due to the number of channels required and the necessity for matching at large radial distances. This work is in progress.

Over the last few years most work in this field has been on the hydrogen atom and it is clear that each spectral portion of this fundamental system must be thoroughly understood before theoretical progress can be made for the corresponding spectral portion of a nonhydrogenic system. The Fourier transform has been taken of the region near the ionization threshold for hydrogen. Repeating this process for nonhydrogenic atoms would help clarify how the multi-electronic core alters the excitation of states corresponding to the frequency regularities in the spectrum in that region.

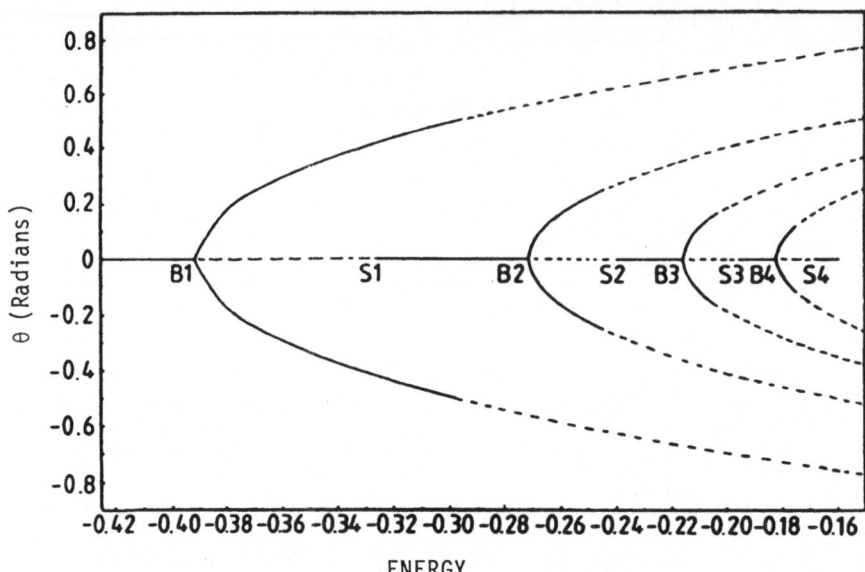

Figure 6   The angle of emission from the origin of periodic trajectories
          as a function of energy. The full line shows where the motion
          is stable whereas the dotted line indicates unstable motion.
          Motion along the z-axis correspond to θ = 0, and this becomes
          unstable at E = 0.39.   The orbit B1 in Fig. 4 is then
          created.   The motion along the Z-axis regains stability at S1
          and the process of stability-instability repeats with the
          generation of the orbits B2, B3, etc. shown in Fig. 4.

ACKNOWLEDGEMENT

The work of P. F. O'M. was supported in part by a grant from the
Science and Engineering Council, U.K.

REFERENCES

1. Nayfeh, M. H. and Clark, C. W., eds. Atomic Excitation and
    Recombination in External Fields (Gordon and Breach, 1985).
2. Holle, A., Wiebusch, Main, J., Rottke, H., Hager, B. and Welge,
    K. H., Phys. Rev. Lett. 56, 2594 (1986).
3. Main, J., Wiebusch, G., Holle, A. and Welge, K. H., Phys. Rev.
    Lett. 57, 2789 (1986).
4. Garton, W.R.S. and Tomkins, F. S., Astrophys. J. 158, 839 (1969a).
5. Clark, C. W. and Taylor, K. T., J. Phys. B 15, 1175 (1982).
6. Wintgen, D. and Friedrich, H., Phys. Rev. Lett. 57, 571 (1986).
7. Delande, D. and Gay, J. C., Phys. Rev. Lett. 57, 2006 (1986).
8. Wunner, G., Woelk, U., Zech, I., Zeller, G., Ertl, T., Geyer, F.,
    Schweizer, W. and Ruder, H., Phys. Rev. Lett. 57, 3261 (1986).
9. Al-Laithy, M. A., O'Mahony, P. F. and Taylor, K. T., J. Phys. B 19
    L773 (1986).
10. Wintgen, D. and Friedrich, H., Phys. Rev. A 36, 131 (1987).
11. O'Mahony, P. F. and Taylor, K. T., Phys. Rev. Lett. 57, 2931
    (1986).
12. Seaton, M. J., J. Phys. B 18, 2111 (1985).
13. Rubbmark, I. R. and Borgstrom, S. A., Physica Scr. 18, 196 (1978).
14. Garton, W.R.S. and Tomkins, F. S., Astrophys. J. 158, 1219 (1969b).

15. Post, B. H., Vassen, W., Hogervost, W., Aymar, M. and Robaux, O., J. Phys. B $\underline{18}$, 187 (1985).
16. Garton, W.R.S., Connerade, J. P., Baig, M. A., Hormes, J. and Alexa, B., J. Phys. B $\underline{16}$, 389 (1983).
17. Parkinson, W. H., Reeves, E. M. and Tomkins, F. S., J. Phys. B $\underline{9}$, 157 (1976).
18. Holle, A., Wiebusch, G., Main, J., Welge, K. H., Zeller, G., Wunner, G., Ertl, T. and Ruder, H., Z. Phys. D $\underline{5}$, 279 (1987).
19. Sumetskii, M. Y., Sov. Phys. J.E.T.P. $\underline{56}$, 959 (1982).
20. Wintgen, D., J. Phys. B $\underline{20}$, L511 (1987).
21. Al-Laithy, M. A. and Farmer, C, J. Phys. B, accepted for publication (1987).

ATOMIC DIAMAGNETISM

Dieter Wintgen[1] and Harald Friedrich[2]

[1]Fakultät für Physik, Universität Freiburg, Hermann-
Herder-Straße 3, D-7800 Freiburg, West Germany
[2]Inst. für Theor. Astrophysik, Universität Tübingen
Auf der Morgenstelle 12c, D-7400 Tübingen, West Germany

ABSTRACT

The hydrogen atom in a uniform magnetic field is a remarkably rich example of a simple but real physical system displaying all the features which are currently causing excitement in the classical and quantum mechanical study of non-integrable systems. The analysis of short range and long range correlations in quantum spectra has led to a deeper understanding of the quantal manifestation of classical chaos, and the appreciation of the role of isolated unstable periodic classical orbits has led to a breakthrough in our understanding of the quasi-Landau phenomenon.

I. H-ATOM IN A UNIFORM MAGNETIC FIELD

The problem of an atom in a uniform magnetic field has recently become one of the most widely studied atomic systems, and it is the subject of almost half of the talks at this conference.[1-10]

The Hamiltonian for a hydrogen atom in a uniform magnetic field is (in cylindrical coordinates and Rydberg units)

$$H = p_\rho^2 + p_z^2 + \ell_z^2/\rho^2 + \frac{1}{4} \gamma^2 \rho^2 - 2\left(\rho^2 + z^2\right)^{-1/2}, \qquad (1)$$

where z (ρ) is the coordinate parallel (perpendicular) to the field and γ is the magnetic field strength measured in units of $B_0 = 2.35 \times 10^5$ T. The trivial paramagnetic term $\gamma \ell_z$ is omitted in Eq. 1. For one electron Rydberg states of more general atoms the potential in (1) is modified near the origin, but this is not important for most of the effects discussed below.

The Hamiltonian (1) is an extremely "simple" example of a non-inte-grable conservative system in two coordinates, and there are several reasons for its great popularity. Because of its simplicity it is amenable to both classical and quantum mechanical ab initio calculations with no approximations bar computer rounding errors. However, the system is complex enough to show such effects as the transition from regular, quasi-periodic motion to completely chaotic classical behaviour (section III) as well as a transition from a quasi-separable to a com-pletely irregular quantum spectrum (section II and III). Furthermore, the Hamiltonian (1) describes a real physical system which can be and has been prepared in the laboratory, so that a state for state comparison between calculated and observed spectra is possible (see section IV).

The study of atoms in a magnetic field received a major thrust in 1969 when Garton and Tomkins discovered almost equidistant peaks in the photoabsorption cross sections of magnetized barium atoms near thresh-old.[11] These so called quasi-Landau peaks, which are separated roughly by 1.5 times the cyclotron energy around threshold, should not be interpreted as individual quantum states but rather as a periodic clustering of states constituting a modulation of the quantum mechanical level spectra and cross sections. Recently the discovery of whole series of such modulations and of their correlation to unstable periodic orbits in the classically chaotic regime has led to a major breathrough in our understanding of the quasi-Landau phenomenon (see section V).

II. APPROXIMATE SEPARABILITY

In 1980/81 Zimmerman et al.[12] and Delande and Gay[13] observed exponentially small anticrossings between (calculated) energy levels of the Hamiltonian (1) and this was interpreted as evidence for the existence of an additional approximate symmetry associated with an approximately conserved quantum number K (in addition to the exactly conserved azimuthal quantum number m and the z-parity π). It was in fact subsequently shown[14-16] that, within a __single__ n-manifold of eigenstates of the field free hydrogen atom, diamagnetic interaction proportional to ρ^2 can, to first order in γ^2, be expressed in terms of the invariant combination

$$\Sigma = 4A^2 - 5A_z^2 \tag{2}$$

of components of the Runge-Lenz-Vector. Quantized eigenstates of the first order invariant (2) can be assigned to two different categories depending on the sign of the eigenvalues σ: rotator states for $\sigma > 0$ and vibrator states for $\sigma > 0$. In this way it was possible to understand some features of the weak-field spectrum (ℓ-mixing regime), e.g. the approximate doublet degeneracy of vibrator states belonging to the same azimuthal quantum number m but to different z-parity π or the different localization in configuration space of wave functions of rotator and vibrator states. However, as emphasized by Delande and Gay,[16] this did not explain the exponentially small anticrossings which were observed between states belonging to adjacent (i.e. different) n-manifolds. Up until now it has not been possible to find a more general expression (valid for higher orders of γ^2) for the operator representing an approximate symmetry, but it has been shown by construction[17] that there is a representation of the Hamiltonian which is approximately separable in a range of energies and field strengths which includes the observed small anticrossings.

This construction is based on transforming the Schrödinger Equation governed by the Hamiltonian (1) via the introduction of dilated semi-parabolic coordinates

$$\mu = (-E)^{1/4} (r + z)^{1/2}, \quad \nu = (-E)^{1/4} (r - z)^{1/2} \tag{3}$$

giving the Schrödinger Equation

$$\left\{ -\Delta_\mu - \Delta_\nu + (\mu^2 + \nu^2) + \left(\frac{\gamma}{2E}\right)^2 \mu^2 \nu^2 (\mu^2 + \nu^2) \right\} \Psi = 4/\sqrt{-E} \; \Psi. \tag{4}$$

Δ_μ and Δ_ν are the radial parts of two-dimensional Laplacians with fixed azimuthal quantum number m.[18] Eq. (4) describes two harmonic oscillators which are coupled by the diamagnetic interaction $\mu^2 \nu^2 (\mu^2 + \nu^2)$ with an "effective strength" proportional to $(\gamma/E)^2$. The γE-plane can be scanned for solutions by diagonalizing Eq. (4) and tuning the effective strength parameter γ/E.

Rewriting the Schrödinger Equation into the form (4) not only provides an extremely efficient algorithm for calculating eigenstates and eigenvalues of (1) in the bound state region,[18] but it also enables the construction of an approximately separable representation of the Hamiltonian which is valid beyond the perturbative low field regime.

Therefore the diamagnetic interaction proportional to $\mu^2 \nu^2 (\mu^2 + \nu^2)$ in each shell of degenerate oscillator eigenstates is pre-diagonalized. Each such shell corresponds to a degenerate n-manifold of the field free hydrogen atom. Following the usual convention the resulting pre-diagonalized states are labelled by the index K which varies from 0 to N−|m|−1 where N = N_μ + N_ν + |m| + 1 is the total oscillator quantum number of the shell and is in fact equal to n. K runs over even (odd) integers in m^π subspaces with even (odd) z-parity π and the highest energy state in each shell carries the label K = 0. Because of the degeneracy of the basis states within each shell, the mixing induced in this pre-diagonalization is independent of E and γ; it yields the eigenstates of the adiabatic invariant (2) in the low field regime, but the present construction of the label K remains valid in the whole range of field strengths and (negative) energies.

Figure 1 shows the resulting eigenvalues of the states originating from the n = 29 to n = 31 manifolds in the m^π = 0^+ subspace for magnetic fields up to about 6 T when diagonalized in subspaces of fixed K. All level crossings are exact in this figure because K is conserved.

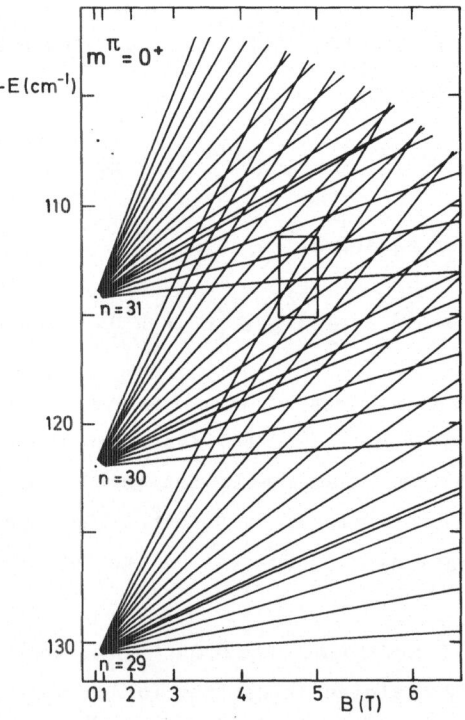

Figure 1. Spectrum of the hydrogen atom obtained by the method of K projection. The area enclosed by a box is shown on an expanded scale in Fig. 2.

The exact solutions of the Schrödinger Equation are obtained by diagonalizing the residual interaction between different K-subspaces. For the states shown in the box of Fig. 1 the results are compared with an exact diagonalization in Fig. 2. Clearly for the states shown, mixing induced by the residual interaction is quite small and the quantum number K defined via prediagonalization is very well conserved.

The approximate separability expressed in the well conserved label K is however not valid in the entire E-γ plane. The method of construction of K breaks down as we approach the zero-field threshold, because the effective residual interaction scales with $(\gamma/E)^2$ and becomes singular at E = 0. A detailed study of the spectrum also reveals a breakdown of approximate separability in the form of increased level repulsion and a decreased probability for small anticrossings as we approach E = 0 from below.

The transition from the approximately separable regime to the strongly non-separable regime is very complex. In the transition region there are some states which, for certain ranges of γ cross all other states nearly almost exactly while undergoing strong level repulsions in between. One interesting feature is that the approximate parity-doublet degeneracy of vibrator states is upheld far beyond the regime of approximate separability.

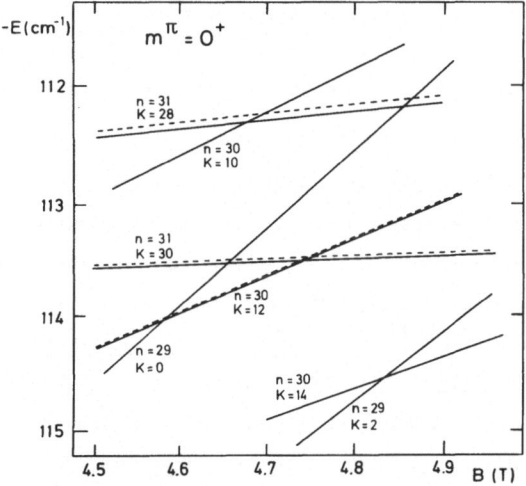

Figure 2. Spectrum of the hydrogen atom obtained by the method of K projection (broken lines) and by exact numerical solution of the Schrödinger Equation (full lines).

It should be mentioned, that the method of generating bound-state spectra along lines of constant E/γ can be formulated more generally to include also the case $0 < E/\gamma < I$, where $I = |m| + 1$ is the real ionization threshold value.[18]

The key point for finding an approximately separable representation of the Hamiltonian (1) was to formulte an eigenvalue equation for fixed values of the effective field strength parameter γ/E by taking advantage of scaling properties of the Schrödinger Equation. These scaling properties are purely quantum mechanical and differ from scaling properties of the classical system (see next chapter). Thus, the use of semi-parabolic coordinates is not essential in this context, although Eq. (4) reflects the quantum scaling in a canonical way. An analogous construction of the label K can be obtained by using, e.g. Sturmian functions in spherical coordinates.[19]

The conjecture[12,13] that the observed approximate level crossings are due to approximate separability is correct,[17] but not necessarily compelling. Classical regular motion in a non-separable system leads to dynamical localization in phase space in general (see, e.g. Ref. 3) and this may also cause very small anticrossings between states localized in different regions of phase space. For example, approximate separability is not strictly fulfilled for vibrator states, which are appreciable coupled by $\Delta K = 2$ chains. However, avoided crossings between vibrator states are also very small if ΔK is large enough. Another example seems to be the hydrogen atom in combined electric and magnetic fields, for which also systematic near-degeneracies were observed.[20]

III. TRANSITION FROM REGULARITY TO IRREGULARITY, CLASSICALLY AND QUANTUM MECHANICALLY

The study of the quantum mechanics and the classical dynamics of simple non-separable systems has evolved into a major field in physics and considerable attention is being given to the question of how the fact that a system is classically chaotic manifests itself in the quantum mechanical observables.[6,21-25] Statistical properties of the quantum mechanical spectra frequently studied in this context are the nearest neighbor spacing distribution (NNS) or the spectral rigidity Δ_3. Guided by the prediction of random matrix theories (RMT) and by several empirical results for model systems, it is now widely agreed that a completely regular quantum system is characterized e.g. by a Poisson distribution of the NNS while a completely irregular time-reversal invariant system corresponds to a Wigner-type distribution. In contrast to these limiting cases, the transition region where regular and irregular dynamics coexist is not well understood. Heuristic[26] or semiclassical formulae[27,28] for the NNS in this region have been discussed, but there is as yet no convincing theory.

The classical dynamics of te Hamiltonian (1) have been studied by several authors in recent years.[29-33] Before discussing the classical dynamics, it is important to reflect on a <u>scaling property</u> of the Hamiltonian (1). With the transformation

$$(\tilde{\rho}, \tilde{z}) = \gamma^{2/3} (\rho, z); \quad (\tilde{p}_\rho, \tilde{p}_z) = \gamma^{-1/3} (p_\rho, p_z) \tag{5}$$

we obtain a scaled Hamiltonian

$$\tilde{H} = \tilde{p}_\rho^2 + \tilde{p}_z^2 + \tilde{\ell}_z^2/\tilde{\rho}^2 + \frac{1}{4} \tilde{\rho}^2 - 2\left(\tilde{\rho}^2 + \tilde{z}^2\right)^{-1/2} = \gamma^{-2/3} H \tag{6}$$

which no longer depends on the field strength γ. Also, the scaled angular momentum $\tilde{\ell}_z = \gamma^{1/3}\ell_z$ is negligibly small under typical laboratory conditions. The scaling property (5), (6) shows that the classical dynamics of the system does not depend on E and γ independently, but only on the scaled energy

$$\hat{E} = \epsilon = E/\gamma^{2/3}. \tag{7}$$

One main result of the classical studies is a quantitative description of the transition from regular to irregular motion.[31] For scaled energies less than ~ -1.2 the whole phase space is filled with regular trajectories confined to invariant tori. As the energy is increased the fraction of phase space in which the motion is irregular increases rather sharply around $\epsilon = -0.7$ and reaches practically 100% at $\epsilon = -0.25$.

In the second half of 1986 three papers[34-36] appeared, in which the statistical properties of the quantum spectra of (1) were studied and compared with the predictions of RMT. In Fig. 3 we show typical NNS histograms for teh classically chaotic region at E = 0 (a), and for the classically regular region. In both cases the results agree well with the expectations based on RMT. Note that, although the classical equations of motion are not time-reversal invariant, one expects a <u>Wigner distribution</u> for the irregular case.[6] Similar results were obtained in Refs. 35 and 36.

For a more detailed analysis of the transition region we have studied spectra at constant values of the scaled energy (7)[37] which, except for E = 0, are slightly more difficult to obtain than the spectra at constant E/γ which were used for fig. 3(a),(b). Fig. 4 shows the histograms obtained for four different values of the scaled energy between -0.8 and -0.2. The solid lines in Fig. 4 are fits to the

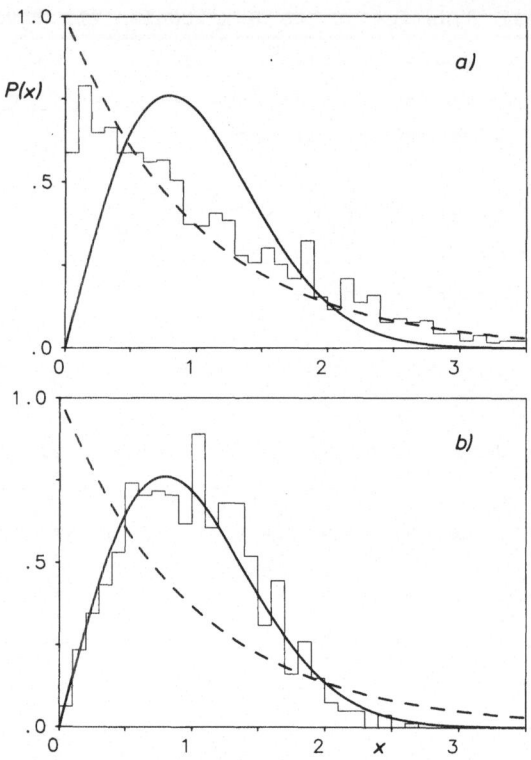

Figure 3. NNS histograms for level spectra (a) in the approximately
separable region, and (b) at the zero-field threshold E = 0,
together with the Wigner (smooth solid lines) and Poisson
distributions (dashed lines). A total of 1823 [810] spacings
is analyzed in panel (a) [(b)].

heuristic formula of Brody[26] and q_b are the corresponding values of the
interpolation parameter which is zero for a Poisson distribution and
unity for a Wigner distribution. The dashed curves in Fig. 4 show fits
to the histograms using the semiclassically derived formula of Berry and
Robnik,[27] characterized by the parameter q_a.

Fig. 5 shows how the q-values thus extracted from the quantum level
spectra vary as a function of scaled energy and compare them with the
curve of Harada and Hasegawa[31] for the fraction q of irregular phase
space. Although the q-values derived from the quantum spectra are not
defined sharply, Fig. 5 does show quantitative correspondence between the
increase in the irregular fraction of classical phase space and the
turnover from a Poisson-type to a Wigner-type NNS distribution.

One feature of the histograms in Fig. 4 is the minimum at small
separations which is better reproduced by the heuristic formula of Brody
than by the Berry-Robnik formula. The origin of this minimum is the fact
that in an almost but not exactly integrable system even a very small

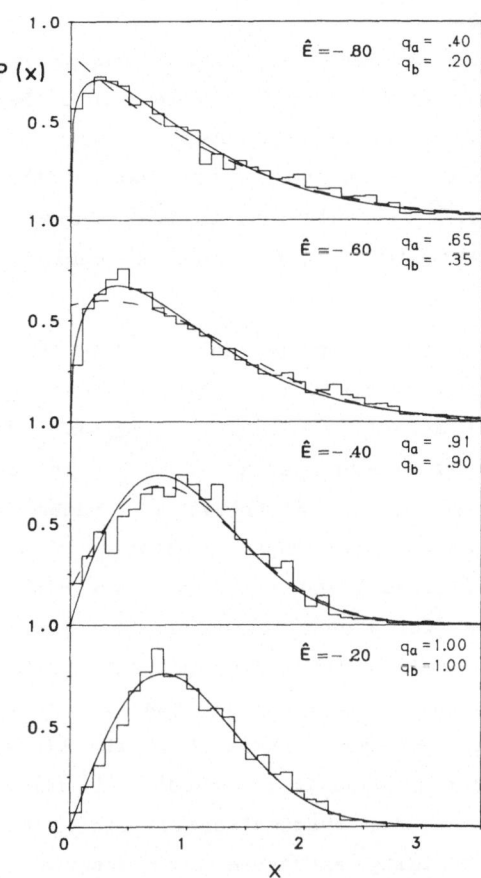

Figure 4. Distributions of NNS calculated at different values of the scaled energy ε. The solid lines show the fits with the Brody formula nd the dashed lines show the fits given by the semiclassical formula of Berry and Robnik. The q-values denote the corresponding fit parameters.

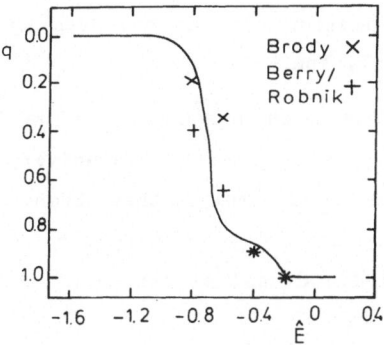

Figure 5. Fraction q of available phase space in which the classical motion is regular, plotted as a function of the scaled energy. The straight (diagonal) crosses are the q-values obtained by fitting the formula of Berry/Robnik (Brody) to the quantum NNS distributions.

residual interaction will cause small level repulsion and make the probability for exact degeneracies (i.e. of vanishing NNS) vanish. This feature is not contained in the original formula of Berry and Robnik, but improved formulae have since been proposed both by Robnik,[38] and by Hasegawa et al[39] who follow up a stochastic differential equations discussed by Yukawa.[40] Calculations to test the applicability of these formulae in the intermediate regime between regularity and irregularity are in progress.

Finally Fig. 6 shows some new and improved results for statistical quantities of the quantum spectra at the scaled energy of −0.2, where the classical system is already completely irregular. The figure shows the NNS histogram, the cumulative spacing distribution, the spectral rigidity Δ_3, and higher moments of the distribution: number variance Σ_2, skew γ_1 and excess γ_2 together with the expectation based on RMT for the completely regular (Poisson) and completely irregular (Gaussian Orthogonal Ensemble = GOE) case (for a description of these fluctuation measures see, e.g. Bohigas et al. in Ref. 21). The spectral rigidity, as already noted in similar calculations in Refs. 35 36, qualitatively agrees with the GOE curve in this case. However, beyond this the present results show two distinct features not previously observed: (i) the spectral rigidity of the calculated spectra lies systematically lower than the GOE prediction, (ii) the spectral rigidity extracted from spectra in m^π-subspaces with negative z-parity is systematically lower than in m^π-subspaces with positive z-parity π. Discrepancy (i) reflects the nonuniversal deviation from the GOE case predicted by Berry[41] and becomes smaller when going to higher excitations in the spectra.

The results for the higher moments of the distributions agree with the GOE predictions quite well except for the variance Σ_2 which shows a systematic discrepancy similar to that observed in the spectral rigidity, but even for comparatively low L.

Finally, we conclude with a warning. The fact, that fluctuation properties of "chaotic" quantum spectra are universal and can be simulated by random matrices does not imply, that these fluctuations carry no information. In fact, if analyzed properly, they contain very much information on the periodic classical orbits of the system. This will be the main subject of chapter V.

In summary, the present results show that the quantum spectra in the completely irregular regime follow the predictions of RMT quite well, but there are sufficient interesting deviations to encourage further more detailed investigations.

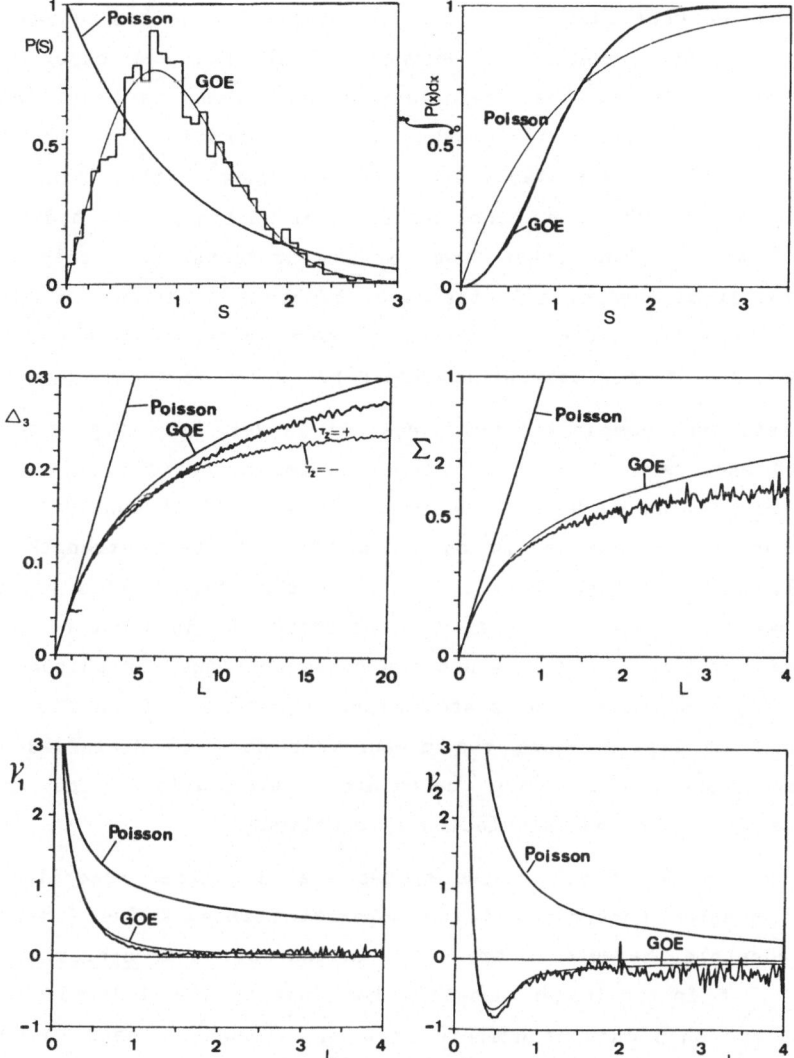

Figure 6. The figure shows various fluctuation measures calculated from eith different spectra for $\varepsilon = -0.2$ $(m = 0,1,2,3,\ \pi = +/-)$: NNS distribution, cumulative spacing distribution, spectral rigidity Δ_3, number variance Σ_2, skew γ_1 and excess γ_2. A total of 3000 calculated energy levels have entered the analysis. Also shown are the Poisson and GOE predictions. The analysis for the spectral rigidity is shown for each z-parity π separately.

IV. COMPARISON WITH EXPERIMENTAL SPECTRA

One of the strongest motivations for studying one electron atoms in a uniform magnetic field is that these atoms are real and, in contrast to the many popular two-dimensional models (billiards, Henon-Heiles problem, various coupled oscillators...), can be studied in the laboratory. Until recently, a direct comparison between calculation and experiment was hampered by the fact, that calculations were done for simplicity for hydrogen, while most measurements were performed, for experimental reasons, with alkali and alkali earth atoms. These difficulties have now been overcome on the one hand through calculations for hydrogen-like atoms[42-44] and on the other hand by measurements on highly resolved spectra for atomic hydrogen in fields up to about 6 T[45] and this has made a direct state for state comparison between calculation and experiment beyond the more or less trivial low field perturbative regime possible.

A first such comparison was presented in Ref. 46. Fig. 7 shows the calculated spectra in the $m^\pi = 0^+$ subspace for magnetic fields up to 7 T and for energies up to 90 cm^{-1} below the zero-field threshold. From the regular pattern of the levels as a function of field strength and the negligibly small anticrossings it is clear that Fig. 7 lies entirely in the approximately separable region, even though it does extend well into the n-mixing regime. Fig. 8 shows the cross sections for photoabsorption from the $2p_0$ state into the states along a cut at 6 T in Fig. 7. The upper panel of Fig. 8 shows the measurement done in Bielefeld and the lower panel shows the values calculated numerically. The agreement between measurements and calculation is excellent.

At the bottom of Fig. 8 the states are classified according to the principal quantum number n and the separation index K (see chapter II). This is possible, because we are still in the regime of <u>approximate separability</u>. It is no longer possible to classify the individual quantum states by two such quantum numbers if we go somewhat closer to threshold and leave the regime of approximate separability. This is illustrated in the left hand part of Fig. 9 which shows the calculated levels for fields around 6 T and energies of 51-64 cm^{-1} below the zero-field threshold.

The right hand part of Fig. 9 shows a photoabsorption spectrum taken at a field strength near 6 T and the level sequence to the left of the measured spectrum is the calculated spectrum at 5.96 T. Again the agreement between calculation and experiment is excellent.

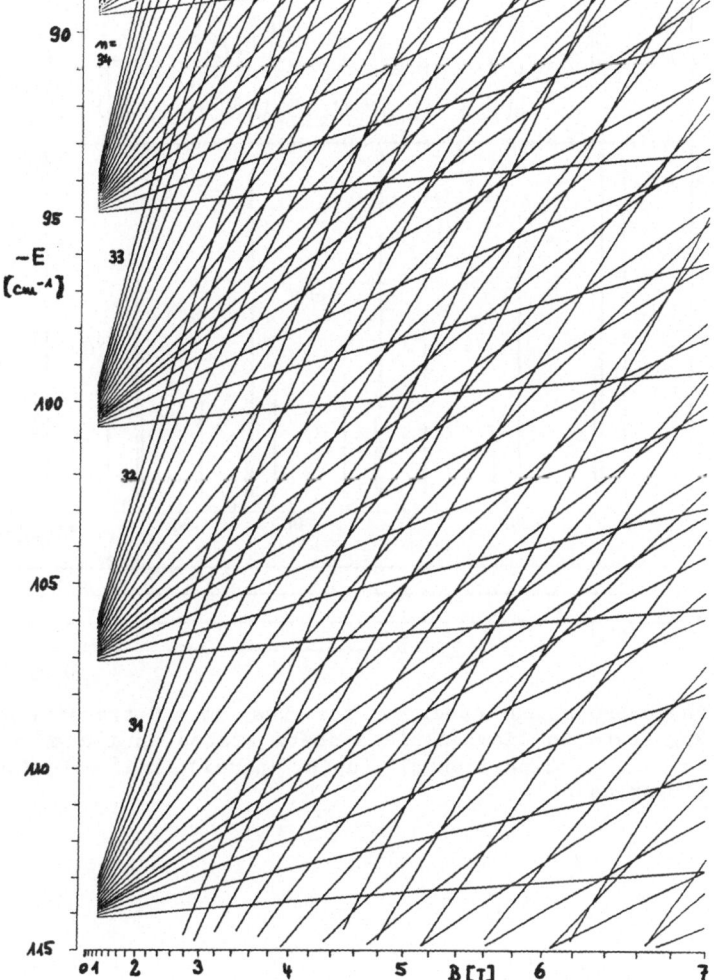

Figure 7. Section of the spectrum of states in the $m^\pi = 0^+$ subspace.

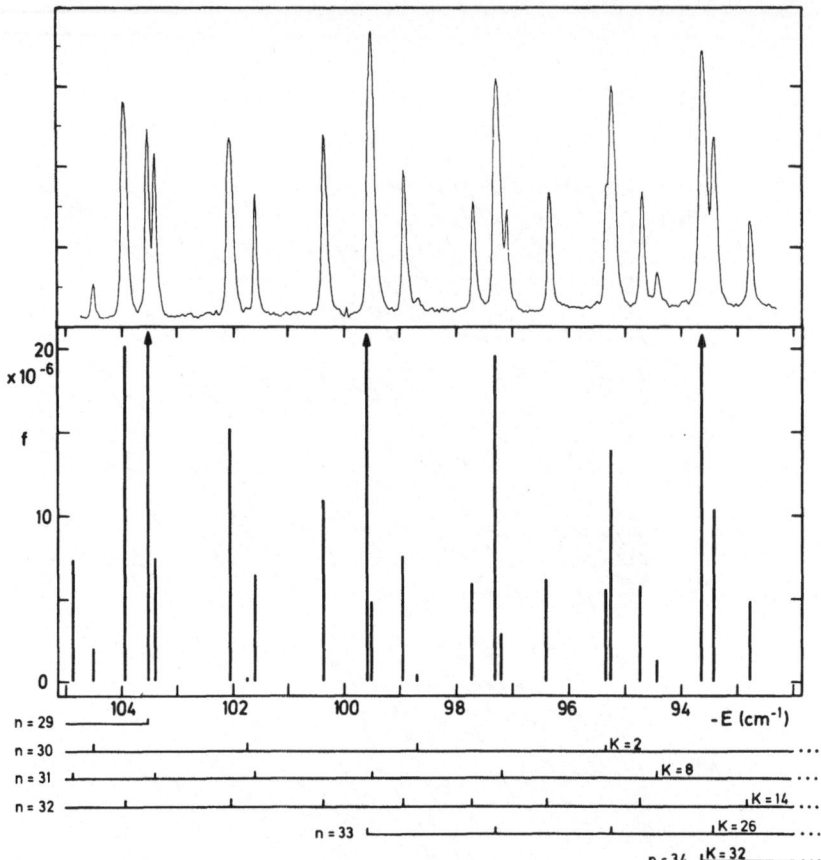

Figure 8. Photoabsorption spectrum for the $\Delta m = 0$ transitions from the $2p_0$ state to diamagnetic Rydberg states in the $m^\pi = 0^+$ subspace at 6 T. Experimental (upper panel) and calculated spectra (lower panel).

Because of the many levels involved and their different dependence on the field strength, the level sequence in any spectrum below the perturbative low-field regime depends very sensitively on the precise strength of the magnetic field. In the situation illustrated in Fig. 9 it was in fact the comparison between calculation and experiment which was used for an accurate determination of the experimental field strength.

More recently further and more detailed calculations of photoabsorption spectra have been done. At 5.96 T the comparison has been extended upto 20 cm^{-1} below the zero-field threshold in Ref. 47 and calculations up to -5 cm^{-1} have been reported at this conference.[2,48] In all cases the agreement between calculation and experiment is excellent.

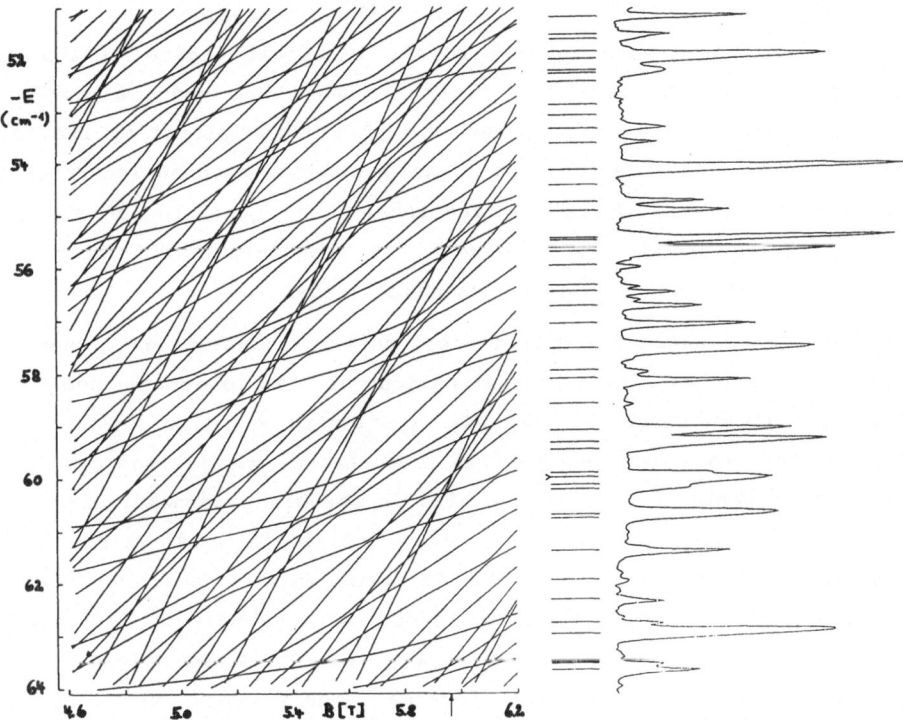

Figure 9.A section of the spectrum in the $m^\pi = 0^+$ subspace (left panel). The middle part shows the spectrum at B = 5.96 T, which should be compared with the experimental spectrum (right panel).

Thus the $m^\pi = 0^+$ spectrum of the hydrogen atom in a magnetic field at field strengths near 6 T has now been followed state for state up to energies E __ -5 cm^{-1}. (Note that the real ionization threshold at 6 T lies near E = +6 cm^{-1}.) Up to this energy the spectrum contains roughly 700 states and extends well into the irregular regime, where a classification according to quantum numbers other than m and π_z is meaningless. To our knowledge this is the first example of a physically real non-separable system whose spectrum has been measured _and_ calculated to such high excitations.

V. UNSTABLE PERIODIC ORBITS AND QUASI-LANDAU MODULATIONS

The discovery of almost equidistant peaks in the photoabsorption spectra of atoms in a magnetic field[11] played a major role in generating the widespread attention given to the problem of magnetized atoms in recent years. Near the ionization threshold, the separtion of these "quasi-Landau" peaks is about 1.5 times the separation of free electrons in a magnetic field. On the basis of semiclassical investigations and of quantum calculations up to a few $\hbar\omega_c$ below E = 0 the quasi-Landau peaks were widely interpreted[49-56] as resonant photoabsorption into individual quantum states which were related to the classical orbits in the plane perpendicular to the magnetic field. Based on quantum mechanical calculations around E = 0 (at somewhat higher than laboratory fields) we showed[57] in 1985 that the physical ideas underlying such an interpretation were, at best, severe oversimplifications. The quasi-Landau region is highly irregular and it is in general not possible to identify simple sequences of individual bound (or resonant) states.

On the other hand, the experimental data on "quasi-Landau resonances" do not at all imply an identification of the observed peaks with individual quantum states. What the experiments actually show are modulations of the cross sections and the quasi-Landau modulations can in fact be related to periodic classical orbits in the plane perpendicular to the field without invoking the existence of individual quasi-Landau quantum states as shown by Reinhardt.[58] In this model peaks in observable quantities such as cross sections occur as a result of constructive interference due to recurrence of a wave packet moving along a closed trajectory. Constructive interference occurs when the "resonance condition"

$$S(E) = \frac{1}{h} \oint \vec{p} \cdot d\vec{r} = n \tag{8}$$

if fulfilled. For unstable periodic orbits Eq. (8) is not a quantization condition and does not define individual quantized states, but it may define the positions of modulation peaks in quantities such as cross sections or spectral densities.

A more quantitative description of how periodic classical orbits influence quantum mechanical observables is given by the semiclassical theory of Gutzwiller, Ballian, Block and Berry (for a Review see Ref. 59). The main result of this theory is that e.g. the quantum spectral density $n(E)$,

$$n(E) = \Sigma_i \ \delta(E - E_i), \tag{9}$$

can be written as a sum over all periodic orbits of the system:

$$n(e) - \overline{n}(E) = \Sigma_r \ \Sigma_j \ a_{rj}(E) \ \cos\{2\pi j(S_r(E) - \alpha_r)\}, \tag{10}$$

where $\overline{n}(E)$ is a smooth mean level density. In (10) r labels all primitive periodic orbits and j runs from 1 to infinity corresponding to multiple traversals of the same primitive orbit. $S_r(E)$ is the action along the primitive period orbit and α_r is an appropriate phase correction.

The beauty of Eq. (10) is that it applies irrespective of whether the system is regular and the orbits stable or irregular and the orbits unstable. The information about <u>stability</u> of an orbit is contained in the amplitude factors a_{rj}. For stable orbits these factors oscillate or behave as a power law in j, depending on whether or not the (stable) orbit is isolated.[59] The summation over all j in (10) leads to a spike in the level density $n(E)$ when $S_r - \alpha_r$ is an integer. For unstable orbits the factors a_{rj} decrease exponentially in j and hence only the leading terms survive in the summation over j. Thus the "resonance condition" leads not to individual quantum states but to modulation peaks which become sharper with increasing stability.

The interest in the quasi-Landau phenomenon received a great boost by the experimental discovery of further modulations related to classical orbits lying outside the plane perpendicular to the magnetic field.[1,56,60,61] The correlation to classical orbits is best studied by calculating the Fourier transforms of the cross sections or spectral densities. Because of the scaling property (5), (6) of the classical Hamiltonian, the resonance condition (8) becomes

$$\gamma^{-1/3} \, S(\varepsilon) = n, \tag{11}$$

where the scaled action

$$S(\varepsilon) = \frac{1}{h} \oint \vec{p} \cdot d\vec{r} \tag{12}$$

now depends only on the scaled energy ε. From (11) it is clear that modulation peaks at given ε are equidistant on a scale linear in $\gamma^{-1/3}$, so modulations are best revealed by studying physical quantities at fixed ε as a function of $\gamma^{-1/3}$. The Fourier transforms then depend on a quantity conjugate to $\gamma^{-1/3}$; we call it $\gamma^{+1/3}$ for simplicity.

In order to study the correspondence between classical orbits and modulations of physical quantities we have performed extensive quantum mechanical calculations of cross sections and made a systematic search for periodic orbits.[62] For the study of photoabsorption transition from a low lying state ($n = 2$) into a high lying Rydberg state, only orbits passing through the origin are relevant.[62] Figs. 10, 11 show such periodic orbits. Fig. 10 shows the series of orbits also identified in Refs. 45, 56, 60 and in a similar theoretical study by Du and Delos.[63]

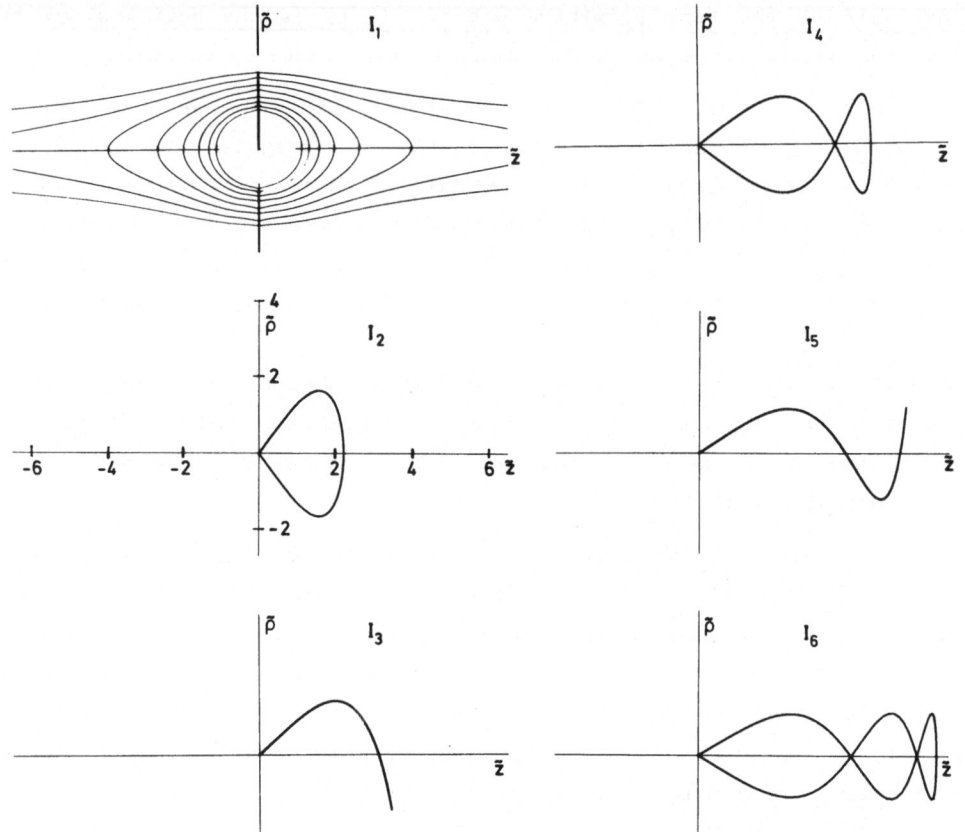

Figure 10. Leading members of a series (I) of classical periodic orbits at E = 0. The diagram showing the first number I_1 of the series also shows the equipotential lines.

Typical Fourier transformed photoabsorption spectra are shown in Fig. 12 for two m^π subspaces and various scaled energies. A very detailed discussion on the quantitative correspondence between the actions of classical orbits and the modulation peaks in the Fourier transformed cross sections can be found in Ref. 62. Let us just mention at this point that there is no peak in Fig. 12 which cannot be correlated with a periodic orbit passing through the origin. The distribution of periodic orbits with scaled actions $S_i < 4$ is not so dense that this is a trivial observation.

If we study not the photoabsorption spectra but the fluctuating part $N - \langle N \rangle$ of the cumulative level density N, $N(E) = \int^E n(E')dE'$, then all periodic orbits and not only those passing through the origin must be taken into account.[64] Fig. 13 shows the power spectra for various m^π subspaces at the scaled energy $\varepsilon = -0.4$ and it does indeed show a few

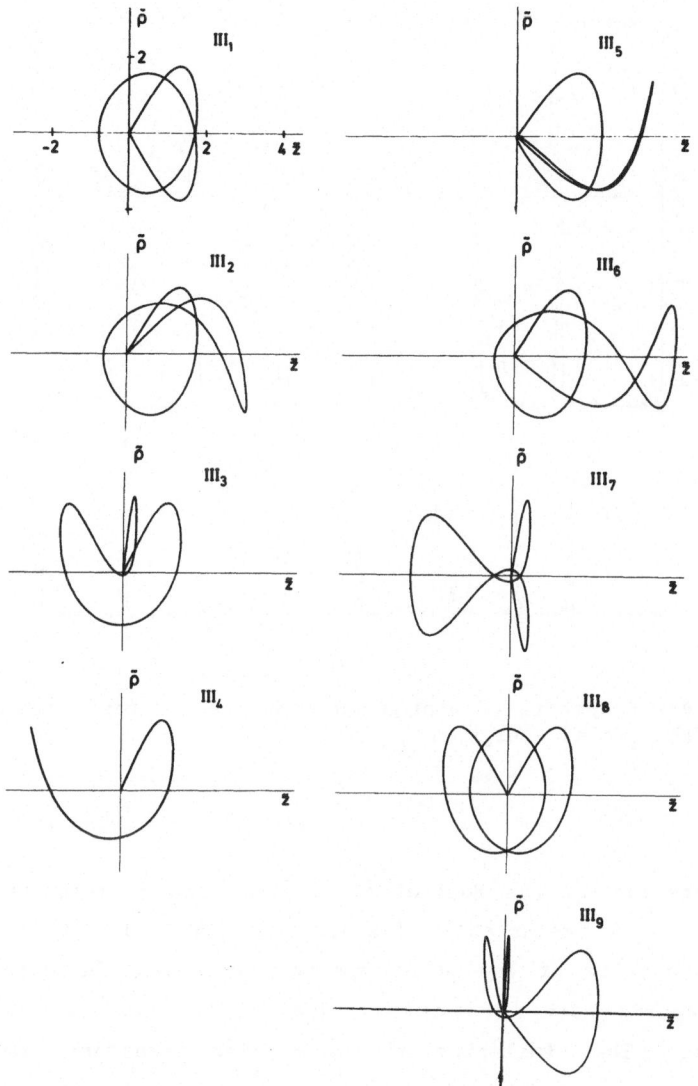

Figure 11.   Further periodic orbits at E = 0 (here called series III).

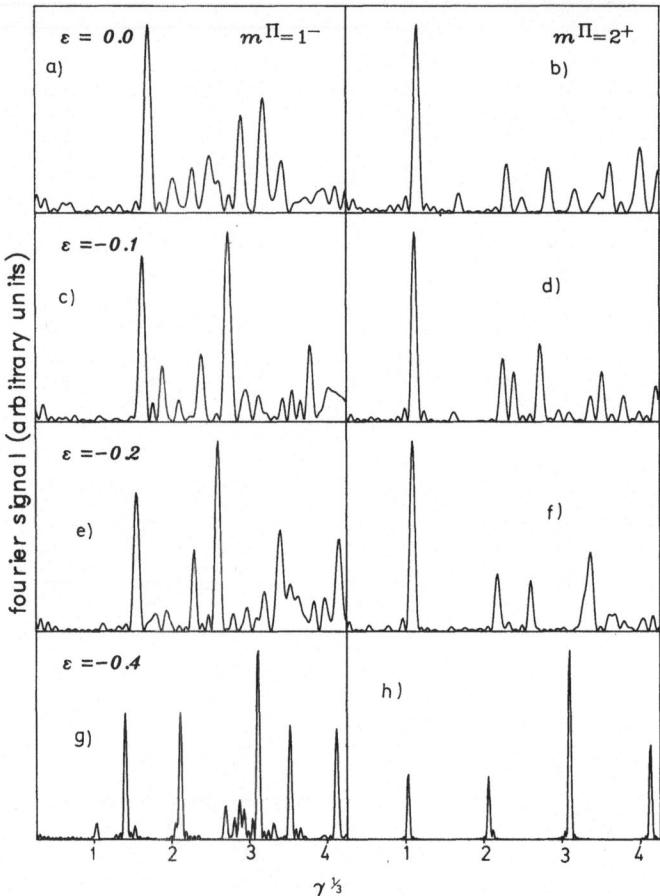

Figure 12. Power spectra of photoabsorption spectra for various values of the scaled energy ε.

peaks not present in the Fourier transformed photo cross sections (see Fig. 12g,h). A remarkable feature of Fig. 13 is the virtual indistinguishability of the power spectra, although the details of the original spectra are of course very different in the various m^π subspaces. The similarity of the Fourier transforms is even more striking than in the Fourier transformed photoabsorption spectra (see Fig. 12), because in the latter, the relative heights of the peaks are strongly influenced by the geometry of initial and final states in the transition concerned.

In our studies so far we have established the correspondence between the positions of peaks in Fourier transformed spectra and the scaled actions of periodic orbits, but we have not quantitatively accounted for the heights of the peaks. As is clear from the discussion in connection

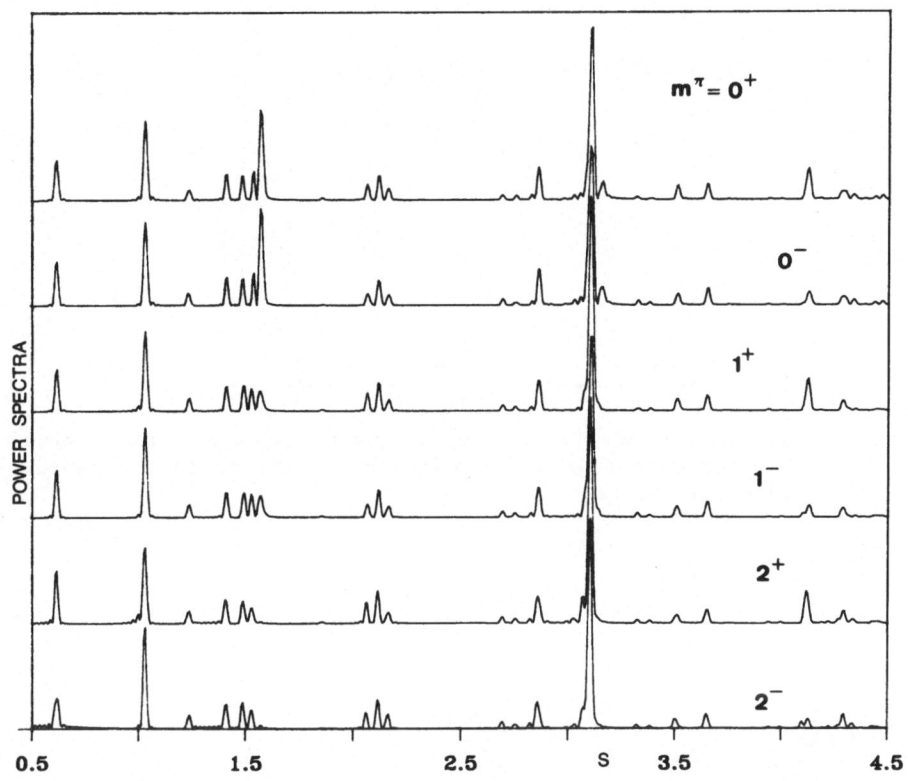

Figure 13. Power spectra of the fluctuating part of the cumulative level density $N(E)$ for different m^π subspaces at $\varepsilon = -0.4$. 750 levels have been included in each Fourier transform.

with Eq. (9) the heights of the peaks are strongly influenced by the degree of instability of the periodic orbits involved, a fact which has been taken into account by Du and Delos.[10,63] A quantitative measure of the instability of a classical orbit is given by the <u>Liapunov exponent</u> which gives the ratio of expnential divergence of neighboring trajectories. First calculations of Liapunov exponents for periodic trajectories in the magnetized hydrogen atom are presented in Ref. 65 for the straight line orbits parallel and perpendicular to the field and in the contributions[2,66] to this conference for other orbits of the series in Fig. 10. Fig. 14 shows the product period times Liapunov exponent for the periodic orbit parallel to the magnetic field as a function of the scaled energy. Note that there are alternating intervals of stability and instability. Moving in the directionof increasing ε the points at which the orbit become unstable are actually points of bifurcation where the orbits $I_2, I_3, I_4, \ldots$ of Fig. 10 are "born."

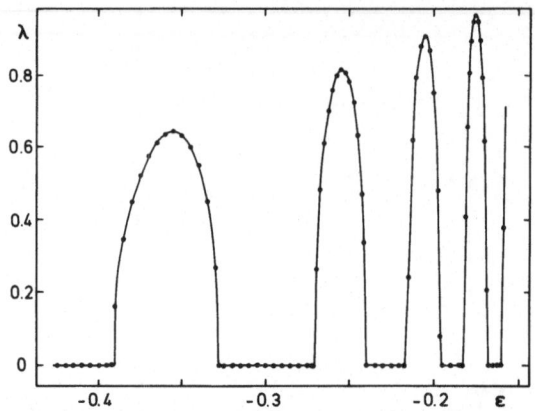

Figure 14. Dependence of the product Liapunov exponent times period on the scaled energy ε for the orbit parallel to the field. (ε in atomic units.)

VI. SUMMARY

The problem of a single electron atom in a uniform magnetic field has become one of the most popular prototype models for studying the classical and quantum mechanical behaviour of simple non-separable systems.

The work done in the last year or two has led to a major breakthrough in our understanding of atomic diamagnetism and in particular of the quasi-Landau phenomenon. We now realize that the peaks observed by Garton and Tomkins in 1969 are a beautiful example of the quantum manifestation of unstable periodic orbits embedded in a classically chaotic regime.

During the conference at Egham attention was drawn to a thesis written by R. A. Pullen under the supervision of A. R. Edmonds in 1981. In this work[30,67] many of the ideas and results discussed in chapter V were already formulated. This includes the scaling of the classical Hamiltonian, the discovery of the "new quasi-Landau modulations" beyond the original "Garton-Tomkins resonances" and the recognition of the importance of periodic orbits outside the plane perpendicular to the field. It also includes a reference to Eq. (9) which is the key to understanding the quasi-Landau phenomenon. If this work had become known earlier many other papers need not have been written and we probably would have reached the present breakthrough in our general understanding of the quasi-Landau phenomenon a few years earlier.

REFERENCES

1. K. H. Welge et al., these proceedings.

2. G. Wunner et al., these proceedings.

3. D. Delande and J.C. Gay, these proceedings.

4. H. Rinneberg et al., these proceedings.

5. U. Fano, these proceedings.

6. M. Robnik, these proceedings.

7. J. Pinard et al., these proceedings.

8. H. Hasegawa, these proceedings.

9. P. F. O'Mahony et al., these proceedings.

10. J. B. Delos et al., these proceedings.

11. W.R.S. Garton and F. S. Tomkins, Astrophys. J. $\underline{158}$ (1969) 839.

12. M. L. Zimmerman, M. M. Kash and D. Kleppner, Phys. Rev. Lett. $\underline{45}$ (1980) 1092.

13. D. Delande and J. C. Gay, Phys. Lett. $\underline{82A}$ (1981) 393.

14. E. A. Solov'ev, JETP Lett. $\underline{34}$ (1981) 265.

15. D. R. Herrick, Phys. Rev. $\underline{A26}$ (1982) 323.

16. D. Delande and J. C. Gay, J. Phys. $\underline{B17}$ (1984) L335.

17. D. Wintgen and H. Friedrich, J. Phys. $\underline{B19}$ (1986) 1261.

18. D. Wintgen and H. Friedrich, J. Phys. $\underline{B19}$ (1986) 991.

19. D. Wintgen, H. Marxer, and J. S. Briggs, J. Phys. $\underline{A20}$ (1987) at press.

20. K. Richter, D. Wintgen and J. S. Briggs, J. Phys. $\underline{B20}$ (1987) at press.

21. T. H. Seligman and H. Nishioka (Eds.) Quantum Chaos and Statistical Nuclear Physics, Lect. Not. Phys. $\underline{263}$, Springer-Verlag, Berlin.

22. E. Haller, H. Köppel and L. S. Cederbaum, Phys. Rev. Lett. $\underline{52}$ (1984) 215.

23. T. H. Seligman, J.J.M. Verbaarshot and M. R. Zirnbauer, Phys. Rev. Lett. $\underline{53}$ (1984) 215.

24. H. D. Meyer, E. Haller, H. Köppel and L. S. Cederbaum, J. Phys. $\underline{A17}$ (1984) L813.

25. T. Zimmermann, H.-D. Meyer, H. Köppel and L. S. Cederbaum, Phys. Rev. $\underline{A33}$ (1986) 4334.

26. T. A. Brody Lett. Nouvo Cim. $\underline{7}$ (1973) 482.

27. M. V. Berry and M. Robnik, J. Phys. $\underline{A17}$ (1984) 2413.

28. T. H. Seligman and J.J.M. Verbaarschot, J. Phys. $\underline{A18}$ (1985) 2227.

29. M. Robnik, J. Phys. $\underline{A14}$ (1981) 3195.

30. R. A. Pullen, D. Phil. thesis, Imperial College, 1981 (unpublished).

31. A. Harada and H. Hasegawa, J.Phys. $\underline{A16}$ (1983) L259.

32. H. Hasegawa, S. Adachi and A. Harada, J. Phys. A16 (1983) L503.

33. J. B. Delos, S. K. Knudson and D. W. Noid, Phys. Rev. A30 (1984) 1208.

34. D. Wintgen and H. Friedrich, Phys. Rev. Lett. 57 (1986) 571.

35. D. Delande and J. C. Gay, Phys. Rev. Lett. 57 (1986) 2006.

36. G. Wunner, U. Woelk, I. Zech, G. Zeller, T. Ertl, F. Geyer, W. Schweizer and H. Ruder, Phys. Rev. Lett. 57 (1986) 3261.

37. D. Wintgen and H. Friedrich, Phys. Rev. A35 (1987) 1464.

38. M. Robnik, J. Phys. A20 (1987) L495.

39. H. Hasegawa, H. J. Mikeska and H. Frahm, submitted to Phys. Rev. Lett.

40. T. Yukawa, Phys. Rev. Lett. 54 (1985) 1883; Phys. Lett. 116A (1986) 227.

41. M. V. Berry, Proc. Roy. Soc. London, Ser. A400 (1985) 229.

42. P. F. O'Mahony and K. T. Taylor, J. Phys. B19 (1986) L65.

43. P. Cacciani, E. Luc-Koenig, J. Pinard, C. Thomas and S. Liberman, J. Phys. B19 (1986) L519.

44. P. F. O'Mahony and K. T. Taylor, Phys. Rev. Lett. 57 (1986) 2931.

45. A. Holle, G. Wiebusch, J. Main, H. Rottke, B. Hager and K. H. Welge, Phys. Rev. Lett. 56 (1986) 2594.

46. D. Wintgen, A. Holle, G. Wiebusch, J. Main, H. Friedrich and K. H. Welge, J. Phys. B19 (1986) L557.

47. A. Holle, G. Wiebusch, J. Main, K. H. Welge, G. Zeller, G. Wunner, T. Ertl and H. Ruder, Z. Phys. D5 (1987) 279.

48. G. Zeller, U. Woelk, G. Wunner and H. Ruder, poster-contribution to this conference.

49. A. R. Edmonds, J. Physique Paris Colloq. 31 (1970) C4-71.

50. A.R.P. Rau, Phys. Rev. A16 (1977) 613.

51. U. Fano, Phys. Rev. A22 (1980) 2260.

52. J. C. Castro, M. L. Zimmerman, R. G. Hulet, D. Kleppner and R. R. Freeman, Phys. Rev. Lett. 45 (1980) 1780.

53. D. Delande, C. Chardonnet, F. Biraben and J. C. Gay, J. Physique Paris Colloq. 43 (1982) C2-97.

54. J.A.C. Gallas, F. Gerck and R. F. O'Connell, Phys. Rev. Lett. 50 (1983) 324.

55. C. W. Clark, in Atomic Excitations and Recombination in External Fields, M. H. Nayfeh and C. W. Clark eds., Gordon and Breach, London, 1985.

56. M. A. Al-Laithy, P. F. O'Mahony and K. T. Taylor, J. Phys. B19 (1986) L773.

57. D. Wintgen and H. Friedrich, J. Phys. B19 (1986) L99.

58. W. P. Reinhardt, J. Phys. B16 (1983) L635.

59. M. V. Berry, in Chaotic Behaviour of Deterministic Systems, eds. G. Iooss et al., North Holland, Amsterdam 1983, p. 171.

60. J. Main G. Wiebusch, A. Holle and K. H. Welge, Phys. Rev. Lett. 57 (1986) 2789.

61. A.R.P. Rau, Nature 325 (1986) 577.

62. D. Wintgen and H. Friedrich, Phys. Rev. A36 (1987) 131.

63. M. L. Du and J. B. Delos, Phys. Rev. Lett. 58 (1987) 1731.

64. D. Wintgen, Phys. Rev. Lett. 58 (1987) 1589.

65. D. Wintgen, J. Phys. B20 (1987) L511.

66. W. Schweizer, H. Friedrich, R. Niemeier, G. Wunner and H. Ruder, poster-contribution to this conference.

67. A. R. Edmonds and R. A. Pullen, Preprint Imperial College, London (1980) unpublished.

OSCILLATIONS IN ABSORPTION SPECTRA OF ATOMS IN MAGNETIC FIELDS

J. B. Delos[*]and M. L. Du

Joint Institute for Laboratory Astrophysics
University of Colorado and National Bureau of Standards
Boulder, Colorado 80309-0440

It is a pleasure to give recognition to M.R.C. McDowell and his many contributions to atomic physics over the years, and also to honor the senior delegates to this Conference, W.R.S. Garton, U. Fano and A. Edmonds, who made the first discoveries that eventually led to the work reported here.

SYSTEM

We consider the near-threshold absorption spectrum of an atom in a magnetic field. The atom considered is hydrogen, and it is placed in a magnetic field of about 6 Tesla. The electron is initially in a $2p_z$ state, and it is then excited by a laser into states with energies very close to the ionization threshold. The phenomena we discuss occur also in other atoms, so while hydrogen provides the simplest case, there is otherwise nothing special about it.

The nuclear motion is ignorable, so the system Hamiltonian is (in atomic units)

$$H = \frac{1}{2} p^2 - \frac{1}{r} + \lambda \rho^2.$$ (1)

The first two terms are the standard Hydrogenic Hamiltonian, and the last is the diagmagnetic term. (The paramagnetic term is simply an additive constant.) Because of the cylindrical symmetry of the system, the

---
[*]JILA Visiting Fellow, 1986-87, on leave from College of William and Mary, Williamsburg, VA 23185.

ϕ-dependence of the wave functions separates, and so only the motion in ρ and z (or r and θ) has to be considered. The kinetic energy term can be written as

$$\frac{1}{2} p^2 = \frac{1}{2} (p_\rho^2 + p_z^2) + L_z^2/2\rho^2$$

and in the case considered here, $L_z = 0$.

OBSERVATIONS

The near-threshold absorption spectrum of various atoms in magnetic fields was first measured by Garton and Tomkins.[1] They found that the absorption as a function of photon energy is oscillatory (Fig. 1). More recently, in a remarkable series of experiments,[2] the Bielefeld group has made corresponding measurements on the hydrogen atom at much higher resolution (Fig. 2a). At first glance, the spectrum is a mess — it looks as if there were a wire loose in the detector. In fact, however, hidden in that spectrum is an unexpected type of order. It becomes apparent when the Fourier transform of the spectrum is computed, converting from energy E to time T as the independent variable:

$$\tilde{Abs}(T) = \int_{E_1}^{E_2} Abs(E) \exp(iET/\hbar)dE. \tag{2}$$

(Here Abs(E) means the measured photon absorption rate in arbitrary units.) The resulting transformed-absorption signal $\tilde{Abs}(T)$ shows a set of sharp peaks (Fig. 2b). This means that the absorption signal Abs(E) is dominantly a superposition of sinusoidal oscillations

Abs(E) = smooth background

$$+ \sum_k C_k \sin(T_k E/\hbar + \Delta_k). \tag{3}$$

The index k labels a peak in Fig. 2b, and T_k is the value of the time-variable at the kth peak. We call C_k and Δ_k respectively the kth spectral-oscillation amplitude and phase.

We seek a theory of these oscillations. We would like to know the physical origin of the oscillations, and we would like to be able to calculate T_k, C_k and Δ_k starting from the Hamiltonian of the system.

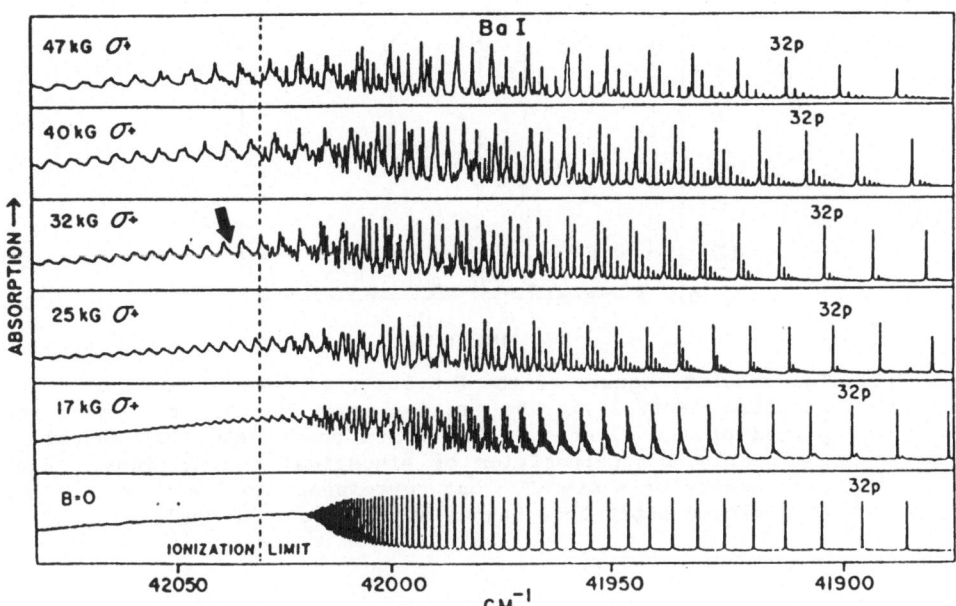

Fig. 1   Absorption spectrum of the barium atom close to the ionization threshold (Ref. 1).   Photon–energy increases toward the left.   The arrow on the left points to oscillations that are visible in the spectrum in the classically-chaotic regime.   This paper summarizes the theory of such oscillations.

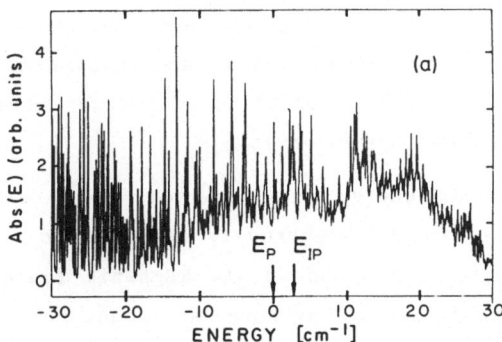

Fig. 2(a)   The absorption spectrum of the hydrogen atom in a magnetic field B = 5.96 T.   Photon energy increases to the right. The zero-field and actual ionization thresholds are indicated.   The initial state was $2p_z$, and the final states have $L_z = m\hbar = 0$.   At this higher resolution, the oscillations seen in Fig. 1 are not apparent to our eyes. They are still there, however.

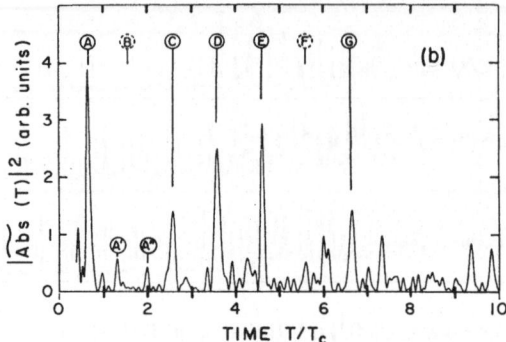

Fig. 2(b) The absolute square of the Fourier-transform of the absorption
spectrum (Ref. 2). The unit of time is T_c, the cyclotron
period of the electron. The sharp peaks show that Abs(E) is
dominantly a superposition of sinusoidal oscillations. Each
peak occurs at a time T_k that corresponds to the time duration
of a closed orbit that begins and ends at the nucleus.

REMARKS

Two facts are particularly relevant.

(1) In the near-threshold states, the diamagnetic and Coulomb terms
are of comparable importance, so the two degrees of freedom are strongly
coupled. Neiter n nor ℓ are good quantum numbers, no conserved
quantities are known (except energy), and there is no known way of
assigning quantum numbers to the eigenstates. Moreover, classical
trajectories of an electron moving in these fields are chaotic. The
classical limit of quantum mechanics for chaotic systems is not very well
understood, and, in particular, there is no theory connecting individual
eigenstates to individual trajectories.

(2) Individual energy levels are not resolved in these
experiments. The absorption spectrum is measured to finite resolution,
but near the ionization threshold in the magnetic field, the density of
states approaches infinity. Therefore we cannot in any case directly
compute all of the energy levels. We would prefer to obtain a theory
that describes the spectrum at finite resolution without requiring
computation of individual levels.

The most important step toward the interpretation of this phenomenon
was taken by Edmonds.[3] It was shown that the time T_k associated with
each oscillation is equal to the time duration of a periodic (or possibly
a closed) orbit of an electron in the combined Coulomb and diamagnetic

fields. The Hamiltonian (1) obviously admits a periodic orbit lying in the line z = 0, and the period of this orbit turns out to be equal to 2/3 of the cyclotron period T_c, in precise agreement with the oscillations observed by Garton and Tomkins. (In 1970, Edmonds had available only the early low-resolution data. However some unpublished data from that period already suggested the presence of additional oscillations, and Edmonds began to search for other periodic orbits as well.)

Fano[4] was also quick to recognize the fundamental importance of this phenomenon, and his discussions of it have provided much stimulation to theorists. Like Edmonds, he used a WKB approximation to describe the wave function on the line z = 0. But he strongly emphasized that a two-dimensional wave function cannot be correlated only with a one-dimensional orbit: it is essential to obtain a description of the wave function in directions transverse to the orbit. He suggested some possible approaches to the solution of this problem, but those approaches have not led to quantitative results.

And so the fundamental problem remained. Why do periodic classical orbits manifest themselves as oscillations in the quantum absorption spectrum?

Our description of this phenomenon is based upon theories created by Balian and Block,[5] Gutzwiller[6] and Berry and Tabor.[7] They examined the (smoothed) density of states of a quantum system as a function of energy, $\bar{n}(E)$. They showed that oscillatory fluctuations in n(E) are correlated with periodic classical orbits in the system:

$$\bar{n}(E) = n_0(E) + \sum_k C_k \sin\left[\int T_k(E')dE'/\hbar + \delta_k \right] . \tag{4}$$

Here $n_0(E)$ is a smooth background, related to the volume in phase space, and $T_k(E')$ is again the period of a periodic orbit (typically such periods are slowly-varying functions of energy).

The close similarity between Eq. (4) and Eq. (3) tells us immediately that a very similar approach must provide an explanation of the observed oscillations. Some modifications are needed, however, because the absorption experiments measure an average oscillator strength rather than an average density-of-states.

THEORETICAL FRAMEWORK

The theory begins from the Green's function for a quantum system. The Green's function $G_E(\vec{q}, \vec{q}_0)$ is the wave at $\vec{q}$ that arises when a steady

101

source of outgoing waves of energy E is placed at $\vec{q}_0$. The Green's function satisfies

$$(E - H)G_E(\vec{q},\vec{q}_0) = \delta(\vec{q} - \vec{q}_0) \tag{5}$$

subject to an outgoing-wave boundary condition.

There is a simple semiclassical approximation to the Green's function, and it is very similar to the semiclassical approximation to other kinds of wave functions. Let us imagine that the source at $\vec{q}_0$ produces a steady stream of classical particles, and that these particles then move symmetrically outward from the source, following trajectories of energy E governed by the Hamiltonian H. Suppose there is just one trajectory of energy E that goes exactly from $\vec{q}_0$ to $\vec{q}$. Then the semiclassical approximation to $G_E(\vec{q},\vec{q}_0)$ is given by

$$G_E(\vec{q},\vec{q}_0) \simeq c \, A_E(\vec{q},\vec{q}_0) \, \exp i[S_E(\vec{q},\vec{q}_0)/\hbar + \eta]. \tag{6}$$

Here $S_E(\vec{q},\vec{q}_0)$ is the action $\int \vec{p}(t) \cdot d\vec{q}/dt \, dt$ on the classical path from $\vec{q}_0$ to $\vec{q}$. $A_E(\vec{q},\vec{q}_0)$ is an amplitude, equal to the square=root of the density of classical particles at $\vec{q}$. (The classical density is inversely related to the speed of the particles at $\vec{q}$, and also inversely related to the divergence of neighboring trajectories from the path that connects $\vec{q}_0$ to $\vec{q}$.) Finally, η is an additional phase shift that results from caustics (turning points) encountered by the trajectory.

Now suppose that several distinct classical orbits of energy E propagate from $\vec{q}_0$ to $\vec{q}$. In this case the semiclassical approximation to the Green's function is

$$G_E(q,q_0) = c \sum_k A_E^{(k)}(q,q_0) \, \exp i\left[S_E^{(k)}(q,q_0)/\hbar + \eta^{(k)}\right]. \tag{7}$$

Here k labels the orbits at energy E that go from $\vec{q}_0$ to $\vec{q}$, and $S_E^{(k)}(\vec{q},\vec{q}_0)$ and $A_E^{(k)}(\vec{q},\vec{q}_0)$ are classical phases and amplitudes associated with the kth orbit.

How does $G_E(\vec{q},\vec{q}_0)$ depend upon the energy E? To answer this, let us examine an individual term in Eq. (7). As an energy E changes a bit, the (k^{th}) trajectory connecting $\vec{q}_0$ to $\vec{q}$ changes slightly, and the value of the amplitude $A_E^{(k)}(\vec{q},\vec{q}_0)$ also changes. But the important change is in the phase $S_E^{(k)}(\vec{q},\vec{q}_0)$. From a well-known theorem in classical mechanics, the rate-of-change of $S_E^{(k)}(\vec{q},\vec{q}_0)$ with E is equal to the time duration of the (k^{th}) orbit from $\vec{q}_0$ to $\vec{q}$:

$$\frac{\partial S_E^{(k)}(\vec{q},\vec{q}_0)}{\partial E} = T_k(E). \tag{8}$$

Hence as the energy changes, the imaginary part of $G_E(\vec{q},\vec{q}_0)$ changes as

$$\mathrm{Im}\ G_E(\vec{q},\vec{q}_0) =$$

$$c \sum_k A_E^{(k)}(\vec{q},\vec{q}_0)\ \sin\Big[\int_{E_0}^{E} T_k(E')\ dE'/\hbar + S_{E_0}^{(k)}(\vec{q},\vec{q}_0)/\hbar + \eta^{(k)}\Big] \tag{9}$$

or (for small changes in E)

$$\mathrm{Im}\ G_E(\vec{q},\vec{q}_0) \simeq c \sum_k A^{(k)}(\vec{q},\vec{q}_0)\ \sin\Big\{\big[T_k E + S_{E_0}^{(k)}(\vec{q},\vec{q}_0)\big]/\hbar + \eta^{(k)}\Big\}. \tag{10}$$

Equation (10) is starting to look like the empirical Eq. (3). However there is a very important difference. In typical bounded two-dimensional systems, since the trajectories wander forever through a finite region of space, at each energy there is an infinite number of orbits connecting $\vec{q}_0$ to $\vec{q}$. Each such orbit contributions a sin-wave oscillation to the energy dependence of $G_E(\vec{q},\vec{q}_0)$. The wavelength along the energy axis of each oscillation is equal to $2\pi\hbar/T_k$. Thus short, direct orbits from $\vec{q}_0$ to $\vec{q}$ produce slow oscillations, while long convoluted orbits produce rapid oscillations. If we try to include all of these orbits in Eq. (10), the resulting Green's function may be so wildly oscillatory that no significant results can be obtained.

Now let us recall that the experimental measurements are made at finite resolution, and we hope to obtain also a finite-resolution theory. Suppose we define a "finite-resolution" or "energy-averaged" Green's function $\bar{G}_E(\vec{q},\vec{q}_0)$ as

$$\bar{G}_E(\vec{q},\vec{q}_0) = \int G_{E'}(\vec{q},\vec{q}_0)\ g(E - E')dE' \tag{11}$$

where $g(E - E')$ is some "convolution function" or "window function" of width ΔE. Can we make a semiclassical approximation to $\bar{G}_E(\vec{q},\vec{q}_0)$? If we put Eq. (10) into Eq. (11), then it is obvious that the rapidly-oscillating terms associated with long orbits will average to zero. Only terms with orbits of duration $T_k \lesssim 2\pi\hbar/\Delta E$ will survive. Hence a semiclassical approximation to $\bar{G}_E(\vec{q},\vec{q}_0)$ can be obtained by including in the sum (7) only orbits which travel from $\vec{q}_0$ to $\vec{q}$ in a time less than $T_{max} = 2\pi\hbar/\Delta E$.

Now it is well known that the Green's function $G_E(\vec{q},\vec{q}_0)$ contains all the information about the quantum mechanics of the system. The energy-averaged $\overline{G}_E(\vec{q}, \vec{q}_0)$ contains the corresponding infomation to lower resolution. In particular, a simple formula relates the Green's function to the absorption spectrum, and the energy-averaged Green's function to the finite-resolution absorption spectrum.

Let us define $\overline{Df}(E)$ as the average oscillator strength, which is proportional to the average rate of photon absorption in a finite-resolution measurement. This quantity is related to the finite-resolution Green's function by the formula

$$\overline{Df}(E) = \text{constants Im} \langle D\psi_i | \overline{G}_E | D\psi_i \rangle. \tag{12}$$

In this formula ψ_i is the initial state (the $2p_z$ state), and D is the component of the dipole operator representing the radiation field.

It naturally follows from Eqs. (12) and (10) that $\overline{Df}(E)$ is given by a smooth background plus a sum of oscillatory terms

$$\overline{Df}(E) = \overline{Df}_0(E) + \sum_k C_k \sin(T_k E/\hbar + \Delta_k). \tag{13}$$

The smooth background $Df_0(E)$ turns out to be equal to the oscillator strength in the absence of an external field. Each oscillatory term is associated with a closed orbit that begins and ends in the vicinity of the nucleus, overlapping with the initial state ψ_i.

PHYSICAL PICTURE

The above formulas are closely connected with a simple physical picture of the absorption process (Fig. 3). The electron is initially in the $2p_z$ state, close to the nucleus, where the diamagnetic term is negligible. The laser excites the electron to states close to the ionization threshold, and the electron goes into a near-zero-energy outgoing Coulomb wave. This wave propagates outward in all directions from the nucleus. Its angular dependence is determined by the spherical harmonics in $|D\psi_i\rangle$.

At moderate distances from the nucleus, the propagation of the wave can be described semiclassically. Associated with the spherically out-going wavefronts are a family of radially outgoing trajectories. The wavefronts propagate outward, transverse to the outward-propagating trajectories, with an amplitude determined by the initial angular dependence of the waves, and by the speed and divergence of the trajectories.

The trajectories and their associated waves continue outward to very large distances (~2000 a_0). However they cannot escape in any direction transverse to the B-field, so eventually they are turned back. Some of the trajectories are pulled in such directions that they return before long to the nucleus. Surrounding each such closed orbit is a family of orbits that return to the vicinity of the nucleus. Each such family carries with it a returning wave.

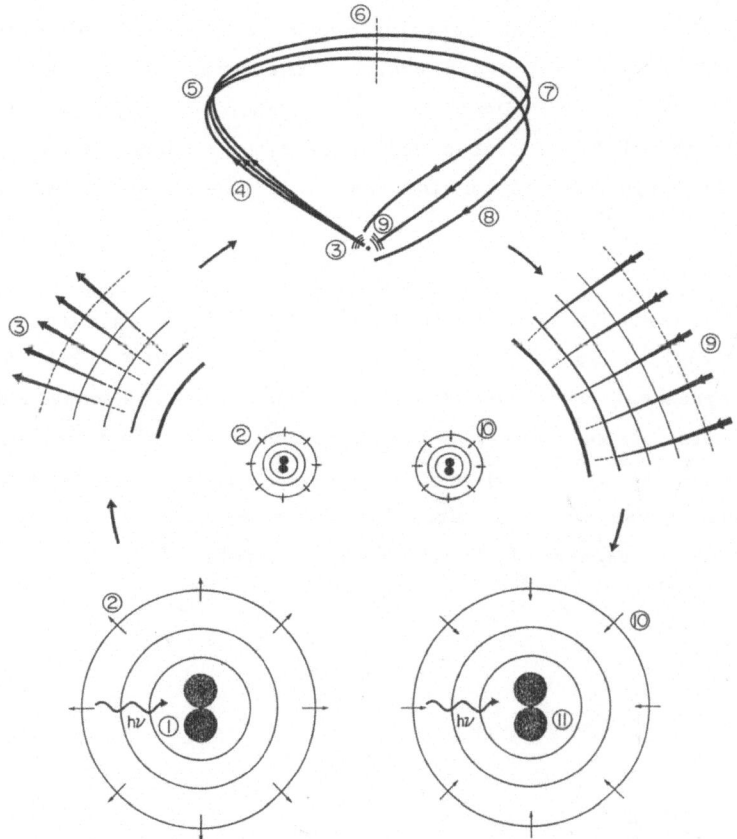

Fig. 3   Physical picture of the absorption process. (1) The atom is initially in the $2p_z$ state, with the oscillating field due to the laser present. (2) The oscillating field produces zero-energy Coulomb waves, which propagate outward in all directions. (3) For distances greater than about 50 a_0, a semiclassical approximation is appropriate, and we can propagate the wave outward by following classical trajectories. (4) A pencil of trajectories propagates outward, encounters a caustic (5), a focus (6), and another caustic (7). This group of trajectories started out in such a direction that it turned around and returned to the atom (8). Around 50 a_0, we describe it as a incoming zero-energy Coulomb wave (9), which continues to propagate inward (10), until it overlaps with the initial $2p_z$ state (11). Interference between steadily-produced outgoing and incoming waves leads to oscillations in the absorption spectrum.

It is the interference between the steadily-produced outgoing waves
and the steadily-returning waves that produces the observed oscillations
in the spectrum. As the energy changes, the action integral along each
closed orbit changes according to Eq. (8), and the relative phase of
outgoing and returning waves varies in the same way, producing an
interference pattern described by Eq. (13).

The amplitude of the kth spectral oscillation depends upon the
amplitude of the kth group of returning waves, and that in turn depends
upon (1) the amplitude of outgoing waves in the initial direction of
motion of the kth closed orbit (which comes from the angular dependence
of $|D\psi_i\rangle$), and (2) the divergence of neighboring trajectories from the
central closed orbit. The absolute phase Δ_k of each spectral oscillation
depends mainly upon the action integral along the closed orbit at $E = 0$.

RESULTS

Spectral-oscillation amplitudes C_k have been calculated for the
process $H(2p_z) + h\nu \rightarrow H(E \sim 0, m = 0)$ at $B = 5.96$ T (Fig. 4). Very
pleasing agreement with the Bielefeld experiments is obtained for the
closed orbits of relatively short duration (T_k/cyclotron period $\lesssim 6$).
For longer orbits there are some discrepancies, but at the moment we do
not consider these to be interesting. The experiment has already been
improved, and if necessary the theory can be refined.

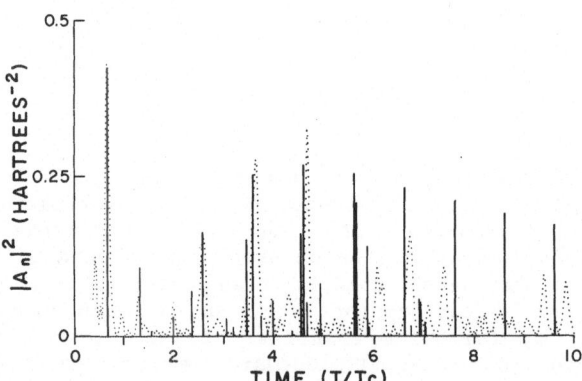

Fig. 4   Fourier transform of the absorption spectrum. The vertical
lines are our spectral-oscillation amplitudes $|C_k|^2$ computed at
$E = 0$. Each amplitude $|C_k|^2$ is placed on the time axis at T_k,
the duration of the orbit. This is compared with the Fourier
transform of the measured absorption spectrum as reported in
Ref. 2.

CONCLUSIONS

A theory is now available which quantitatively accounts for the oscillations observed in the absorption spectrum of a hydrogen atom in a magnetic field. The theory gives information about the finite-resolution spectrum from examination of classical trajectories of the system. (As presently applied, the theory says nothing about individual discrete or quasi-discrete states of the system.)

A more general remark is relevant to quantum chaology (the study of the quantum properties of classically chaotic systems). Even in this classically chaotic system, when the trajectories are examined on a limited time scale, up to some maximum time T_{max}, a kind of order is retained. Surrounding each closed orbit is a family of orbits traveling in step with the closed orbit, though gradually moving away from it. (For any finite time t, there is a smooth relationship between $\vec{q}(t)$ and the initial conditions of the orbit.)

The surprise is that this finite-time order is manifested in the finite-resoultion spectrum. For, associated with each such family of orbits is a propagating wave, which is as orderly as the family of orbits. This wave produces oscillations which are visible in the absorption spectrum if it is measured to a resolution $\Delta E \sim 2\pi\hbar/T_{max}$.

ACKNOWLEDGEMENTS

This work was supported by the National Science Foundation. It was carried out while JBD was a Visiting Fellow at JILA, and we both acknowledge the hospitality of the Institute.

REFERENCES

1. W.R.S. Garton and F. S. Tomkins, Ap. J. 158, 839 (1969); K. T. Lu, W.R.S. Garton and F. S. Tomkins, Proc. Roy. Soc. A 362, 421 (1978).

2. A. Holle, G. Wiebusch, J. Main, B. Hager, H. Rottke and K. H. Welge, Phys. Rev. Lett. 56, 2594 (1986); J. Main, G. Wiebusch, A. Holle and K. H. Welge, ibid. 57, 2789 (1986).

3. A. R. Edmonds, J. Phys. 31, C4-71 (1970).

4. U. Fano, J. Phys. B 13, L519 (1980); Phys. Rev. A 22, 2660 (1980).

5. R. Balian and C. Bloch, Ann. Phys. 60, 401 (1970); 64, 271 (1971); 69, 76 (1972); 85, 514 (1974).

6. M. C. Gutzwiller, J. Math. Phys. 8, 1979 (1967); 10, 1004 (1969); 11, 1791 (1970); 12, 343 (1971).

7. M. V. Berry and M. Tabor, Proc. Roy. Soc. A 349, 101 (1976).

SPECTRA OF THE STRONGLY MAGNETIZED HYDROGEN ATOM

CLOSE TO THE IONIZATION THRESHOLD

G. Zeller, U. Woelk, G. Wunner, and H. Ruder

Lehrstuhl für Theoretische Astrophysik
Universität Tübingen
D-7400 Tübingen, Federal Republic of Germany

We determine numerically energies and wavefunctions for the highly excited hydrogen atom in magnetic fields of about 6 Tesla and compute the spectra of Balmer transitions up to energies of $\sim -10 \text{ cm}^{-1}$ below the field-free ionization limit. Fourier transforms of the spectra show how the structural characteristics of the spectra change with small variations of the magnetic field strength.

The hydrogen atom in th presence of a homogeneous magnetic field (in z-direction) is represented by its Hamiltonian

$$H = p^2 - \frac{2}{r} + 2\beta L_z + \beta^2(x^2 + y^2)$$

(in atomic units, $\beta = B/B_0$ with $B_0 = 2(\alpha m_e c)^2/(e\hbar) \cong 4.70 \cdot 10^5$ T), which reduces after separating the trivial azimuthal dependence to that of a nonintegrable system with two degrees of freedom. To solve this Schrödinger equation we expanded the wavefunctions in terms of spherical harmonics,

$$\Psi_m \sim \sum_\ell g_{n,\ell}(r) \, Y_{\ell,m}(\theta,\phi),$$

(with fixed magnetic quantum number m) and the radial functions $g_{n,\ell}(r)$ in the complete, orthonormal set of functions

$$g_{n,\ell}^{(\zeta)}(r) = \zeta^{\frac{3}{2}} \sqrt{\frac{n!}{(n + 2\ell + 2)!}} \, e^{\frac{-\zeta r}{2}} (\zeta r)^\ell \, L_n^{(2\ell+2)}(\zeta r),$$

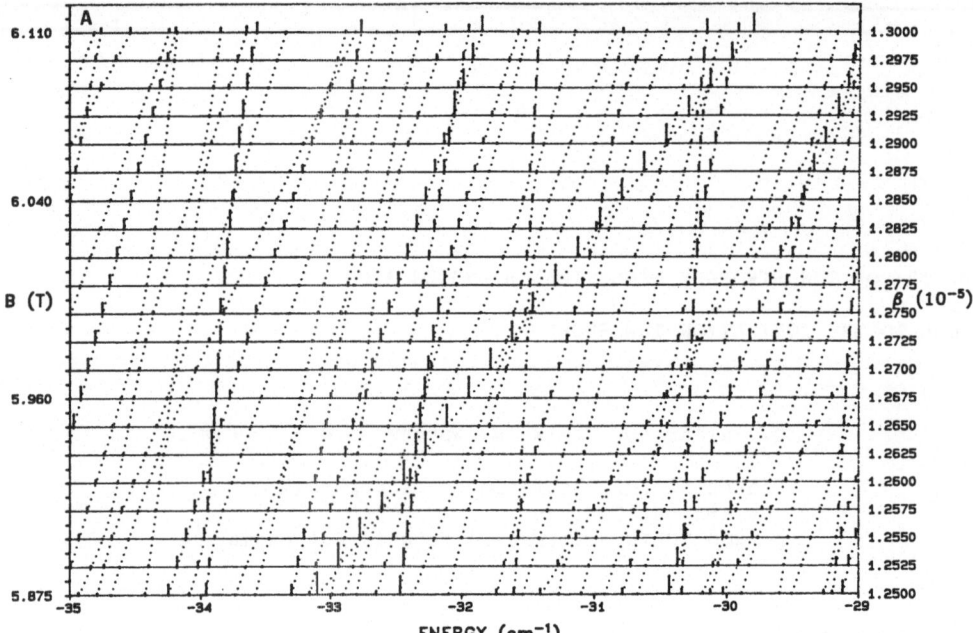

Figure 1a. Theoretical photoabsorption spectra of $\Delta m = 0$ transitions from $2p_0$ to even parity final Rydberg states with energies between -35 cm^{-1} and -29 cm^{-1} for different magnetic field strengths.

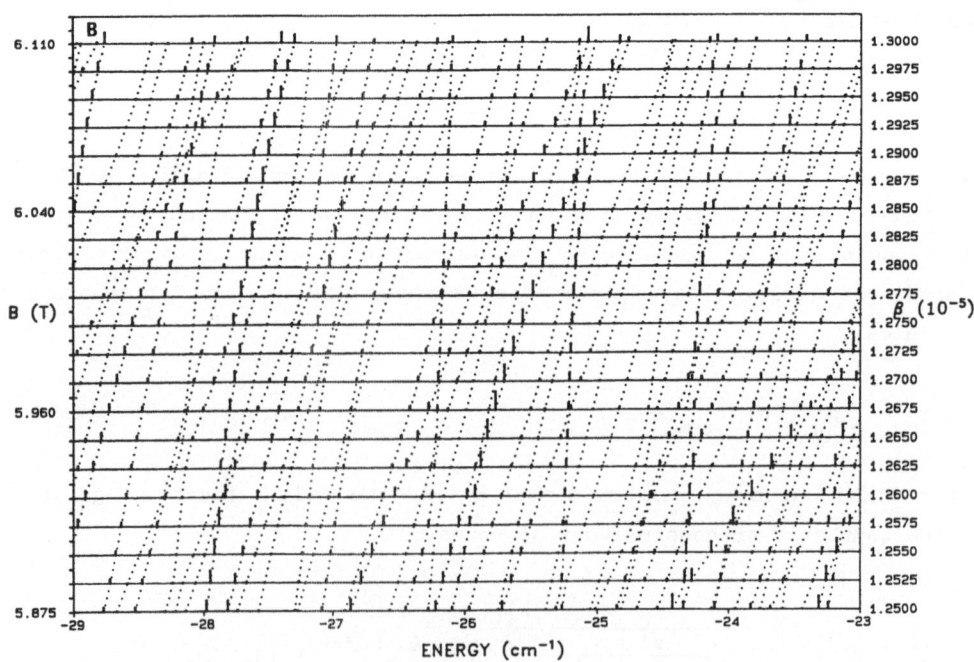

Figure 1b. Spectra between -29 cm^{-1} and -23 cm^{-1}.

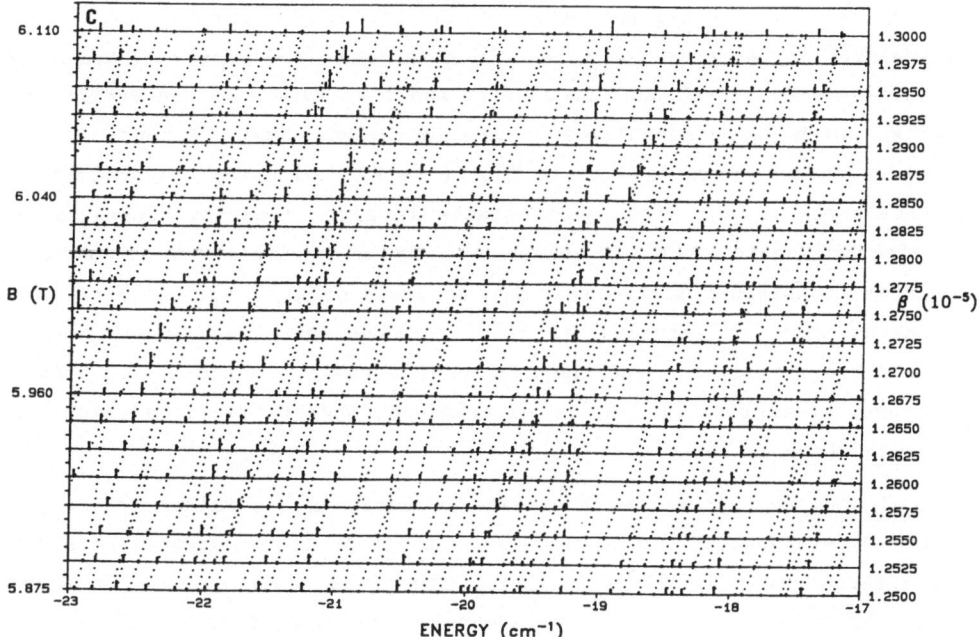

Figure 1c. Spectra between -23 cm^{-1} and -17 cm^{-1}.

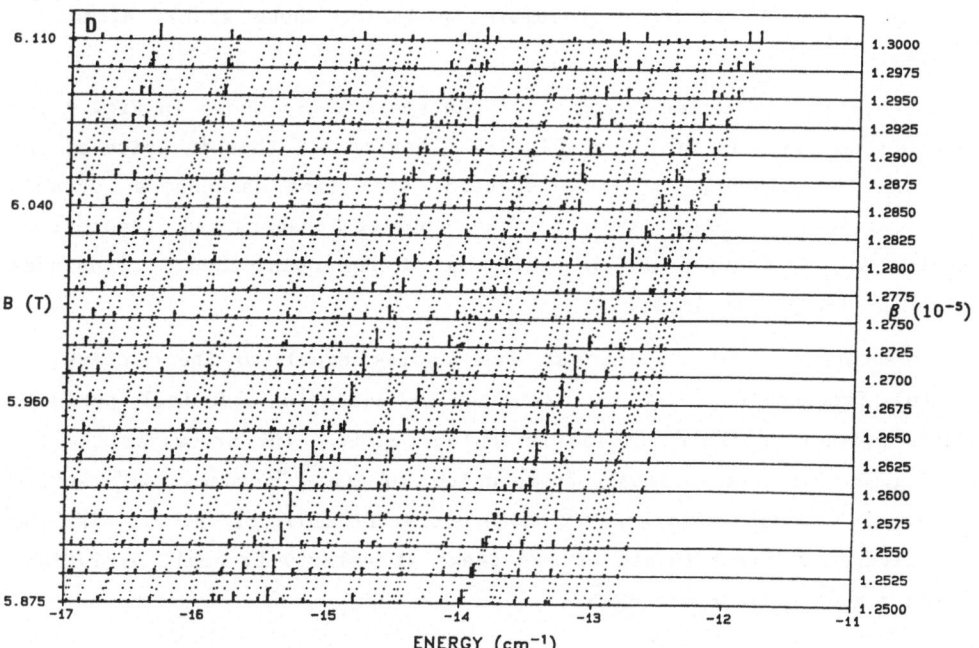

Figure 1d. Spectra between -17 cm^{-1} and -11 cm^{-1}.

where ζ denotes a scale parameter of inverse length and the $L_n^{(2\ell+2)}$ are generalized Laguerre polynomials. In this basis the matrix elements can be expressed in closed analytical form and they form a banded Hamiltonian matrix, which was diagonalized by standard algorithms. In our calculations we worked up to 6400 basis functions for computing eigenvalues and eigenvectorss of the first ~ 700 states in given m-parity subspaces and in fixed magnetic field strengths β. So we obtained the energies of spectral transitions with a relative accuracy of 10^{-6} and the oscillator strengths with an accuracy of 10^{-3}. Convergence was established by varying the parameter ζ and the size of the basis. A comparison between our theoretical results and the experimental spectra yields an excellent agreement (see Holle et al.[1]).

Figures 1a-d show the theoretical spectra $\Delta m = 0$ Balmer transitions to $m = 0$, even-parity final states as a function of the magnetic field in the interval from 5.95 T to 6.05 T. The plots cover a range of energy from -35 cm^{-1} to -11 cm^{-1} where the transition from regularity to irregularity occurs. In these figures we find an extreme sensitivity of the oscillator strengths, and thus of observable spectra, with respect to small variations of an external parameter, viz. the magnetic field strength. While in the range of Fig. 1a, where a small part of the transitions is still regular, the oscillator strengths are nearly preserved passing through avoided crossings, we see in the totally irregular range of Fig. 1d that some strong lines disappear in the crossings, and it becomes impossible to follow these states along the increasing field.

Examples of Fourier analyses of the spectra are shown in Fig. 2 for four of the magnetic fields of Fig. 1. They reveal that in spite of the seeming irregularity of the spectra the line sequences contain periodicities. These peaks in the Fourier transformed theoretical spectra are indicative of the build-up of the quasi-Landau resonances below the ionization limit.

The resonances, which appear at frequencies equal to their classical orbital frequencies, can be assigned to unstable classical periodic orbits present in the otherwise irregular regime. The arrows at the top of Fig. 2 indicate orbital frequencies of the "fundamental" series discussed by Main et al.[2] A noteworthy finding of Fig. 2 is the sensitivity of the strength of the peaks at relatively small variations of the magnetic field strength.

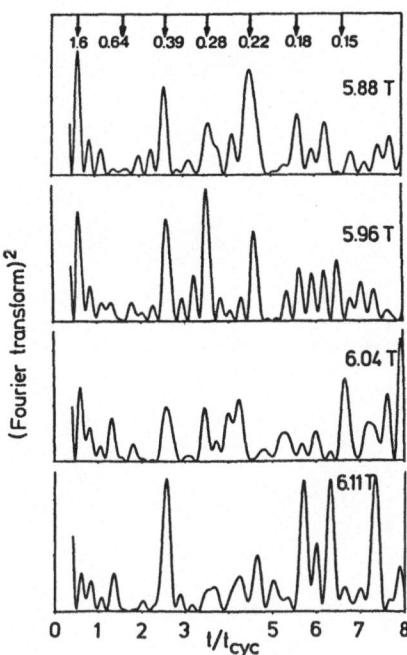

Figure 2. Fourier spectra of the theoretical photoabsorption spectra with m = 0 in the range of −35 cm^{-1} to −13 cm^{-1}.

REFERENCES

1. A. Holle, G. Wiebusch, J. Main, K. H. Welge, G. Zeller, G. Wunner, T. Ertl, H. Ruder, Z. Phys. D5, 279, (1987)

2. J. Main, G. Wiebusch, A. Holle, K. H. Welge, Phys. Rev. Lett. 57, 2789, (1986).

ACCURATE EIGENENERGIES FOR THE MAGNETIZED HYDROGEN ATOM

Paulo C. Rech, Márcia R. Gallas and Jason A.C. Gallas

Universidade Federal de Santa Catarina
Departamento de Física
88049 Florianópolis, Brazil

The year of 1986 was extremely fruitful in results for the quadratic Zeeman effect. We mention three examples: first, quasi-Landau resonances having a closer spacing than 1.5 $\hbar\omega$ near the zero-energy threshold have been reported experimentally by the Bielefeld group[1,2] and explained theoretically.[3-6] The second example is the investigations connected with the manifestation of quantum chaos, i.e. how does classical deterministic chaos manifest itself in the quantum spectrum of energies.[4,7,8] Connected with this is the very interesting work of Wintgen[9] showing the existence of long-range correlations in the quantum spectrum. These contributions are particularly interesting because in contrast to previous studies of model Hamiltonians, they are based on a real physical system which can be investigated in the laboratory. As observed by Professor Friedrich in his lecture at this Conference, the magnetized hydrogen atom is becoming the system _par excellence_ to investigate quantum chaos. The third example is the beautiful results of O'Mahony and Taylor on the quadratic Zeeman effect for nonhydrogenic systems.[10,11] The central point in almost all the aforementioned theoretical works was the calculation of the energy spectrum of a magnetized atom.

The purpose of this paper is to present a single variational function capable of producing very accurate approximations to the eigenenergies of a hydrogen atom in a uniform magnetic field. In fact, to the best of our knowledge, the trial function presented here is the only _single function_ capable of consistently generating eigenenergies with at least 5 significant digits in the whole parameter range $0 \leq \gamma < \infty$, where γ is the magnetic field strength in units of 2.35×10^5 T. Our function is based on an expansion in parabolic coordinates proposed by one of us,[12] using a basis combining characteristics of both the Coulomb and Landau regimes. Calling

$$
\begin{aligned}
f(x,y) = {} & 1 + x_1 c(x+y) + x_2 c^2 xy + x_3 c^2(x^2 + y^2) + x_4 c^3 xy(x+y) \\
& + x_5 c^3(x^3+y^3) + x_6 c^4 x^2 y^2 + x_7 c^4 xy(x^2+y^2) + x_8 c^4(x^4+y^4) \\
& + x_9 c^5(x^5+y^5) + x_{10} c^5 xy(x^3+y^3) + x_{11} c^5 x^2 y^2(x+y) \\
& + x_{12} c^6 x^2 y^2(x^2+y^2) + x_{13} c^6 xy(x^4+y^4) + x_{14} c^6 x^3 y^3,
\end{aligned}
$$

the trial function is given by $\psi(x,y) = f(x,y) \exp[-\frac{1}{2} a(x+y+acxy)]$, where $x \equiv \xi$ and $y \equiv \eta$ are the usual parabolic coordinates and a trivial normalization constant has been omitted. The great advantage of this trial function is having the exponential dependence appropriate to the symmetry of the problem.

Table 1. Comparison of binding energies (in a.u.) for a magnetized hydrogen atom. E_{RWHR} are the energies of Rösner, Wunner, Herold and Ruder.[14] The number in parenthesis indicates the number of configurations needed to guarantee their quoted digits. E_N are energies obtained by us using N configurations of the basis proposed by Gallas.[12]

γ	E_{RWHR}	E_6	E_{12}	E_{15}
1	0.831169(7)	0.83117	0.83117	0.831169
2	1.022214(11)	1.02220	1.02221	1.022214
3	1.164533(11)	1.16451	1.16453	1.164533
4	1.280798(12)	1.28077	1.28080	1.280798
10	1.747797(19)	1.74774	1.74780	1.747797
20	2.215398(20)	2.21533	2.21539	2.215398
40	2.801029(24)	2.80094	2.80102	2.801028
100	3.78905 (12)	3.78952	3.78977	3.789799
200	4.72655 (12)	4.72626	4.72709	4.727134
300	5.36030 (12)	5.35920	5.36074	5.360799
1000	7.66205 (12)	7.65562	7.66231	7.662388
2000	9.30448 (12)	9.29149	9.30462	9.304706
10000		14.09563	14.14038	14.140629
20000		16.63834	16.70516	16.705632

Peculiarities of the basis being used were already discussed by us in the literature.[12,13]

Using the above function a generalized 15×15 matrix eigenvalue problem was solved for the eigenenergies. The last column in Table 1 presented the results of such a calculation. In this table our results are compared with the results of the very detailed 1984 eigenfunction expansion of Rösner, Wunner, Herold and Ruder.[14] Their results are the most accurate available in the literature covering the whole range $0 \leq \gamma < \infty$, and were obtained by expanding the eigenfunction either in a Coulomb basis (for $\gamma \leq 40$) or in a Landau basis (for $\gamma > 40$). The number in parenthesis after their energies indicates the number of expansion terms that they needed to guarantee the quoted digits in the energy. In their calculations it was important to obtain first <u>starting functions</u> for the computation (thereby needing to solve numerically an eigenvalue equation), then to solve the actual problem, and, finally, to study the convergence of the procedure as a function of the terms in the spherical <u>and</u> the cylindrical basis. In contrast, our method involves a search for the minimum of $E = E(a,c)$ as a function of the two non-linear variational parameters a and c. The procedure involves a trivial and stable diagonalization of a 15×15 matrix. Altogether, our calculation involves 16 variational parameters (14 linear + 2 non-linear) but, obviously, 14 of them are automatically determined by the diagonalization procedure. To give an idea of convergence properties within the basis being used, Table 1 presents two further columns obtained by considering (i) a 6×6 matrix formed only with terms involving x_1, x_2, x_4, x_6 and x_7 in the above equation, and (ii) the 12×12 matrix obtained by neglecting x_{12}, x_{13} and x_{14} in the trial function.

From the energies presented in Table 1 one can assess the accuracy of the calculation of Rösner et al.[14] One sees that the region roughly between $10 \leq \gamma \leq 200$ is a "difficult" region for their calculations: up to about $\gamma=40$ they had to greatly increase the size of the basis in order to maintain the same number of significant digits in the eigenvalues. For $\gamma=100$ their value obtained using 12 configurations is not as good as ours obtained with just 6 configurations (involving 7 variational parameters). At this point it would be useful to know how to compare the effort needed to obtain eigenvalues of the same accuracy by both methods. This question is difficult to answer but it seems safe to say that for equal number of terms in the expansions our method is much easier to implement than that of Herold et al. Therefore we believe our results involving 12 configurations to be much simpler to obtain than the corresponding ones obtained by Herold et al., also from 12 configurations. It is also worth mentioning that the non-relativistic model on which all calculations are based imposes a limit on the strength of the magnetic field for the energies to remain meaningful. As mentioned in Ref. 12, a limit on the magnetic field strength for the model to remain reliable is obtained by estimating the field required to produce an energy difference of mc^2 between neighbouring Landau levels. This defines the threshold $\gamma_{th} \cong 18700$, which is therefore the upper limit for the variation of γ. That is why our table includes results up to $\gamma \cong \gamma_{th}$. A further noteworthy point is the existence of calculations based on high-field expansions as, for example, that of Baye and Vincke.[15] Using at least 18 variational parameters they reported $E_B = 7.662405$ and 9.30475 for $\gamma = 1000$ and 2000, respectively. Being results of a specific high-field calculation, these 2 energies are better than the corresponding ones of Rösner et al. They are also better than the more accurate results being reported here. However it is important to realize that they use a larger basis and that their calculations are totally unable to deal with the Coulomb limit.

In summary, we presented binding enegies E_6, E_{12} and E_{15} for a magnetized hydrogen atom, based on variational calculations involving 7, 13, and 16 variational parameters, respectively. These energies cover the whole range of magnetic field strengths, from the Coulomb to the Landau limit. Our energies were compared with the most accurate results presently available in the literature, namely with the results of Rösner et al.[14] Our results obtained from the 13-parameter calculation are as good as the ones obtained by Rösner et al. The energies obtained from the 16-parameter calculation agree to at least 7 significant digits for γ less than about 40, and are superior for all other γ. We have also investigated the excited states obtaining similar results. A detailed report of our findings will be presented elsewhere.

PCR and MRG are predoctoral fellows. JACG is a research fellow of the Brazilian Research Council (CNPq). The work of our group is partially supported by the Brazilian agencies CNPq, FINEP and CAPES.

REFERENCES

1. A. Holle, G. Wiebusch, J. Main, B. Hager, H. Rottke and K. H. Welge, Diamagnetism of the hydrogen atom in the quasi-Landau regime, Phys. Rev. Lett. 56, 2594 (1986).

2. J. Main, G. Wiebusch, A. Holle and K. H. Welge, New quasi-Landau structure of highly excited atoms: the hydrogen atom, Phys. Rev. Lett. 57, 2789 (1986).

3 D. Wintgen, A. Holle, G. Wiebusch, J. Main, H. Friedrich and K. H. Welge, Precision measurements and exact quantum mechanical calculations for diamagnetic Rydberg states in hydrogen, J. Phys. B 19, L557 (1986).

4. D. Wintgen and H. Friedrich, Regularity and irregularity in spectra of the magnetized hydrogen atom, Phys. Rev. Lett. 57 571 (1986).

5. M. A. Al-Laithy, P. F. O'Mahony and K. T. Taylor, Quantization of new periodic orbits for the hydrogen atom in a magnetic field, J. Phys. B 19, L773 (1986).

6. D. Wintgen and H. Friedrich, Correspondence of unstable periodic orbits and quasi-Landau modulations, Phys. Rev. A 36, 131 (1987).

7. D. Delande and J. C. Gay, Quantum chaos and statistical properties of energy levels: numerical study of the hydrogen atom in a magnetic field, Phys. Rev. Lett. 57, 2006 (1986), errata 57, 2877 (1986).

8. G. Wunner, U. Woelk, I. Zech, G. Zeller, T. ERtl, F. Geyer, W. Schweitzer and H. Ruder, Rydberg atoms in uniform magnetic fields: uncovering the transition from regularity to irregularity in a quantum system, Phys. Rev. Lett. 57, 3261 (1986).

9. D. Wintgen, Connection between long-range correlations in quantum spectra and classical periodic orbits, Phys. Rev. lett. 58, 1589 (1987).

10. P. F. O'Mahony and K. T. Taylor, the quadratic Zeeman effect in caesium: departures from hydrogenic behaviour, J. Phys. B 19, L65 (1986).

11. P. F. O'Mahony and K. T. Taylor, Quadratic Zeeman effect for nonhydrogenic systems: application to the Sr and Ba atoms, Phys. Rev. lett. 57, 2931 (1986).

12. J.A.C. Gallas, Zeeman diamagnetism in hydrogen: on variational Ansätze for arbitrary field strengths, J. Phys. B 18, 2199 (1985).

13. P. C. Rech, M. R. Gallas, and J.A.C. Gallas, Zeeman diamagnetism in hydrogen at arbitrary field strengths, J. Phys. B 19, L215 (1986).

14. W. Rösner, G. Wunner, H. Herold and H. Ruder, Hydrogen atoms in arbitrary magnetic fields: I. Energy levels and wave-functions, J. Phys. B 17, 29 (1984).

15. D. Baye and M. Vincke, A simple variational basis for the study of hydrogen atoms in strong magnetic fields, J. Phys. B 17, L631 (1984).

LIAPUNOV EXPONENTS FOR THE DIAMAGNETIC KEPLER PROBLEM AND THE TRANSITION FROM REGULAR TO IRREGULAR MOTION

W. Schweizer,[*] H. Friedrich, G. Wunner, R. Niemier and H. Ruder

Univeristät Tübingen, Lehrstuhl f. Theoretische Astrophysik, D 7400 Tübingen, FRG
[*]Univeristy of London, Royal Holloway and Bedford New College, Department of Mathematics, Egham, TW20 0EX, GB

ABSTRACT

For the diamagnetic Kepler problem the transition from regular to irregular motion is observed by calculating Liapunov exponents of periodic orbits over a large energy scale.

I. INTRODUCTION

One of the most important and unsolved questions in physics is the relationship between classical mechanics and quantum mechnaics for irregular or chaotic systems.[1] As the magnetized hydrogen atom is complex enough to display the essential features of a chaotic system, and as it is a real system which has been prepared in the laboratory[2] it is an ideal physical example for studying this relationship.

Garton and Tomkins[3] discovered unexpected oscillatory structure in the photoabsorption cross sectionsof Rydberg atoms in a magnetic field. Together with a general description of the classical motion of the diamagnetic Kepler problem, Edmonds and Pullen[4] showed that this structure is correlated with the periodic orbit in the plane perpendicular to the magnetic field. Besides these quasi-Landau modulations further modulations have recently been measured[2,5] and have been related to periodic orbits outside the plane perpendicular to teh field.[5-8] Therefore an observable manifestation of these classically unstable periodic orbits in quantum mechanics is the occurence of many

---

[*]Permanent address: Universität Tübingen, Lehrstuhl f. Theoretische Astrophysik, D7400 Tübingen, FRG

superimposed modulations in the photoabsorption cross section.

Whether or not neighboring trajectories separate exponentially with time (unstable or stable) is indicated by the maximal Liapunov exponent.[9] Wintgen[10] calculated Liapunov exponents in the anharmonic oscillator picture of the diamagnetic Kepler problem for the trajectories parallel and perpendicular to the magnetic field. For Hamiltonian systems, Meyer[11] has given an operational definitionof Liapunov exponents which is both mathematically rigorous and amenable to practical calculation. Using his definition we calculated Liapunov exponents for periodic orbits of the diamagnetic Kepler problem over a large energy scale.

II. METHODS

The Hamiltonian

The Hamiltonian of the problem in cylindrical coordinates and atomic units is

$$H = \frac{1}{2} (p_z^2 + p_\rho^2) + \frac{m^2}{2\rho^2} - \frac{1}{r} + \frac{1}{2} \beta^2 \rho^2 \tag{1}$$

where β is the magnetic field strength in units of $B = 4.7 \times 10^5 T$ and the trivial linear Zeeman term is omitted. This Hamiltonian is simplified employing a scaling law used by Robnik.[12]

$$H(\vec{p}, \vec{r}; m; \beta) = \beta^{2/3} H(\beta^{-1/3} \vec{p}, \beta^{2/3} \vec{r}; \beta^{1/3}m; \beta = 1), \quad t \to \beta t \tag{2}$$

The scaled angular momentum $\beta^{1/3}m$ is negligibly small under laboratory conditions and will be ignored inthe following. If we introduce scaled semiparabolic coordinates

$$\mu = \beta^{1/3}(r + z)^{1/2} \qquad \nu = \beta^{1/3}(r - z)^{1/2} \tag{3}$$

and the canonical conjugate momenta

$$p_\mu - \frac{d\mu}{d\tau} \qquad\qquad p_\nu = \frac{d\nu}{d\tau} \tag{4}$$

corresponding to the scaled time τ given by

$$dt = (\mu^2 + \nu^2) \, d\tau \tag{5}$$

then the equation of motion for fixed scaled energy ε are equivalent to the equations generated by the Hamiltonian

$$K = \frac{1}{2}(p_\mu^2 + p_\nu^2) - \varepsilon(\mu^2 + \nu^2) + \frac{1}{2}\mu^2\nu^2(\mu^2 + \nu^2) \equiv 2.$$ (6)

For numerical calculations this Hamiltonian is much easier to deal with then the original scaled Hamiltonian (2), whose potential is singular at the origin.

The Liapunov exponents

Meyer[11] calculated the Liapunov exponent of a trajectory which passes through a phase space point $\vec{\gamma}(t = 0)$ of a Hamiltonian system via the stability matrix M(t', t) defined by

$$M_{ij}(t', t) = \frac{\partial \vec{\gamma}_i(t')}{\partial \vec{\gamma}_j(t)}.$$ (7)

The Transformation induced by M is canonical, and therefore M is a symplectic 2n × 2n matrix[13] for a system of n degrees of freedom. In an approbriate order the eigenvalues of M fulfill

$$d_{i+n} = d_i^{-1} \quad 1 < i < n$$ (8)

and the eigenvalues MM^t, where t denotes the transposed matrix,

$$0 < d_1^2 < \ldots\ldots < d_n^2 < 1.$$ (9)

The maximum Liapunov exponent λ_∞ is then given by[11,14]

$$\lambda(t) = \frac{1}{t} \log \|M(t, 0)\|, \quad \lambda_\infty = \lim_{t\to\infty} \lambda(t)$$ (10)

where $\|M(t,0)\|$ denotes the matrixnorm[14] given by the maximum eigenvalue of MM^t and $\lambda(t)$ the (unphysical) Liapunov function. From the definition of the stability matrix M it follows that

$$M(t + t', 0) = M(t + t', t) \cdot M(t, 0)$$ (11)

which allows us to calulate the full stability matrix step by step by matrix multiplication.

For periodic orbits the Liapunov exponent becomes particularly easy to calculate. If T is the period of the orbit, then $M(n \cdot T, 0) = M(t, 0)^n$ and

$$\|M(T, 0)^n\| = \rho^n \text{ for } n \to \infty, \text{ where } \rho = \|M(T, 0)\|, \tag{12}$$

which can be derived by diagonalization. But note, that in general $\|M(T, 0)^n\| \neq \|M(T, 0)\|^n$ for any finite integer value n as the Liapunov function is basis dependent in contrast to the Liapunov exponent.

The calculation

In practical calculations we integrate the equations of motion derived from the Hamiltonian (6) by iterative Taylor expansions for fairly long times. As emphasized by Meyer[11] such a method is very powerful in situations where high derivatives can be caluclated analytically via the iterative use of the canonical equations of motion. The stability matrix is evaluated at each integration step along a given trajectory and the Liapunov function (11) is obtained as the commulative matrix product

$$M(T, 0) = M(T, t_n) \, M(t_n, t_{n-1}) \, \ldots\ldots\ldots \, M(t_1, 0) \quad \text{with}$$

$$T > t_n > t_{n-1} > \ldots > t_1 > 0 \tag{13}$$

Both the scaled "Kepler time" and the scaled "semiparabolic time" τ (5) are stored during integration, and the corresponding scaled Liapunov exponents are given by (10) and scales as

$$\lambda_\infty\big|_\varepsilon = \beta^{-1} \cdot \lambda_\infty\big|_{\beta^{2/3}\varepsilon = E} \cdot \tag{14}$$

Clearly physical characteristics, such as the question of stability of a trajectory, cannot depend onthe choice of the representation; however, the quantities which measure those characteristics, such as the positive Liapunov exponent of a trajectory in an irregular part of phase space, may have very different absolute values and even energy dependence inthe different representations.

III) THE RESULTS

Because of lack of space we restrict the following results to the most important periodic orbits. Further results will be published elsewhere.

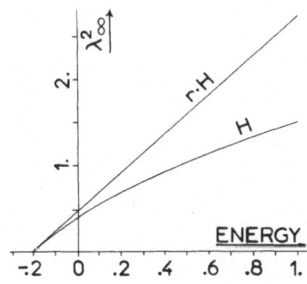

Figure 1.   Scaled energy and squared Liapunov exponent λ_∞ dependence for the trajectory perpendicular to the field in the anharmonic oscillator representation (r · H) and in the Kepler representation (H).

The results for the periodic orbit in the plane perpendicular to the magnetic field are shown in figure 1.   This orbit is stable up to $\varepsilon = -0.202$ in agreement with the value published by Delande and Gay.[15] The graph r · H is calculated with respect to the scaled "semiparabolic time" τ and can be fitted by a square-root law

$$\lambda_\infty^2\big|_{semi} = 2.30 \cdot (\varepsilon + 0.202) \qquad (15)$$

as previously published by Wintgen.[10] The other graph is calculated with respect to the scaled real physical time t and shows the Liapunov exponent dependence onthe chosen framework as mention above.

The scaled energy and Liapunov exponent λ_∞ dependence of the trajectory parallel to the field is shown by the shaded graph in figure 2 and illustrates the succesion of stable and unstable regions.   The scaled energy values where this orbit becomes unstable and again stable can be parametrized by

$$-\varepsilon_{i,j} = (\beta_{o,j} + \beta_{i,j})^{-2/3} \qquad (16)$$

where (j - 1), j ⩾ 2 labels the points of transition from stable to unstable with respect to increasing energy for i = 1, and for i = 2 from unstable to stable.   This equation was proposed by Main et al.[7] for the transition stable/unstable.   The values

$$\beta_{o,j} = \frac{1}{2}(j - 2), \quad \beta_{1,j} = j, \quad \text{and} \quad \beta_{2,j} = \frac{3}{4} + j$$

give a good fit to the true calculated results (see table I), which are in agreement with the previously published ones of Wintgen.[10]

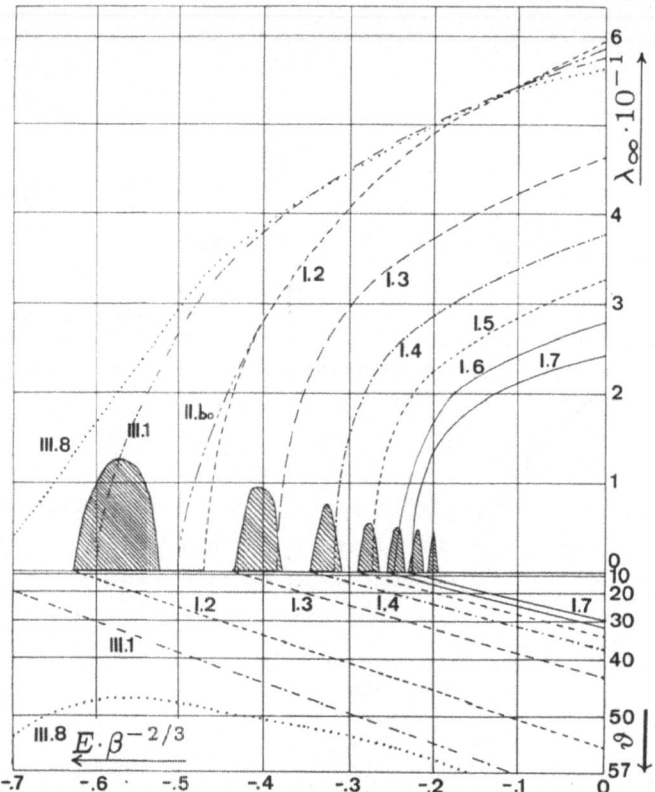

Figure 2. The figure in the upper part shows the scaled energy and Liapunov exponent λ_∞ dependence of some periodic orbits, which are labeled as in Ref. 8. The shaded graph belongs to the trajectory parallel to the field. In the lower part of the figure the initial angle (0 in the field direction) of the periodic orbits is drawn as a function of the scaled energy. These graphs show that the periodic orbits of the main series commence stable at the transition from stable to unstable regions of the field-parallel trajectory.

At each critical value $\varepsilon_{i,j}$ a bifurcation process take place. More specifically, the new periodic orbits, mentioned above, arise at the energy values $\varepsilon_{i,j}$. These orbits are stable from the energy value $\varepsilon_{i,j}$ up to some

Table I

Scaled transition energies $\varepsilon_{i,j}$ of Fig. 2, parameters of Eqs. (16) and (18) with corresponding orbits I_n.

n	I_n	$-a$	$-\varepsilon_{1,n}$	$-\varepsilon_{2,n}$	$\beta_{0,n}$	$\beta_{1,n}$	$\beta_{2,n}$
2	I_2	0.436	0.622	0.515	0	2	$2\frac{3}{4}$
3	I_3	0.352	0.433	0.378	$\frac{1}{2}$	3	$3\frac{3}{4}$
4	I_4	0.347	0.342	0.310	1	4	$4\frac{3}{4}$
5	I_5	0.338	0.289	0.265	$1\frac{1}{2}$	5	$5\frac{3}{4}$
6	I_6	0.313	0.254	0.238	2	6	$6\frac{3}{4}$
7	I_7	0.345	0.226	0.213	$2\frac{1}{2}$	7	$7\frac{3}{4}$
8	–	–	0.200	0.195	3	8	$8\frac{3}{4}$
–	III_1	0.475	0.773	–	–	–	–

higher energy where instability occurs. This can be seen (figure 2) by the energy gap between their place of commencement and the first occurence of non-zero Liapunov exponents. In the lower part of figure 2 we draw the initial angle θ of these periodic orbits, defined by[8]

$$\theta = \lim_{t \to 0} \arctan \frac{\rho(t)}{z(t)}, \tag{17}$$

as a function of the scaled energy, in such a way that it shows the phenomenological asymptotic square-root law

$$\lim_{\varepsilon \to \varepsilon_{1,n}} \left\{ \lim_{t \to 0} \left[\frac{p_\mu(t)}{p_\nu(t)} \right]^2 \right\} = a \cdot (\varepsilon - \varepsilon_{1,n}). \tag{18}$$

The upper part shows their scaled energy and Liapunov exponent λ_∞ dependence calculated in the Kepler representation.

W. S. acknowledges support from the Deutsche Forschungsgemeinschaft and would like to thank Dr. K. T. Taylor for his hospitality and encouragment at the RHBNC.

REFERENCES

1.) M. Berry, "Chaotic Behaviour of Deterministic Systems," North Holland, Amsterdam, 1983 p. 172.

2.) A. Holle, G. Wiebusch, J. main, B. Hager, K. H. Welge, Phys. Rev. Lett. 56, 1986, 2594.

3.) W.R.S. Garton, F. S. Tomkins, Astrophys. J. 158, 1969, 839.

4.) A. R. Edmonds, R. A. Pullen, Imperial College; ICTP/79-80/28.

5.) J. Main, G. Wiebusch, A. Holle, K. H. Welge, Phys. Rev. Lett. 57, 1986, 2789.

6.) M. A. Al-Laithy, P. F. O'Mahony, K. T. Taylor, J. Phys. B19, 1986, L773.

7.) J. Main, A. Holle, G. Wiebusch, K. H. Welge, Z. Phys. D 1987 (at press).

8.) D. Wintgen, H. Friedrich, Phys. Rev. A 1987 (at press).

9.) A. J. Lichtenberg, M. A. Liebermann, "Regular and Stochastic Motion," Springer, New York 1983.

10.) D. Wintgen, J. Phys. B20, 1987, L511.

11.) H. D. Meyer, J. Chem. Phys. 84, 1986, 3147.

12.) M. Robnik, J. Phys. A14, 1981, 3195.

13.) V. I. Arnold, "Mathematical Methods of Classical Mechanics," Springer, New York 1978.

14.) V. I. Oseledec, Trans. Mosc. Math. Soc. 19, 1968, 197.

15.) D. Delande, I. C. Gay, Phys. Rev. Lett. 57, 1986, 2006.

CROSS SECTION FOR POTENTIAL SCATTERING IN THE PRESENCE

OF A MAGNETIC FIELD AND LIMIT FOR B → 0

Silvano Nuzzo and Michelangelo Zarcone

Instituto di Fisica dell'Università
Via Archirafi, 36
90123 Palermo, Italia

In the recent past astrophysical objects such as pulsars and white dwarfs, suspected to possess huge magnetic fields, higher than 10^2 T and up to 10^9 T, have stimulated increasing interest in the study of electron scattering assisted by such strong fields.[1] This Note is intended to provide a few comments on the peculiarities of the scattering cross section, in order to understand how the scattering process is altered by the magnetic field and how the cross section evolves towards the field free one as the field goes to zero.

In the presence of a quantising magnetic field the electron motion is free along the direction of the field and becomes confined in the transverse direction. In cartesian coordinates and in the Landau gauge the wave-function for an electron in the presence of a magnetic field is given by[2]

$$\Psi = C \exp[i(p_z z - p_x x)\hbar] \quad \exp\{-\eta^2/2\} H_n(\eta) \tag{1}$$

where C is the normalization constant, $H_n(\eta)$ are Hermite polynomials with $\eta = (y - y_o)\rho_o$, $y_o = -p_x \rho_o^2/\hbar$ and $\rho_o = \sqrt{(c\hbar/eB)}$ is the cyclotron radius.

The correspponding energy eigenvalues are

$$E = E_\perp + E_z = (n + 1/2)\hbar\omega_c + p_z^2/2m \tag{2}$$

where $\omega_c = eB/mc$ is the cyclotron fequency.

In the S-Matrix approach the cross section is obtained dividing the probability transition per unit time P_{fi} by the incident current density

J_z and summing over all the final states (n_f, p_{xf}, p_{zf}) and over the degenerate initial states p_{xi}:

$$\sigma_T = \sum_{n_f, p_{xf}, p_{zf}} \sum_{p_{xi}} \frac{P_{fi}}{J_z} \tag{3}$$

where

$$P_{fi} = \frac{2\pi}{\hbar} \left| \langle \Psi_f |V| \Psi_i \rangle \right|^2 \delta(E_f - E_i) \tag{4}$$

and the current density of the incident particle summed over p_{xi} is

$$J_z = \frac{1}{2\pi\rho_o^2} \frac{1}{L_z} \frac{P_{zi}}{m} \tag{5}$$

To transform the sum over p_z into an integral over E, we introduce the density of final states along z

$$\sum_{p_{zf}} \rightarrow \int \rho(E) \, dE$$

with

$$\rho(E) = \frac{L_z}{2\pi} \frac{m}{k_{zf}} \tag{6}$$

For a screened Coulomb potential of the form

$$V(\underline{r}) = V_o \exp(-\alpha/r)$$

the total scattering cross section results[3]

$$\sigma_T = \frac{m^2}{\hbar^4} \sum_{n_f} \frac{1}{k_{zi} \, k_{zf}(n_f)} \frac{1}{2\pi\rho_o^2} \left| M^B_{fi} \right|^2 \tag{7}$$

130

where

$$|M_{fi}^B|^2 = 4 \pi^2 \rho_o^4 V_o^2 \frac{n_i!}{n_f!} \int_o^\infty ds \frac{\exp(-s)}{(s+\zeta)^2} s^{n_f-n_i} |L_{n_i}^{n_f-n_i}(s)|^2 \qquad (8)$$

In Eqs. (10) and (11)

$$\zeta = \frac{\rho_o^2}{2} \left\{ \frac{(p_{zi}-p_{zf})^2}{\hbar^2} + \alpha^2 \right\} \qquad (9)$$

$L_{n_i}^{n_f-n_i}$ are associated Laguerre polynomials and

$$k_{zf} = \pm \left[k_{zi}^2 + \frac{2m}{\hbar^2} (n_i-n_f)\hbar\omega_c \right]^{1/2} \qquad (10)$$

the positive values corresponding to transmission and the negative ones to reflection.

The cross section (7) presents two important features:
  i)   the one dimensional final state density introduces a dependence over $1/k_{zf}(n_f)$,
  ii) the bound electron motion in the plane perpendicular to the field gives a dependence over $1/(2\pi\rho_o^2)$.

Properties (i) and (ii) are essentially due to the form of the energy spectrum, Eq. (2), consisting of a combination of a discrete and a continuous spectrum, relative to the motion in the plane perpendicular and parallel to the field.

A similar behaviour arises in all the physical situations where the energy spectrum of the scattering particles has the same form of Eq. (2) as, for instance, in the case of field-free potential scattering of a charged particle when we impose a finite costraint to the motion in the plane (xy). In this case we have

$$\sigma_T^{FF} = \frac{m^2}{\hbar^4} \sum_{n_x n_y} \frac{1}{k_{zi} k_{zf}} \frac{1}{L_x L_y} |V(\underline{k}_f - \underline{k}_i)|^2 \qquad (11)$$

where the sum is over the discrete states of a particle in a two dimensional box $(L_x L_y)$ and $V(\underline{k}_f - \underline{k}_i)$ is the Fourier transform of the potential. In both cases, the cross sections Eq. (7) and Eq. (11) are found to be proportional to k_{zf}^{-1} (apart from the additional k_{zf} dependence of the matrix element).

When the initial kinetic energy (along z) matches exactly the energy difference between the initial and final Landau levels, k_{zf} goes to zero and the cross section undergoes a resonant growth. To remove the unphysical infinity of the cross section at the resonances one can introduce in the theory a natural width of the Landau levels and go over the F.B.A. in the calculation of the matrix elements when k_{zf} gets smaller.

Another interesting point deserving some attention is the analysis of the above cross section in the limit of vanishing magnetic field. From Eq. (7) we see that as B goes to zero the apparent vanishing contribution to the cross section of each term of the sum over $n_f m$ due to the factor $(\rho_0^{-1} \propto B)$, is compensated by the increasing number of final Landau levels to be included in the sum, going as $(n_{f,Max} \propto B^{-1})$.

For a fixed energy $E_\perp$, Eq. (4) implies that for B going to zero the number of allowed final Landau levels becomes infinite. Consequently the sum over n_f can be transformed into an integral

$$\sum_{n_f} \rightarrow \int dn_f = \frac{\hbar}{m \, \omega_c} \int k_{zf} \, dk_{zf} \tag{12}$$

and the cross section becomes

$$\sigma_T = \frac{m^2}{\hbar^4} \frac{1}{k_{zi}} \frac{1}{2\pi} \int |M_{fi}^B|^2 \, dk_{zf} \tag{13}$$

Considering the case when $n_i = 0$ we express k_{zi} and k_{zf} in terms of k and a scattering angle θ, in complete analogy with the field free case as $k_{zi} = k$; $k_{zf} = k \cos \theta$ getting

$$\sigma_T = \frac{m^2}{\hbar^4} \frac{1}{2\pi} \int |M_{fi}^B|^2 \sin\theta \, d\theta \tag{14}$$

This expression reduces to the field free cross section provided that

$$\lim_{B \to 0} |M_{fi}^B|^2 = |V(\underline{k}_f = \underline{k}_i)|^2 \tag{15}$$

The validity of the above expression has been proved analytically and numerically.[4] Moreover for an energy of the incident electrons of above 100 eV the identity (15) is verified for a magnetic field as high as B = 10 T.

We can conclude saying that the potential scattering assisted by a magnetic field shows many interesting new features with respect to the field free case, due mainly to the form of the energy spectrum and to the behaviour of the density of final states.

However, for laboratory fields (up to 10 T) for which we have to extend the sum in the cross section over 10^5 Landau levels, the resulting value of the cross section is identical to the field free one.

For very intense magnetic fields (B = 10^5 T), instead, the Landau levels involved in the scattering process are only $n_f = 0$ and/or $n_f = 1$, the matrix element Eq. (8) is very different from the field free one and new effects are present.

References

1. See, for instance, M.R.C. McDowell and M. Zarcone, Scattering in strong magnetic fields, Adv. Mol. Phys., 21:255 (1985), and references therein.
2. L. D. Landau and E. M. Lifshitz, "Quantum Mechanics," Pergamon, Oxford (1965).
3. G. G. Pavlov and D. G. Yakovlev, Coulomb deceleration of fast protons in a strong magnetic field, Sov. Phys. JETP, 43:389 (1976).
4. Detail of the calculations will be reported in a paper now in preparation.

THE HYDROGEN ATOM IN A STRONG MAGNETIC FIELD

Alexander Alijah, John T. Broad[*] and Juergen Hinze

Fakultät für Chemie, Universität Bielefeld, 4800 Bielefeld
*Fakultät für Physik, Universität Freiburg, 7800 Freiburg
Federal Republic of Germany

ABSTRACT

We present a method to calculate photoionization and scattering cross sections for hydrogen atoms in magnetic fields of arbitrary strengths. The method is based on a coordinate transformation from spherical coordinates in the vicinity of the nucleus, where the Coulomb attraction dominates, to cylindrical coordinates in the asymptotic region.

INTRODUCTION

The behaviour of a hydrogen atom in a homogeneous mangetic field B, B∥z, is described by the Schrödinger equation

$$(T + V_B + V_C - E)\Psi = 0 \tag{1}$$

with the magnetic interaction

$$V_B = \frac{1}{2\mu} \left[\frac{e\hbar}{c} (m_\ell + 2m_s)B + \frac{e^2}{4c^2} B^2(x^2+y^2) \right] \tag{2}$$

and the Coulomb attraction

$$V_C = - \frac{e^2}{r} . \tag{3}$$

Since we have neglected the spin-orbit coupling, the projections of the angular momentum, m_ℓ, and the spin, m_s, are conserved separately (Paschen-Back effect). The other quantities being conserved are the z-parity π for a reflection of the wave function at the plane with z=0 and of course the energy E.

Unfortunately, the Hamiltonian is not separable in any coordinate system. Because of the different symmetries of the Coulomb and magnetic forces, we use spherical harmonics in the inner region, where the Coulomb force dominates, and Landau state functions in the outer region to expand the total wavefunction. The resulting systems of coupled differential equations for the radial motion and the motion along the field, respectively, are described by Rösner et al.[1] We integrate these equations numerically with Johnson's log derivate method[2] and then calculate the wavefunction with proper initial conditions near the nucleus.

MATCHING THE INNER AND OUTER SOLUTIONS

The z-components $Z_{N,N}(z)$ of the outer solutions can be calculated by projecting the total wavefunction onto the Landau states $R_{N,m}(\rho,\phi)$ at fixed z. Inserting the spherical basis of the inner region we obtain

$$Z_{N'N}(z) = \sum_{1} < R_{N'm}(\rho,\phi) \mid \sum_{1'} Y_{1'm}(\theta,\phi) \frac{\Omega_{1'1}(r)}{r} >_{\rho,\phi} \alpha_{1N}$$

$$= \sum_{1} \tilde{Z}_{N'1} \alpha_{1N} . \tag{4}$$

An expression for Z' can be derived by differentiating Eq. (4) with respect to z. The matrix α combines the solutions such as to satisfy the asymptotic conditions at large z (see next chapter). We can now solve the equation

$$Z' = R^{(z)}Z \tag{5}$$

for the unknown log derivative matrix in z, $R^{(z)}$. Note that $R^{(z)}$ doesn't depend on the matrix α.

ASYMPTOTIC (LARGE z) BOUNDARY CONDITIONS

The electronic motion in the outer region can be regarded as a superposition of oscillations perpendicular to the magnetic field and the motion in the remaining weak Coulomb field along the magnetic field. The oscillations are represented by the Landau state functions,[3] while the z-motion can be expressed in terms of the three Coulomb functions F, G and Q and the K-matrix

$$Z_{N'N}(z) = F_{N'N} + \sum_{L \, \epsilon \, \text{open}} G_{N'L} K_{LN} + \sum_{L \, \epsilon \, \text{closed}} Q_{N'L} K_{LN}$$

$$= F_{N'N} + \sum_{L} P_{N'L} K_{LN} \tag{6}$$

(**P** may be either **G** or **Q**). Asymptotically these function matrices become diagonal. **F** and **G** oscillate like a sine and cosine, respectively, while **Q** decreases exponentially.[4] Thus, **F** and **G** have non-zero diagonal elements for the open channels, **Q** for the closed channels. From Eqs. (5) and (6) we can now derive a system of linear equations for the elements of the full **K**-matrix

$$\sum_{M} (\sum_{L} R_{N'L}^{(z)} P_{LM} - P'_{LM}) \, K_{MN} = \sum_{L} -R_{N'L} \, F_{LN} + F'_{N'N} \, . \tag{7}$$

Once **K** is known, we solve the equations

$$\sum_{1} \tilde{Z}_{N'1} \, \alpha_{1N} = F_{N'N} + \sum_{L} P_{N'L} K_{LN} \tag{8}$$

for α. Unfortunately, in the asymptotic region (7) has a unique solution only for the open-open and closed-open parts of the **K**-matrix. If we are interested in the scattering information only, we can propagate the log derivative matrix $\mathbf{R^{(z)}}$ into the asymptotic region and calculate the open-open **K**-matrix and the **S**-matrix. Otherwise we have to propagate these solutions from infinity to the matching point. To explain photoionization experiments,[5] we can build up the dipole transition matrix while integrating out the radial functions and then post-multiply it by the matrix α.

FRAME TRANSFORMATION

From Eq. (4) we have

$$\tilde{Z}_{N1'}^{m}(z) = \sum_{1} \int_{0}^{\infty} \rho d\rho \, \phi_{N}^{m}(\rho) \, \theta_{1}^{m}(\cos\theta) \, \frac{\Omega_{11'}(r)}{r} \, . \tag{9}$$

The functions $\phi(\rho)$ and $\theta(\cos\theta)$ are the non-ϕ-dependent parts of the Landau functions and the spherical harmonics, respectively. Since the radial functions are known only numerically, we change the integration variable and integrate over r.

$$\tilde{Z}_{N1'}^{m}(z) = \int_{z}^{\infty} dr \, \phi_{N}^{m}((r^2 - z^2)^{1/2}) \sum_{1} \theta_{1}^{m}(\frac{z}{r}) \, \Omega_{11'}(r) \tag{10}$$

For the z-derivative we obtain

$$\tilde{Z}'^m_{N1}(z) = \int\limits_z^\infty dr\ \phi^m_N((r^2 - z^2)^{1/2}) \sum_l [\frac{z}{r}\ \theta^m_l(\frac{z}{r}) \sum_k R^{(r)}_{1k}\Omega_{kl'}(r)$$

$$+ (1\ \frac{z}{r}\ \theta^m_l(\frac{z}{r}) - (1 - m + 1)\ \theta^m_{l+1}(\frac{z}{r}))\ \frac{\Omega_{11'}(r)}{r}]\ . \tag{11}$$

with

$$L^m_{N1'}(z) = \int\limits_z^\infty dr\ \phi^m_N((r^2-z^2)^{1/2}) \sum_l [\frac{r}{z}\ \theta^m_l(\frac{z}{r}) \sum_k R^{(r)}_{1k}\Omega_{k'1'}(r) +$$

$$\frac{1}{r}\ \frac{d\theta^m_l(\frac{z}{r})}{d\ \frac{z}{r}}\ \Omega_{11'}(r)]\ . \tag{12}$$

While integrating the functions Ω and $R^{(r)}$ outward, we form the sums for these integrals simultaneously using the trapezoidal rule. When convergence is achieved, the integration stops.

SOLVING THE DIFFERENTIAL EQUATIONS

The second order matrix equation

$$F'' + U\ F = 0 \tag{13}$$

can be split into two first order matrix equations in many ways. If we define the logarithmic derivative matrix R

$$R = F'\ F^{-1} \tag{14}$$

we obtain a Ricatti equation for R,

$$R' + U + R\ R = 0\ . \tag{15}$$

This equation can be integrated numerically with Johnson's log derivative method,[2] which is inherently stable with respect to closed channels.

Then we have to solve the equation

$$F' = R\ F\ . \tag{16}$$

The matrix R can be diagonalized in a sector j, so that the solutions in this sector are known analytically. F can be propagated by the matching condition between two sectors

$$F(x_{j+1} - \frac{h}{2}) = F(x_j + \frac{h}{2})\ . \tag{17}$$

138

However, this method is stable only for the step size h $\ll$ x$_j$. If this condition is not satisfied, we use the formula

$$F(x_j + h) = (1 + hR - \frac{h^2}{2} U) F(x_j) + O(h^3) . \tag{18}$$

Because of the stability of Johnson's algorithm, the combination of the three methods above is much more satisfactory than integrating (14) directly.

ACKNOWLEDGEMENTS

The authors acknowledge the support of the Deutsche Forschungs-gemeinschaft, Sonderforschungsbereich 216.

REFERENCES

1. W. Rösner, G. Wunner, H. Herold and H. Ruder, J. Phys. B 17, 29 (1984).
2. B. R. Johnson, J. Comp. Phys. 13, 445 (1973).
3. L. D. Landau and E. M. Lifshitz, Quantenmechanik, (Berlin, 1979).
4. P. G. Burke and W. D. Robb, Adv. At. Molec. Phys. 11, 143 (1975).
5. A. Holle, G. Wiebusch, J. Main, B. Hager, H. Rottke and K. H. Welge, Phys. Rev. Lett. 56, 2595 (1986); J. Main, G. Wiebusch, A. Holle and K. H. Welge, Phys. Rev. Lett. 57, 2789 (1986).

THE HYDROGEN ATOMS IN STRONG ELECTRIC FIELDS

R. J. Damburg

Institute of Physics
Latvian USSR Academy of Sciences
229021 Riga, Salaspils, U.S.S.R.

INTRODUCTION

Considerable theoretical and experimental efforts have been recently undertaken to study the non-relativistic hydrogen atom in static electric fields.[1,2] These efforts have been quite successful and, as a result, there is even an opinion that the above problem may be now considered as solved.[3] But in our opinion there is at least one aspect of the theory which was not studied in detail and is relevant to the experimental studies.[4,5]

We begin from the known facts. When the electric field F is small, the problem can be solved both numerically and asymptotically.[1] Results obtained by both methods agree in the regions where they should agree. This gives us confidence that everything is understood properly. The experimental results for the Stark energies E_0 and the level widths Γ at low fields F are in excellent agreement with theory for many cases: different states and different fields. One point, however, should be always kept in mind: the very notion of the smallness of the electric field F depends essentially on the atomic state, or, in other words, on the parabolic quantum numbers n_1, n_2, m.

Experimental studies of the photoionization of atoms in the presence of external static electric fields provide new information on Stark effect. Such experiments were started with Rb more than ten years ago by Freeman and Economou.[6] Two years ago experimental results on the photoionization of a hydrogen atom in static electric fields were published by Glab and Nayfeh[4] and independently by Rottke and Welge.[5] Below the zero-field ionization limit E = 0 in the photoionization spectrum sharp peaks which correspond to high Stark states were found in both experiments. Reliable assignments for the sharp peaks were made by comparison with theoretical calculations.[4,5] As energy E increases, fairly broad resonances appear just below $E = 0$. These persist into the region of positive energies. It seems that one characteristic feature of the photoionization spectrum, i.e. small oscillations which accompany sharp peaks just below $E = 0$ and oscillatory structure of broad peaks above $E = 0$, was not discussed before. We attempt to explain qualitatively these peculiarities of experimental data by the indication on the existence of anomalous Stark states in the strong electric field.

THEORY

Let us first consider Schrödinger equation

$$\left(\frac{1}{2}\Delta + \frac{1}{r} - Fz + E\right)\psi = 0 \tag{1}$$

It can be separated in semi-parabolic coordinates $x = \mu\nu\cos\phi$, $y = \mu\nu\cos\phi$, $z = \frac{1}{2}(\mu^2 - \nu^2)$.

Defining the wave function as the product

$$\psi = \frac{1}{2}(\mu\nu)^{-1/2}M(\mu)N(\nu)e^{\pm im\phi} \tag{2}$$

and substituting (2) into (1), we obtain, for $M(\mu)$ and $N(\nu)$ the equations

$$\left(\frac{d^2}{d\mu^2} + \frac{1 - 4m^2}{4\mu^2} + 2E\mu^2 + Z_1 - F\mu^4\right)M(\mu) = 0 \tag{3}$$

$$\left(\frac{d^2}{d\nu^2} + \frac{1 - 4m^2}{4\nu^2} + 2E\nu^2 + Z_2 + F\nu^4\right)N(\nu) = 0 \tag{4}$$

$$Z_1 + Z_2 = 4 \tag{5}$$

Boundary conditions for $M(\mu)$ and $N(\nu)$ are the following

$$M(\mu) \sim \mu^{m+1/2}; \quad N(\nu) \sim \nu^{m+1/2}$$
$$\mu \to 0 \qquad\qquad \nu \to 0 \tag{6}$$

$$M(\mu) \underset{\mu \to \infty}{\sim} \frac{a}{\mu}\exp\left(-\frac{\sqrt{F}\,\mu^3}{3} + \frac{E}{\sqrt{F}}\mu\right)$$

$$N(\nu) \underset{\nu \to \infty}{\sim} \frac{b}{\nu}\sin\left(\frac{\sqrt{F}\,\nu^3}{3} + \frac{E}{\sqrt{F}}\nu + \phi\right) \tag{7}$$

The physical meaning of Eqs. (3) and (4) is clear. When F is small, then for some internal region, where μ and ν are also small, perturbations $F\mu^4$ and $F\nu^4$ can be dropped and thus Eqs. (3)–(5) simply describe the hydrogen atom. Therefore, hydrogenic states at $F = 0$ are chosen as the starting point for the solution of the Stark problem. Practically, in numerical solution of Eqs. (3)–(5) we are taking values given by perturbation theory valid for small F as input data for E and Z_2 (or Z_1). Then we obtain the Stark energy E_o and level width Γ.[1] Some difficulties arise in the case with large n_1 and small n_2 and m, where retaining only a few terms of perturbation expansion gives a poor representation of E.

There is, however, a question. Is it possible that in the Stark problem there are quasistationary states different from low field Stark states, and which consequently do not directly originate from hydrogenic states at $F = 0$? If such quasistationary states exist, we should indicate a different procedure of solving Eqs. (3)–(5). But first we show that such states do exist.

When F is small, the level width Γ can be written as[1]

$$\Gamma = \frac{(4R)^{2n_2 + m + 1}}{n^3 n_2! (n_2 + m)!} e^{-2/5\,R} \left(1 + 0(F)\right) \qquad (8)$$

where

$$R = \frac{(-2E_o)^{3/2}}{F}$$

As can be seen from formula (8), the most stable among the states with the same principle quantum number $n = n_1 + n_2 + m + 1$ is the highest one with $n_1 = n - 1$, $n_2 = m = 0$.

Though formula (8) is not valid for $E_o > 0$, the conclusion that the above states with the highest possible values of $n_1 (n_2 = m = 0)$ can survive at positive E is not absolutely senseless, at least as a first guess. This conclusion was made in Ref. [7].

Sharp peaks which are seen in the photoionization spectra below $E = 0$ in Refs. [4,5] belong to the states with quantum numbers $n_1 = n - 1$, or with $n_1 = n - 2$, $n_2 = 1$.

Since we have experimental data on different Stark energies close to $E = 0$, confirmed by theoretical calculations, we can satisfy our curiosity and look at potential $U(\nu)$ of Eq. (4) responsible for quasistationary states in the Stark problem.

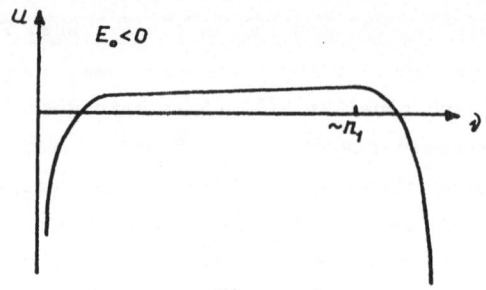

Figure 1

$$U(\nu) = -\frac{1}{4\nu^2} - 2E_o\nu^2 - F\nu^4$$

The forms of potential $U(\nu)$ are shown schematically in Fig. 1 for negative E_o and in Fig. 2 for positive E_o. Values of E_o in the vicinity of $E = 0$ are small. We show later why the term $F\nu^4$ becomes important only at large distances $\nu \sim n_1 \gg 1$. It is known that the potential $- 1/4\nu^2$ cannot support stationary states. Stationary states can be supported by the potential $-2E_o\nu^2$ (if E_o is negative, $F = 0$) and this is the reason why the radius of the atom is determined by the energy. When E_o is positive, no term in potential $U(\nu)$ can support stationary states. But quasistationary states exist for $E > 0$ and we see these on the experimental curves. The explanation for this phenomena was given from the point of view of classical mechanics by Rau.[8] Then Rau and Lu[9] extended this explanation to quantum theory finding some analogy with the autoioni-

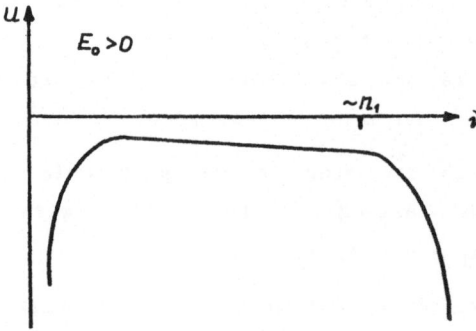

Figure 2

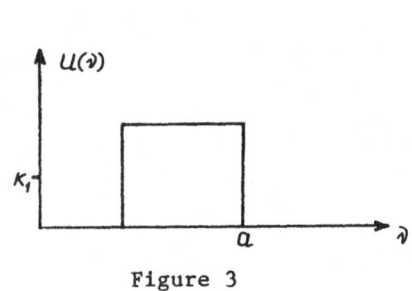

Figure 3

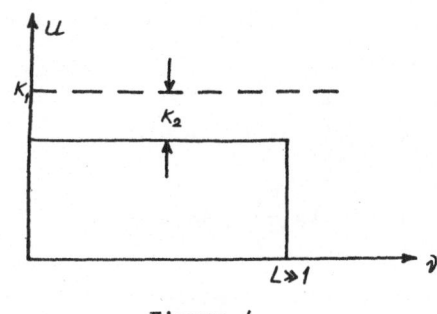

Figure 4

zation. Quantum mechanical consideration of the photoionization of the hydrogen atom in a static electric field for E > 0 was also given in papers [7,10]. But our approach to this problem is different. We compare the potential U(ν) for Z_2 for the Stark effect with model potentials and then make conclusions.

Specific Model of Potentials

It is clear that quasistationary states can exist for the potential presented in Fig. 3. The potential shown in Fig. 4 is unusual for atomic and molecular physics. We can find the solution of the Schrödinger equation for this case easily.

$$\phi(\nu) = \sin K_2 \nu \qquad \nu < L \qquad (9)$$
$$\phi(\nu) = A\sin(K_1 \nu + \phi) \qquad \nu > L \qquad (10)$$

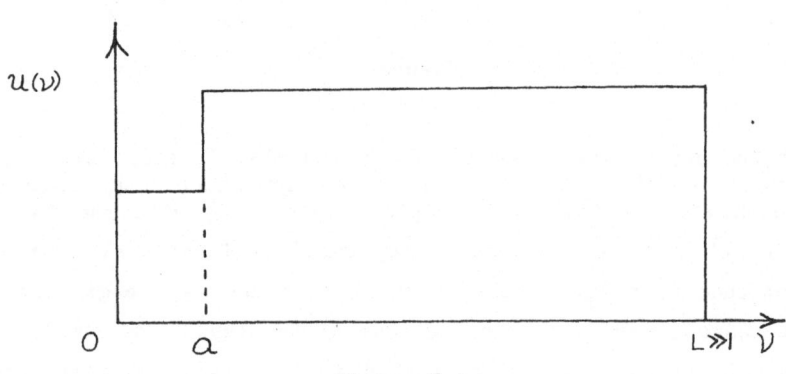

Figure 5

A and ϕ can be determined by matching (9) and (10) and their derivatives at $\nu = L$

$$A = \left(\frac{K_2^2}{K_1^2} \cos^2 K_2 L + \sin K_2 L\right)^{1/2} \tag{11}$$

$$\phi = \text{arctg}\left(\frac{K_1}{K_2} + gK_2 L\right) - K_1 L \tag{12}$$

The amplitude A has minima at $K_2 L = n\pi$, where n is the integer.

$$A_{\min} = \frac{n\pi}{K_1 L}$$

The amplitude $A_{\min}$ increases slowly with increasing K_2 when L is large.

In both cases shown in Fig. 3 and Fig. 4 the wave nature of quantum mechanics is responsible for the existence of the quasistationary states. The first case where we have the phenomenon of the penetration of the particle through the potential barrier is more typical for atomic physics. The second case which can be characterized as the reflection of the particle from the boundary or from the edge is more typical for solid states physics.

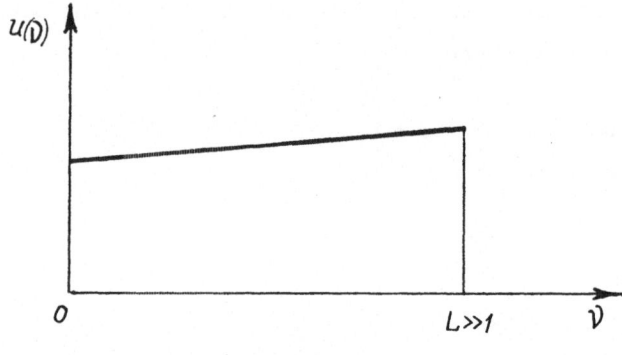

Figure 6

For the potentials shown in Fig. 5 and Fig. 6, there are two possible types of quasistationary states. The lowest of them could be very sharp, with the amplitude $A_{\min}$ rapidly increasing with the increase of the "energy." Then there can be many small oscillations. The oscillation amplitude (not $A_{\min}$) will be small from the very beginning, but it would decrease slowly with the increase of "energy." We can look now at experimental curves (Fig. 7) presented in Ref. [4]. We see that, when $E < 0$, sharp peaks in the vicinity $E = 0$ are accompanied by small ones in full agreement with the above prediction.

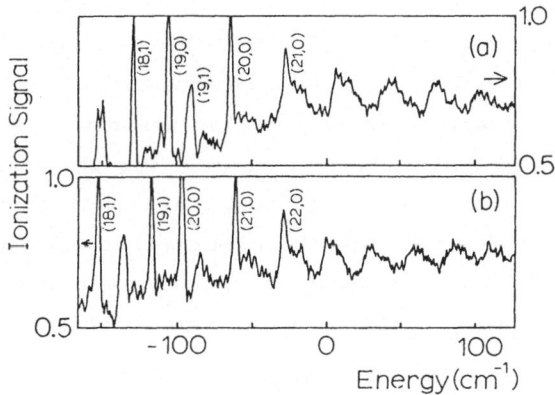

Figure 7

Case E > 0. In case E > 0 can be understood by looking at Fig. 2. It is clear that sharp peaks here are impossible there exist only small oscillations. The resonance values of Z_2 can be both positive and negative. Quasistationary values of E_0 are only slightly dependent on Z_2 since $n_1 \gg 1$. In the photoionization spectra we can see the picture where small peaks stick together around centers with different n_1 and so it looks like widely separated resonances covered by small bumps.

Now we should explain why the term $-F\nu^4$ in the potential $U(\nu)$ of Eq. (5) is small up to the large values of $\nu \sim n_1 \gg 1$ for the states with quantum numbers $n_1 = n - 1$ or may be $n_1 = n - 2$, $n_2 = 1$ in the vicinity $E = 0$.

Consider the perturbation expansion for Stark energy

$$E_0 = \sum_{K=0} a_K(n_1, n_2, m)F^K = -\frac{1}{2n^2} + \frac{3}{2}n(n_1 - n_2)F - \ldots \tag{13}$$

When n_1 is large, but n_2 and m are small, immediate application of (13) would give a very poor value of E_0. We can say even that approximation (13) becomes worse with decreasing F. This statement seems strange and controversial and, therefore, should be explained. Series (13) are asymptotic and, therefore, when applied they should be terminated just before the smallest term. The error of calculation of E_0 then would not exceed the first discarded term, i.e. the smallest one. But at the same time the error will be much larger than the value of the level width Γ

and the corresponding ratio would grow with decreasing F. Such a situation does not exist for the similar one-dimensional problem, namely the anharmonic oscillator with a negative anharmonicity $\sim -\alpha x^4$. In the latter case, perturbation series similar to (13) are not alternating. Series (13) with $n_1 > n_2$ are sign-changing because the whole problem is three-dimensional. In the one-dimensional case, perturbation series for eigenvalues are alternating for the anharmonic oscillator with positive anharmonicity $\sim \alpha x^4$. In this case we can obtain exact eigenvalues from asymptotically divergent series by using Pade approximants.

Strictly speaking, for the summation of (13) the Borel method should be used since the eigenvalues are complex. However, the Pade approximation here gives excellent results for E_0 at large n_1. The explanation is as follows. Expansion (13) actually represents the sum of two factorially growing series: one alternating (dominant) and one where all terms are negative. This can be seen from an asymptotic formula in the parameter K for a_K that was previously derived in Ref. [1]. The practical efficiency of the Pade method is a consequence of the fact that the series with nonalternating terms is small compared with the dominant series.

When $n_1 \rightarrow \infty$ instead of as in (13) we can use the approximation

$$E_o = \frac{1}{2n^2} \sum_{K=0} b_K (n_1^4 F)^K + O(n_1^{-3}) \tag{14}$$

Therefore, it is natural to expect in this case

$$E_o = \frac{A(B)}{n^2} \tag{15}$$

$$F = \frac{B}{n_1^4} \tag{16}$$

While the value of B is changing insignificantly for the present consideration from ~ 0.2 to 0.5, A is changing drastically from small negative values ~ -0.2 to small positive values ~ 0.3 passing zero.

Formulae (15) and (16) show why the term $F\nu^4$ becomes dominant in the potential of Eq. (4) only at $\nu \sim n_1 \gg 1$. It is evident that for $E > 0$ the quantum number n_2 has lost its meaning and formula (15) is no longer well defined. But since $n_1 \gg 1$, the uncertainty of using it is not crucial for qualitative conclusions. On the contrary, the quantum number n_1 for all fields F continues to be meaningful and equal to the number of nodes in Eq. (3).

The quantum number n_2 becomes meaningless also for anomalous Stark states for $E < 0$. By looking at the perturbation expansion for Z_2 we can make qualitative conclusions

$$Z_2 = 4\sqrt{-2E_o}\left(n_2 + \frac{1}{2} - \frac{1}{2R}(3n^2 + 3n + 1) - \ldots\right) \tag{17}$$

Expansion (17) and, consequently n_2, are meaningless when the second term in it exceeds the first one. Here it is impossible to improve the situation by using Pade or Borel methods. Examples are given in Table I. The values of F and E_o are taken from experimental data.[4]

Table I. F = 8KV, $n_2 = m = 0$

n_1	E_a, a.u.	A	B	R
20	-0.28×10^{-3}	-0.124	0.25	8.5
21	-0.116×10^{-3}	-0.059	0.3	2.2

We see from (17) and Table I that for $n_1 = 20$, n_2 can acquire values 0,1,2; for n = 21, n_2 can be equal to 0.

CONCLUSION

We have described qualitatively anomalous Stark states, but they can be found also by exact numerical solution of Eqs. (3)-(5). When $E < 0$, the first low field Stark states with large n_1 close to E = 0 should be considered. Anomalous quasistationary states should be searched for in the vicinities of sharp Stark states, with the precaution that the value of n_1 should not change in the process of solution.

We see that our conclusions on anomalous Stark states are in good agreement with experimental data.[4,5] But, on the other hand, the arguments presented here are actually independent of experiments.

REFERENCES

1. Rydberg atoms in electric fields, in: "Rydberg States of Atoms and Molecules," R. F. Stebbings and F. B. Dunning, eds., Cambridge University Press, Cambridge, London, New York (1983).
2. P. M. Koch, Rydberg studies using fast beams, ibid.
3. K. H. Welge and A. Holle, The highly excited hydrogen atom in strong magnetic fields, in: "Electronic and Atomic Collisions," D. C. Lorents, W. F. Meyerhof and F. R. Peterson, eds., North Holland (1986).
4. W. L. Glab and M. H. Nayfeh, Stark-induced resonances in the photoionization of hydrogen, Phys. Rev. A 31:530 (1985).
5. H. Rottke and K. H. Welge, Photoionization of the hydrogen atom near the ionization limit in strong electric fields, Phys. Rev. A 33:301 (1986).
6. R. R. Freeman and N. P. Economou, Electric field dependence of the photoionization cross section of Rb, Phys. Rev. A 20:2356 (1979).
7. V. D. Kondratovich and V. N. Ostrovsky, Resonance structure of the cross section for photoionization of a hydrogen atom in an electric field, Zh. Eksp. Teor. Fiz. 83:1256 (1982).
8. A.R.P. Rau, Rydberg states in electric and magnetic fields near-zero-energy resonances, F. Phys. B 12:L193 (1979).
9. A.R.P. Rau and K. T. Lu, Comments on near-zero-energy resonances in atoms in external field, Phys. Rev. A 21:1057 (1980).
10. E. Luc-Koenig and A. Bachelier, J. Phys. B 13:1769 (1980).

QUANTUM BEATS IN THE ELECTRIC-FIELD MIXING

OF THE n = 2 STATES OF ATOMIC HYDROGEN

J. F. Williams, E. L.Heck and H. Slim[1]

University of Western Australia
Nedlands 6009 Perth, Australia
[1]Now at Department of Physics, University of Durham, UK

INTRODUCTION

The Lyman-alpha "Stark beats" produced by the coherent decay of the perturbed 2s and 2p states has been extensively studied, for example in collisions of electrons with atoms,[1] in heavy particle collisions[2,3] and beam-foil[4,5,6] or beam-gas[7] excitation. The $2s_{1/2}$ states of hydrogen is metastable with a field-free lifetime of 0.14 sec. It has four hyperfine states which will mix in an electric field with twelve possible $2p_{1/2}$ and $2p_{3/2}$ states with lifetimes of 1.6 nsec which can radiatively decay to the $1s_{1/2}$ ground state with the emission of a Lyman-alpha photon. If the mixing field is applied suddenly compared to the response time of the atom, the decay radiation shows quantum beats. The pattern of the beats contains large oscillations near the Stark shifted $2s_{1/2} - 2p_{1/2}$ frequency (about 1,000 MHz) and the envelope of these oscillations is modulated at the hyperfine splitting of the $2s_{1/2}$ state (about 180 MHz) and each peak contains rapid oscillations near the $2s_{1/2} - 2p_{1/2}$ frequency (about 10,000 MHz). Each frequency is a group of frequencies arising from the Stark and hyperfine splitting of the states. The above types of measurements provide information on the coherence of the excitation process and on the fine and hyperfine beat structure but they are limited in providing only the products of excitation amplitudes averaged over all scattering angles and the initial state amplitudes and their phases in the beam emerging from the foil are not adequately known.

The developments of fast detectors and sub-nanosecond coincidence timing resolution have permitted the simultaneous detection of the scattered electron and radiated photon from the same scattering event.

These have led to the determination of <u>scattering amplitudes</u> and their relative phases, or equivalently the <u>state multipoles</u> or the density matrix, as reviewed most recently by Slevin.[8] In <u>atomic hydrogen</u> these quantities have been measured for the 2P magnetic sublevels.[9] The extension of this approach to study the coherence of the S and P levels was proposed by Blum and Kleinpoppen[10] and partially achieved by Back et al.[11] Recent work by Heck and Williams[12] using a coincidence time resolution of 0.47 nsec and appropriate collision dynamics, has clearly observed quantum beats in the circular polarization <u>Stokes parameter η_2</u>. The time dependence of the η_2 parameter also revealed the changing roles of the S and P components of the superposition state by showing a change of sign from negative to positive after about 3 nsec and by a phase change of about 180° when a negative electric field was used. This talk presents an extension of that work and reports angular and polarization correlations from which the first complete determination of all the elements ofthe density matrix of the n = 2 state has been made.

DESCRIPTION OF THE COINCIDENCE MEASUREMENT

A physical picture of the excited system is obtained with the multipole expansion of the density matrix ρ_{orb} as

$$\rho_{orb} = \sum_{L'LKQ} \langle (L'L)^+_{KQ} \rangle \; T(L'L)_{KQ} \tag{1}$$

where $T(L'L)_{KQ}$ are irreducible tensor operators of rank k and component Q and $\langle T(L'L)^+_{KQ} \rangle$ are the state multipoles of the excited atomic system. The state multipoles of n = 2 excited hydrogen atom have been given in terms of the scattering amplitudes by Blum and Kleinpoppen and are related to the normalized cross sections to the R and J parameters introduced by Morgan and McDowell and, to the R(sp) and J(sp) parameters defined by Wyngaarden and Walters[15] as shown below.

$$\langle T(00)^+_{00} \rangle = \langle |f_{00}|^2 \rangle = \sigma(2S) \qquad\qquad \text{Monopole}$$

$$\langle T(10)^+_{10} \rangle = \langle f_{10} f_{00}^* \rangle$$
$$= [\sigma(2S)\sigma(2P_0)]^{1/2} [R(P_0 S^*) + iJ(P_0 S^*)] \qquad \text{Electric Dipole}$$

$$\langle T(11)^+_{00} \rangle - 1/\sqrt{3}[2\langle |f_{11}|^2 \rangle + \langle |f_{10}|^2 \rangle] = 1/\sqrt{3}\sigma(2P) \qquad \text{Monopole}$$

$$\langle T(11)^+_{11}\rangle \;\; -i\sqrt{2/3}\;\; \mathrm{Im}\langle f_{11}f_{10}^{\,*}\rangle = -i\sqrt{2}\,\sigma(2P)J \qquad\qquad \text{Magnetic}$$

$$\langle T(11)^+_{10}\rangle = 0 \qquad\qquad\qquad\qquad\qquad\qquad\qquad\qquad \text{dipole moment}$$

$$\langle T(11)^+_{20}\rangle = \sqrt{2/3}\,[\langle |f_{11}|^2\rangle - \langle |f_{10}|^2\rangle] = \sqrt{2/3}\,[\sigma(2P_1) - \sigma(2P_0)]$$

$$\langle T(11)^+_{21}\rangle = \sqrt{2}\;\mathrm{Re}\langle f_{10}f_{11}^{\,*}\rangle = -\sqrt{2}\,\sigma(2P)R \qquad\qquad \text{Alignment}$$

$$\langle T(11)^+_{22}\rangle = -\langle |f_{11}|^2\rangle = -\sigma(2P_1) = -0.5(1-\lambda)\,\sigma(2P)$$

with

$$R(P_m S^*) = \mathrm{Re}\langle f_{1m}f_{00}^{\,*}\rangle/[\sigma(2S)\sigma(2Pm)]^{1/2}, \quad m = 0,1 \;\text{ and }\; \lambda = \sigma(2P_0)/\sigma(2P)$$

$$J(P_m S^*) = \mathrm{Im}\langle f_{1m}f_{00}^{\,*}\rangle/[\sigma(2S)\sigma(2P_m)]^{1/2} \;\text{ and }$$

$$\langle T(J'J)^+_{KQ}\rangle = (-1)^{J'-J+Q}\langle T(J'J)^+_{KQ}\rangle^* \qquad\qquad\qquad\qquad (2)$$

The essential new feature of our work is the observation of the time development of the excited atoms which takes place in the presence of an external weak electric field F and the deduction of all the elements of the density matrix of the excited state. As shown by Blum and Kleinpoppen the state ofthe photons emitted at time t and measured in coincidence with the scattered electrons depends on the state of the atoms just after the excitation and on the time development under the internal and external interactions of the excited atomic state until the time of decay. When the excited atomic system is subjected to the influence of a weak electric field, Stark mixing of levels of opposite parity occurs and, although at the time of the decay the selection rules of the dipole transition still apply, the Stark mixing has transformed the S and P states at the instant of excitation, t = 0, into linear superpositions of S and P states at time t. This time evolution of the S and P coherence in the excitation may be observed under selected dynamical coditions. The state of the photons is completely described by the intensity I and the Stokes parameters $I\eta_1$, $I\eta_2$ and $I\eta_3$ which were calculated using the assumption that hyperfine interaction and the $S_{1/2} - P_{3/2}$ ($m_j = 1/2$) mixing could be neglected for a weak external electric field. The Stokes parameters are defined as the difference of two intensities $|(\beta)$ at polarization angles β, that is

$$I = I(LHC) + I(RHC) \qquad\qquad In_1 = I(45°) - I(135°)$$
$$In_2 = I(LHC) - I(RHC) \qquad\qquad In_3 = I(0°) - I(90°) \qquad (3)$$

where right hand circular polarization is defined as a clockwise rotation of the electric vector looking against the direction of propagation of the photon. The relationships of the Stokes parameters to the state multipoles are written below in the form used in the present analysis for the case of the photon detector normal to the scattering plane, that is $\theta_\gamma = 90° = \phi_\gamma$.

$$I(t) = \text{Amp}/3 \ \{B|a(10;t)|^2 + 1/3|a(11;t)|^2 + 5/6 \ \exp(-\gamma_p t)$$
$$- 2/\sqrt{3} \ [\text{Re}\langle t(10)^+ 10\rangle \ \text{Re}a(11;t) \ a(10;t)^*$$
$$- \text{Im}\langle t(10)^+\rangle \ \text{Im}a(11;t)a(10;t)^*]\} \qquad (4)$$

$$In_2(t) = \text{Amp} \ 2/3 \ \{\sqrt{2}\sqrt{3}[\text{Im}\langle t(10)^+ 11\rangle \ \text{Re}a(11;t)a(10;t)^*$$
$$+ \text{Re}\langle t(10)^+_{11}\rangle \text{Im}a(11;t)a(10;t)^*]$$
$$+ 1/3 \ \ \text{Im}\langle t(11)^+_{11}\rangle [|a(11;t)|^2 + 2.5\exp(-\gamma_p t)]\} \qquad (5)$$

$$I(0°;t) = 1/2 \ |(1 + \eta_3)(t) = \text{Amp}/6 \{B|a(10;t|^2 + 1/3|a(11;t)^2$$
$$+ \ (1/3 + \gamma)\exp(-\gamma p t) - 2/\sqrt{3}[\text{Re}\langle t(10)^+_{10}\rangle \text{Re}a(11;t)a(10;t)^* \qquad (6)$$
$$- \text{Im}\langle t(10)^+_{10}\rangle \text{Im}a(11;t)a(10;t^*]\}$$

$$I(45°;t) = 1/2I(1+\eta_1)(t) = \text{Amp}/6\{B|a(10;t|^2+1/3|a(11;t)^2$$
$$+ \ (5/6-\sqrt{2}R)\exp(-\gamma_p t)-2/\sqrt{3}[\text{Re}\langle t(10)^+_{10}\rangle \ \text{Re}a(11;t)a(10;t)^* \qquad (7)$$
$$- \text{Im}\langle t(10)^+_{10}\rangle \text{Im}a(11;t)a(10;t)^*]\}$$

where $\text{Amp} = 2^{16/310} \ C(\omega) \ \sigma(2P)$ and $\langle t(L'L)^+_{KQ}\rangle = \langle T(L'L)^+_{KQ}\rangle\sigma(2P)$,
$\qquad B = \sigma(2S)/\sigma(2P)$, and
$\quad a(LL;t) =$ the time devlopment coefficients given elesewhere.[16]

The measurement problem is how to determine all the multipoles. First consider the case of zero electric field. The J parameter can only be determined by a measurement of the circular polarization out of the scattering plane and if it is measured normal to that plane the following expression shows that there is only a single multipole present.

$$I_{tot} \; \eta_2 = Amp \; 7/9 \; J/\sigma(2P) \qquad (8)$$

The λ and R parameters can be determined from a measurement of either the linear polarizations or angular correlations, however the latter method was chosen because it was independent of the linear and circular polarizers which are used for measurements in the presence of the electric field. These parameters were determined from the expression

$$I_{tot} = Amp/36 \{ 11 + 3\lambda + 3(1 - 3\lambda)\cos^2\theta$$
$$+ 6\sqrt{2}R \; \sin\theta \; \cos\phi - 3(1 - \lambda) \; \sin^2\theta \; \cos2\phi \} \qquad (9)$$

Secondly, in the presence of the electric field, it is more convenient to measure the Stokes parameters which require rotations of the polarisers rather than the whole photon detector. Also it is the form of the time dependence of the Stokes parameters really determines which of the state multipoles can be readily deduced. The expected form was calculated using the state multipoles of Wyngaarden and Walters in the $s_{1/2} - p_{1/2}$ mixing model and is shown in figure 1. When the coincidence resolving time of 0.47 nsec is folded into these predictions, the fine structure oscillations average and the Lamb shift beats with a period of about 0.7 nsec should be observable in the η_2 parameter. Similar calculations at other energies suggested that the quantum beats would be most visible for about 350 eV incident electron energies and an electric field of about 250 V/cm in the circular polarization.

RESULTS

The apparatus and experimental method are unchanged from that reported previously.[12,13] Throughout this work an incident electron energy of 350 eV, a scattering angle of 3° and an electric field of 250 V/cm have been used. The energy loss electrons scattered through an angle θ and detected without spin analysis in coincidence with the radiated photons, whose polarisation is measured when the detector is normal to the scattering plane for polarisation correlations but is not measured when the detector is in the scattering plane for angular correlations. Extensive measurements were made to validate the experimental method and included, for example, showing that the time coincidence signal arose entirely from atomic rather than molecular hydrogen adn that the scattering plane was well defined for such a small scattering angle by observing a reversal of the handedness of the polarisation with reversal of the sign of the scattering angle.

The sum of I(RHC) and I(LHC) gives the total intensity which is plotted in Fig. 2. This data was fitted to equation (4), using the four quantities AMP, $\sigma(2S)/\sigma(2P)$, $\text{Re}\langle T(10)^+_{10}\rangle\rangle/\sigma(2P)$ and $\text{Im}\langle T(10)^+_{10}\rangle\rangle/\sigma(2p)$ as variable. The results shown in Table 1 indicate very large experimental uncertainties, particularly for $\text{Re}\langle T(10)^+_{10}\rangle/(2P)$ and $\text{Im}\langle T(10)^+_{10}\rangle/\sigma(2P)$, which occurs because the contribution of these terms to the total is very small and indicates the small contribution of the s-p coherence to the total intensity.

The normalising factor AMP, so determined for positive and negative fields, were then used to normalize the experimental data for $I\eta_2$ for positive and negative electric fields. For a field parallel to the incident beam, only the circular polarization depends on the $\langle T(L'L)^+_{11}\rangle$ multipoles. A three parameter fit of $\text{Im}\langle T(11)^+_{11}\rangle/\sigma(2P)$, $\text{Re}\langle T(10)^+_{11}\rangle/\sigma(2P)$ and $\text{Im}\langle T(10)^+_{11}\rangle/\sigma(2P)$ of equation (5) was made to the measured values of $I\eta_2$ shown in Figs. 3 and 4. The results shown in Table 1 indicate a good consistency between the positive and negative field data with relatively good experimental uncertainty. Since the photon emission angle appeared in equation (3) as an overall multiplication factor, the identification ofthe three $\langle T(L'L)^+_{11}\rangle$ multipoles occurs primarily through their multiplying time dependent coefficients which emphasizes the need for good timing resolution. The difficulty of these measurements is indicated by the very small value of $|\eta_2$ for the incident electron energy and scattering angle of the present work and the additive errors of I(RHC) and I(LHC). Attempts to fit either I(RHC) or I(LHC), rather than η_2 to the data required a fit of seven parameters of which six are very small. Unacceptably high chisquare values were obtained.

To make the determination of the multipoles complete, values of λ and R are required. It is seen from equation (4) and (7) that the total intensity does not depend on either λ and R when the photons are detected perpendicular to the scattering plane and that the linearly polarized light intensity with polaerization axis at $0°$ depends on λ and that with polarization axis at 45° depends on R. Measurements of the photon intensities for polarization angles of 0°, 90°, 45° and 135° were made for a positive electric field to determine the Stokes parameters. However, it may be seen from the expressions for the Stokes parameters and the polarization intensities that it is simpler in principle to deduce the state multipoles related to λ and R from the intensities $I(\beta)$ rather than from their differences at selected polarization angles. Unfortunately these measurements were not as good as the other measurements above because there was a time walk between the summed data which constituted each of η_1 and η_3 as well as change in the AMP factor.

Table 1

The fitted state multipoles for n = 2 atomic hydrogen are shown for positive, negative and zero electric fields. Theory(a) values are first Born approximation and theory(b) are from Wyngaarden and Walters. The incident electron energy is 350 eV and the electron scattering angle is 3°.

	+250 V/cm	−250 V/cm	Zero Field	Theory(a)	Theory(b)
$\mathrm{Re}\langle T(10)_{10}^{+}\rangle\rangle$	0.169±0.708	0.771±0.879	−	0	0.39
$\mathrm{Im}\langle T(10)_{10}^{+}\rangle\rangle$	0.819±0.616	0.289±0.736	−	1.32	1.15
$\mathrm{Re}\langle T(10)_{11}^{+}\rangle\rangle$	0.868±0.329	0.362±0.306	−	0	1.02
$\mathrm{Im}\langle T(10)_{11}^{+}\rangle\rangle$	1.88±0.303	1.88±0.472	−	3.0	2.6
$\mathrm{Im}\langle T(11)_{11}^{+}\rangle\rangle$	0.362±0.089	0.217±0.082	0.341±0.010	0	0.31
$\langle T(11)_{20}^{+}\rangle\rangle$			6.50+3.96	7.3	6.99
$\langle T(11)_{21}^{+}\rangle\rangle$			−7.26+0.77	−6.89	−6.46
$\langle T(11)_{22}^{+}\rangle\rangle$			−10.68+0.79	−11.08	−10.54
λ	−	−	0.114±0.025		
R	−	−	0.213±0.013		
J			0.010±0.003		
$\sigma(2S)/\sigma(2P)$	0.087±0.014	0.055±0.17	0.030±0.001		
$\sigma(2P)\ a_{0}^{2}/sr$	−	−	24.1±1.1	24.3	23.05
$\sigma(2S)\ a_{0}^{2}/sr$	−	−	0.73±0.053	0.81	0.74

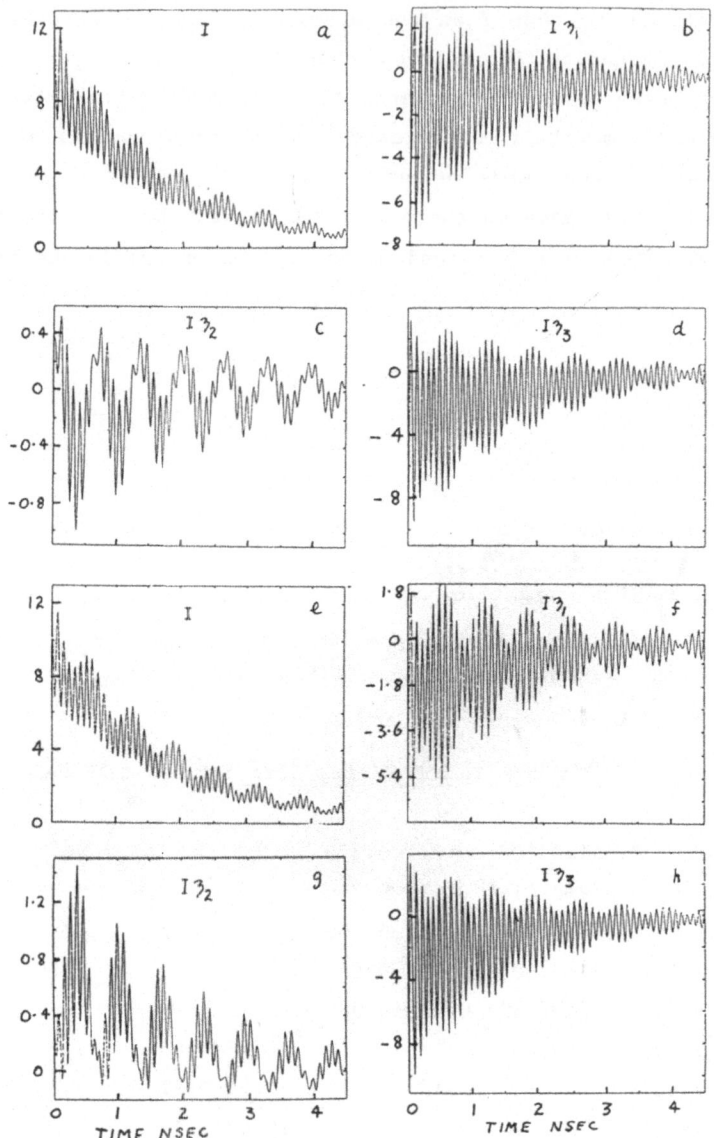

Figure 1 Theoretical values of the Stokes parameters using the state
multipoles of Ref. 15 for $E_0 = 350$ N and $\theta_e = 3°$. The electric
field of 250 V/cm is in the direction of the negative z-axis
for figures (a) to (d) and the positive z-axis for figures (e)
to (h).

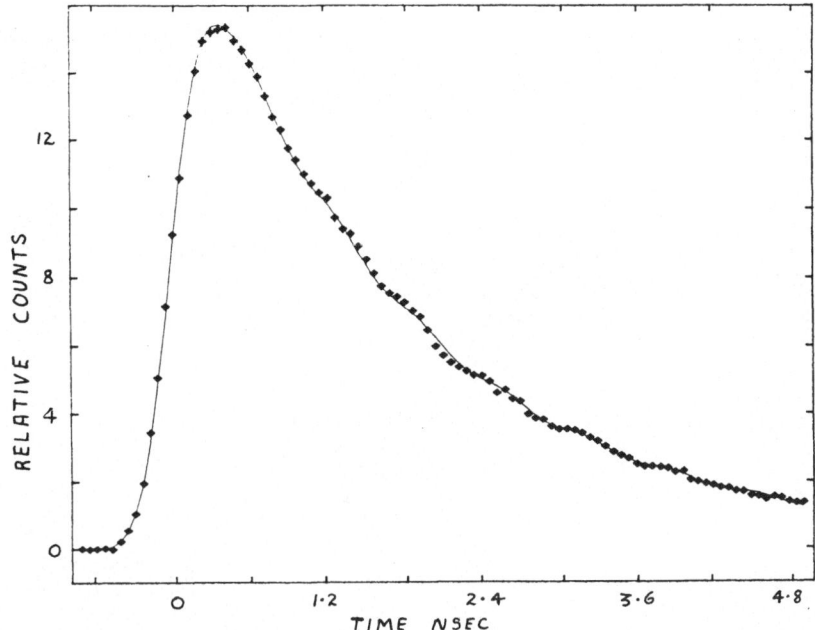

Figure 2   The total photon intensity at 10^{-2} steradians at $(\theta, \phi) = (\pi/2,$
$\pi/2)$, is shown as a function of time.   The electric field is
250 Vcm^{-1}, the incident electron energy 350 eV, the electron
scattering angle is 3°.   The full width half maximum timing
resolution is   0.46 ± 0.01 nsec.   Statistical   error   bars,
of ± one standard deviation, are shown for all data points.
The full line is the optimized least squares fit of eqn.
convoluted with a Gaussian to the summed data using the
theoretical data of Wyngaarden and Walters.

An attempt to make a five parameter fit of AMP, B, λ or R, $\mathrm{Re}\langle t(10)_{10}^{+}\rangle$
and $\mathrm{Im}\langle t(10)_{10}\rangle$ failed to give reliable values, even though the reduced
chisquare value was 3.4, presumable because of the relatively large
statistical errors on the linear polarization total counts and the
relatively small values of λ, R, and B.

The   in-plane   scattered   electron-radiation   photon   angular
correlations were measured without an electric field.   The fitted values
of λ and R shown in Table 1 gave an excellent reduced chisquared value of
0.97 which is reflected in the relatively small experimental uncertainty
for the fitted values.   The procedure for determing the absolute value of
$\sigma(2P)$ and the ratio $\sigma(2S)/\sigma(2P)$ was given by Williams[13] and the results
are given in Table 1.

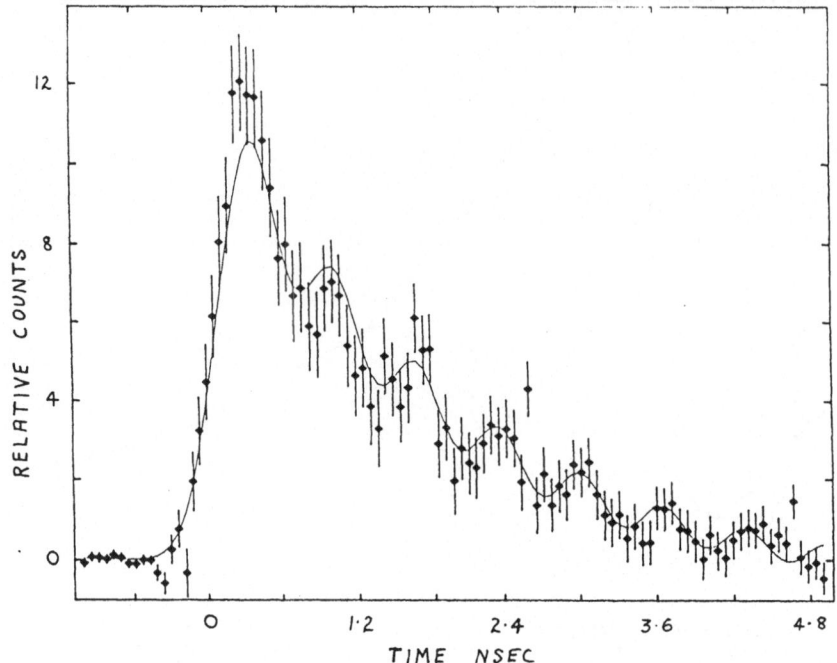

Figure 3 The time development of In_2 is shown for an electric field of 250 V/cm^{-1}. Other parameters are as given for Fig. 2.

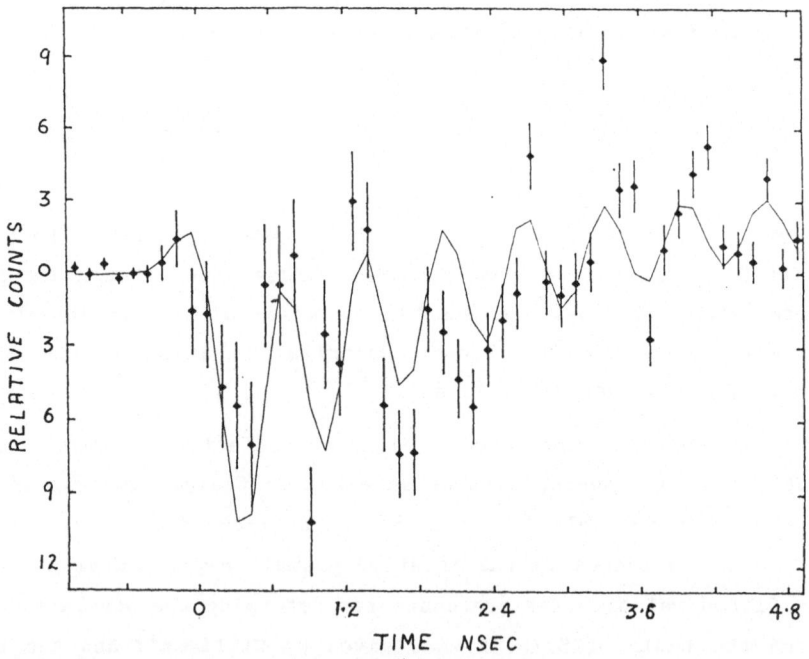

Figure 4 The time development of In_2 is shown for a negative electric field of 250 V/cm^{-1}. Other parameters are given for Fig. 2.

DISCUSSION

Values for all of the state multipoles, as shown in Table 1, have been determined. The most accurate determinations were made for λ, R and J with zero field measurements which is not surprising. The effects of the electric field on the electron projectories have not been studied quantitatively. Only the $\langle T(11)^+_{11} \rangle$ multipole, proportional to the J parameter, has been determined for positive, negative, and zero fields, and this demonstrates rather well the change in experimental uncertainty for the three cases. It was also expected, because of the experimental arrangements for biasing the electric field with respect to the beam energy and analyzer potentials, that the negative fields may give rise to larger experimental uncertainty than positive fields, and this is generally reflected in the measurements. The expectation is also readily seen from the experimental uncertainty in the $I\eta_2$ data of figure 4 with a negative field. Fair values of $\langle T(11)_{11} \rangle$ were obtained in spite of poor statistics because the time dependence ofthe signal $I\eta_2$ was measured and it is the time development coefficients which sort out the multipole from other terms in the fitted form of the Stokes parameter.

The present measurements were limited basically by the large value of $\sigma(2P)$ compared with $\sigma(2S)$ for 350 eV scattering at 3°. Measurements at large angles, where the ratio of $\sigma(2P)/\sigma(2S)$ tends to unity and the ratio of the $\langle T(10)^+_{KQ} \rangle$ multipoles to the $\langle T(11)^+_{2Q} \rangle$ multipoles becomes larger would enhance the beat structure in all the Stokes parameters, better accuracy of the $\langle T(10)_{10} \rangle$ multipoles would be obtained however the differential cross sections become smaller making the total data accumulation times much longer. The inadequacies and strengths of the first Born approximation are clearly seen. The data of Wyngaarden and Walters are in fair agreement with the measured values considering the experimental uncertainties. In conclusion, the measurement of polarization and angular correlations in the presence of Stark mixing of the n = 2 levels has led to the determination of a complete set of the state multipoles.

ACKNOWLEDGEMENTS

This research was supported by the Australian Research Grants Scheme and the University of Western Australia. The authors are grateful to many people for assistance includeing P. Hayes with the operation of the computer and to the departmental workshops with continuing excellent work and maintenance without which this would not have been possible.

REFERENCES

1. A. H. Mahan and S. J. Smith, 1977 Phys. Rev. $\underline{A16}$, 1789.

2. I. A. Sellin, L. Libjeby, S. Mannervik and S. Hultberg, 1970 Phys. Rev. Lett. $\underline{42}$, 570-3.

3. R. Krotkov and J. Stone, 1980 Phys. Rev. $\underline{A22}$, 473-82.

4. A. Gaupp, H. J. Andra and J. Macek, 1974 Phys. Rev. Lett. $\underline{32}$, 1335-7.

5. H. J. Andra, 1974 Phys. Scrip. $\underline{9}$, 257-80.

6. M. J. Alquard and C. W. Drake, 1973 Phys. Rev. $\underline{A8}$, 27-36.

7. A. van Wijngaarden, E. Goh, G.W.F. Drake and P. S. Farago, 1976 J. Phys. $\underline{B9}$, 2017-25.

8. J. Slevin, 1985 Rep. Prog. Phys. $\underline{47}$, 461-512.

9. J. F. Williams, 1986 Aust. J. Phys. $\underline{39}$, 621-32.

10. K. Blum and H. Kleinpoppen, 1977 J. Phys. B: At. Mol. Phys. $\underline{10}$, 3283-95.

11. C. G. Back, S. Watkin, M. Eminyan, K. rubin, J. Slevin and J. M. Woolsey, 1984, J. Phys. B: At. Mol. Phys. $\underline{17}$, 2695-2706.

12. L. Heck and J. F. Williams, 1987 J. Phys. b: At. Mol. Phys. $\underline{20}$, 2871-86.

13. J. F. Williams, 1981, J. Phys. B: At. Mol. Phys. $\underline{14}$, 1197-218.

14. L. A. Morgan and M.R.C. McDowell, 1975, J. Phys. B: At. Mol. Phys. $\underline{8}$, 1073-81.

15. W. I. van Wyngaarden and H.R.J. Walters, 1985, J. Phys. B: At. Mol. Phys. $\underline{18}$, L689-94.

16. J. F. Williams and L. Heck, 1987 (in preparation).

NOBLE GAS ATOMS IN STRONG DC ELECTRIC FIELDS

Rudolf Ch. Ziegelbecker and Laurentius Windholz

Institut für Experimentalphysik
Technische Universität Graz
A-8010 Graz, Austria

INTRODUCTION

The general behavior of atomic levels in a d.c. electric field was already outlined by Bethe,[1] but the Stark effect never reached the same importance as the Zeeman effect. The "state of the art" in investigating atoms and molecules in electric fields 11 years ago is described by Ryde.[2]

Due to the increasing interest in Rydberg atoms and autoionizing states there has been new progress in the theory of the Stark effect again (see e.g. Harmin[3] and Bergeman[4]).

Also experimentally electric field strengths higher than those mentioned by Ryde[2] were reached by Windholz[5] and Jäger et al.[6] when investigating the Stark effect of noble gases. But up to now there were no comprehensive theoretical studies that explained the structure of the splitting, the number of Stark components showing similar shifts, or that permitted to assign quantum numbers to each Stark level observed.

Recently Ziegelbecker[7] investigated the problems mentioned above using a model of maximum simplicity. This model, mainly based on jl coupling (also known as jK coupling[8]), was applied to neutral neon. As a result, the calculated Stark levels were plotted and compared with the experimental curves derived from the visible part of the neon I Stark spectrum, up to the 7s levels. Also, the computed Stark shifts of the upper levels, the line intensities and wavefunction compositions were tabulated. Using these tables, we could identify most of the experimental Stark levels and assign the correct total magnetic quantum number to them. In many cases it was also possible to characterize the observed Stark levels quite well by a nomenclature first proposed by Ziegelbecker[7] and presented here with a little modification. These results will be published in a subsequent paper.[9]

THE COMPUTATION OF THE NEON I STARK EFFECT

The Method

To calculate the Stark effect of an (excited) atom we may solve the eigenvalue equation

$$(H_0 + H_1 F) \, |\psi\rangle = E|\psi\rangle \qquad (1)$$

numerically for a sufficient number of values for the electric field strength F, $H_1 F$ being the d.c. electric field operator. H_0 is the Hamiltonian of the atom in absence of an external field and satisfies

$$H_0 \, |\psi_i^0\rangle = E_i \, |\psi_i^0\rangle \qquad (2)$$

Expanding a $|\psi_j\rangle$ in (1) – corresponding to the eigenvalue E_j – in terms of $|\psi_i^0\rangle$

$$|\psi_j\rangle = \sum_{i=1}^{N} a_{ij} \, |\psi_i^0\rangle, \qquad (3)$$

multiplying by $|\psi_k^0\rangle$ and using (2) leads to a system of coupled equations

$$\sum_{i=1}^{N} [(E_i^0 - E_j)\delta_{ki} + \langle\psi_k^0|H_1|\psi_i^0\rangle F]a_{ij} = 0, \qquad (4)$$

which can be solved numerically for the energy of the jth level, E_j, and the corresponding eigenvector components a_{ij} for any field strength F, provided the E_i^0 and the dipole matrix elements

$$\langle|\psi_k^0|H_1||\psi_i^0\rangle \qquad (5)$$

are known.

Starting with (3), we restrict ourselves to a limited number N of basis functions belonging to the discrete spectrum only. This is no disadvantage, because we want to calculate only the spectroscopically observed quasi–discrete Stark levels. Nevertheless we shall determine that region of the field where quasidiscreteness cannot be assumed any more.

The most exact values for the "unperturbed" energies E_i^0 are certainly those taken from experiment; we used the data of Moore.[10] For energy values not listed by Moore[10] we assumed either degeneracy, or the same quantum defect as observed at another principal quantum number, or hydrogenic values as in the case of the neon I ng-, nh-, ni- ... levels.

The Calculation of the Matrix Elements

For the spectra of singly excited neutral noble gas atoms, jl coupling gives the best theoretical description in terms of a definite coupling scheme. In this scheme first suggested by Racah[11] the orbital angular momentum $\vec{l}_c$ and the spin $\vec{s}_c$ (c indicating the core) are coupled to give an angular momentum $\vec{j}_c$ of the core with a definite value j_c. Since $l_c = 1$ in the case of singly excited noble gas atoms, $j_c = 3/2$ or $1/2$. $\vec{j}_c$ couples to the orbital angular momentum $\vec{l}$ of the excited electron to give an angular momentum $\vec{K}$, and finally the spin $\vec{s}$ of the excited electron is added to give the total angular momentum $\vec{j}$.

The avoid very complex and time consuming programs for calculating (5) we assume pure jl coupling for all (excited) states $||\psi_i^0\rangle$. This permits us to approximte the many particle problem by a two particle (electron-hole) problem in a central potential.[7] In this model a basis function $\langle|\psi^0|$ can be written in the following form

$$\langle\dagger|\psi^0| = \langle j_c \, n \, 1 \, (K) \, J \, M| = \langle R_c(j_c)| \langle R(j_c,n,1,K,J)| \langle j_c \, 1 \, (K) \, J \, M|. \quad (6)$$

The second bra vector is labelled by the jl coupling quantum numbers, n is the principal quantum number of the excited electron. R_c is the radial function of the core, which we consider independent of the state of the excited electron ("frozen core" approximation), while the radial function R of the excited electron may depend on all but the magnetic quantum numbers. In this particle-hole formalism the electric field operator is

$$H_1 F = (-z_c + z)F = (-r_c \cos\theta_c + r \cos\theta) \cdot F. \quad (7)$$

For calculating the nonradial part of (5) we must expand $\langle j_c 1(K) \, J \, M|$ in terms of $\langle l_c m_{l_c}| \langle s_c m_{s_c}| \langle 1_{m_l}| \langle s m_s|$. Because of the frozen core ($l_c = 1$) and the spatial antisymmetry of $\cos\theta_c$ all terms containing $\cos\theta_c$ become zero. As a consequence one finds that a quasi-one-electron formalism using an operator

$$H_1 = z = r \cos\theta \qquad (8)$$

and a wave function

$$\langle\psi^0| = \langle j_c \, n \, 1 \, (K) \, J \, M| = \langle R(j_c,n,1,K,J)| \sum_i c_i \langle j_c m_c|_i \langle 1 \, m_1|_i \langle s \, m_s|_i. \qquad (9)$$

will give the same values for the matrix elements (5) as the two-particle scheme (7) does if the core is purely in the state $l_c = 1$. The c_i are up to 8 expansion coefficients of the spin-orbital part in 76).

Note that z does not act on $\langle j_c m_c|$ in (9). This leads to a factor $\delta_{j_c j_c'}$ in the matrix element preventing any interaction due to the field F between levels with different $j_c(3/2$ or $1/2)$, which approximation is close to reality. However, in practice there is always at least a slight interaction between levels with the same quantum numbers $|M|$ and S(S ... reflection symmetry; see next section), resulting in anticrossings of such levels.

We obtain the radial functions $P = r \cdot R$ by integrating the one electron radial equation inwards from a starting point r_A towards the origin. In order to get a relatively constant number of (about 300) integration steps per antinode of the function, we transform the equation to a new variable $u = \sqrt{r}$ (using Hartree atomic units):

$$\frac{d^2 P(u)}{du^2} = \frac{1}{u}\frac{dP(u)}{du} - P(u)\cdot\left[\frac{41(1+1)}{u^2} - 8u^2(V(u) - E^0)\right] \qquad (10)$$

For E^0 we insert the experimental energy of the level with respect to its related ionization limit. $V = Z_{eff}/r$ is the potential energy function. For $Z_{eff}(u)$ we use a simple analytic expression consisting of a parabolic curve with its maximum at $u = 0$, declining to a turning point at $u = 0.28$ where it is continued by a second parabolic curve whose minimum $Z_{eff} = 1$ is at $u = 1.52$. Z_{eff} is equal to 1 for $u > 1.52$, which corresponds to a core radius of $1.52^2 = 2.31$ atomic units. We use this parametric potential in order to extend (typically $r = 0$ for s- and $r \approx 0.05$ for p-orbitals). Its exact shape is not too critical for the values of the matrix elements, but with the mentioned parameters it agrees quite well with the potential calculated from the Hartree-Fock charge distribution for the ground state of neon I[12] or with the parametric potential used by Feneuille et al.[13]

The integration of the radial equation is carried out simultaneously for all radial functions ns to ng, the radial matrix elements being obtained at the same time. For radial matrix elements containing hydrogenic radial functions only (ng, nh, ...), we use the Gordon formula.[14]

Group Theoretical Considerations

The invariance group of the Hamiltonian of an atom in a static homogeneous electric field has two-dimensional, irreducible representations for $|M| = 1/2, 1, 3/2, 2, ...$ and two inequivalent one-dimensional irreducible representations for $M = 0$, which are labelled $0+$ and $0-$ (or[15]: Σ^+ and Σ^-) and which differ by the reflection symmetry S of the wavefunction with respect to a plane parallel to the z-axis. The eigenvalues of S are $+1$ and -1.

It is known practice to use eigenfunctions $|\gamma; M\rangle$ of the operator J_z as basis functions for calculating (5), which procedure decomposes the Stark matrix [....] in (4) according to M, e.g. $M = ..., -2, -1, 0, 1, 2, ...$ (γ stands for all other quantum numbers including j_c). However, in the case of atoms with integer values of J it is advantageous to use simultaneous eigenfunctions $|\gamma; |M|, S\rangle$ of the operator $|J_z|$ ($|M| = 0, 1, 2, ...$) and of the mentioned reflection operator S_{op}. With these the matrix elements (5) become

$$\langle \gamma'; |M|'S' |z| \gamma; |M|S\rangle = \langle \gamma'; M|z|\gamma; M\rangle \cdot \delta_{|M|,|M|'} \cdot \delta_{S,S'} , \qquad (11)$$

which leads to an additional decomposition of the $|M| = 0$ matrix into two parts of equal size, $S = +1$ and $S = -1$. According to group theory, this is the maximum decomposition of the general Stark problem and we therefore may consider $|M|$ and S to be the natural (conserved) quantum numbers of the Stark effect. In our special case, a further decomposition due to different values of j_c is introduced because of our approximations of jl coupling at $F = 0$ and of a "frozen" wavefunction of the core hole with both $l_c = 1$ and radial part fixed.

Ziegelbecker and Schnizer showed[17] that in the case of singly excited nobel gas atoms every wavefunction $|\gamma; M = 0\rangle$ has a definite reflection symmetry

$$S = (-1)^{1-J}, \qquad (12)$$

and how to construct wavefunctions with definite S ($+1$ and -1) for $|M| > 0$.

Computation of Eigenvalues and Eigenvectors

Equation (4) then is solved numerically, separately for each j_c, for each $|M| \neq 0$ of the absolute total magnetic quantum number and for the two noninteracting systems $|M|$, $S = 0+$ and $0-$ yielding the wanted eigenvalues E_j. In each case this is done for about $i_{max} = 50$ values of the field strength F, distributed according to

$$F_i = F_{max} \cdot \left(\frac{i - 1}{i_{max} - 1}\right)^{1.3} . \tag{13}$$

To obtain an informative representation of the eigenvectors (columns of a_{ij} in (4)) we transform them to a basis $|j_c m_c\rangle |n\ 1\ m_l\rangle |s\ m_s\rangle$, which is possible if we ignore the slight differences which the radial functions belonging to the same set j_c, n, l exhibit because of different values of K and J.

RESULTS

The Rough Structure of the Level Shifts

To get a rough picture of the Stark effect of neon I we ignore the fine structure and calculate the Stark splitting for both values of j_c using an n, l, m_l-basis instead of (9), and energies E^0 obtained by averaging over all levels of a group j_c, n, l with the weight $2J + 1$. With this basis the Stark matrix also decomposes in the described way, with the only exception that $|m_l|$, $S = 0-$ does not exist because $|n, 1, m_l\rangle$ is an elementary basis function.

In Fig. 1 the rough splitting of the neon I levels 6s to 6p is shown as an example. Levels characterized by different $|m_l|$ (indicated by the numbers at the end of each curve) as well as by different j_c ($j_c = 1/2$ indicated by a "prime" in the level specification) do not interact and therefore frequenctly cross each other.

The interaction between levels with equal $|m_l|$ is observable as a "repelling force," or better "bending away force" between the curves, which – according to perturbation theory – depends on the square of the dipole matrix element (selection rule $\Delta l = \pm 1$ at $F = 0$) divided by the difference in the energies of the considered levels.

For nearly degenerate levels (5d, f, g at low field strengths) we get an almost symmetrical splitting pattern, the size of the shifts of the outermost components appearing to be proportional to $l_+^2 - m_l^2$ as a consequence of the factor $\sqrt{l_+^2 - m_l^2}$ contained in the formula for the angular part of the dipole matrix element.[1] It is largest for $|m_l| = 0$ and zero for $|m_l| = l_+$. l_+ is the larger one of both l.

At higher field strengths the 5d0 (abbreviation for 5d, $|m_1| = 0$) is "repelled" by the 6s0 level leading to a crossing with 5d1 (i.e. 5d, $|m_1| = 1$), and the 5g1 is "repelled" by the 6p1, while 5g0 is "repelled" by the 6p0 <u>and</u> the 7s0 level. The very small shift of 5g4 (4 ... highest possible $|m_1|$) is only due to the interaction with $|m_1| = 4$ levels of higher principal quantum numbers. The same (with $|m_1| = 2$) is true for 5f2 that shifts very little because it is the middle one of an odd number of levels which split in a symmetrical pattern. 5f0 and 5f1 behave like 5f2 for small fields, but for F > 100 kV/cm they are shifted much more than 5f2 because of the influence of 7s and 6p. 5f3 and 5g3 are mainly influenced by each other and split very symmetrically. The 6s' level is only influenced by other "primed" levels.

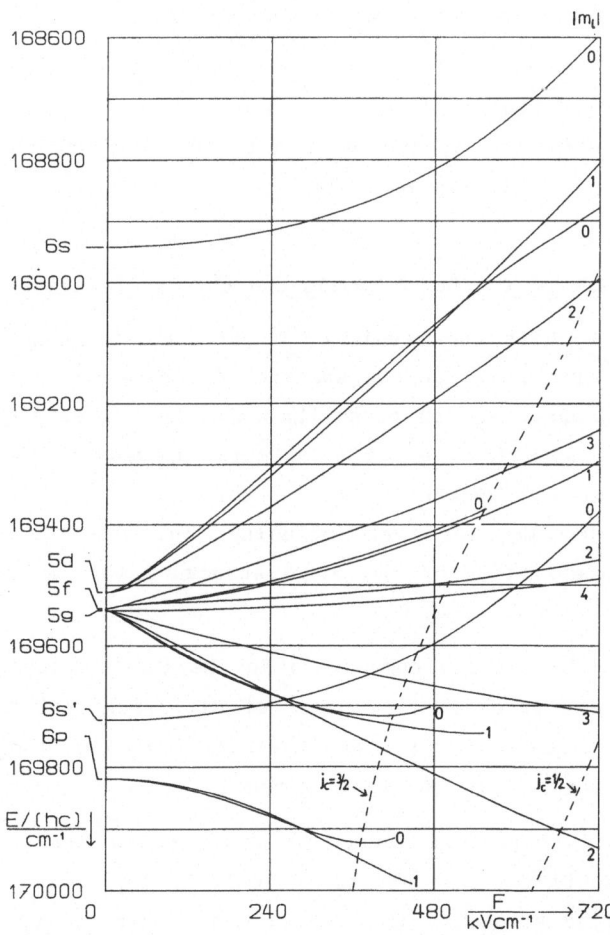

Fig. 1 The rough Stark splitting of the neon I levels 6s to 6p calculated in j_c, n, l, m_1 basis. The value for $|m_1|$ is indicated at the end of each curve. The dashed lines show the electric field strength at which the drawing of a curve is stopped in Fig. 2.

An important remark must be made on the maximum field strength above which individual Stark levels cease to exist: as we increased the size of the basis (in order to achieve more precision) to more than 4 or 5 principal quantum numbers above the level considered, new eigenvalues intruded between the drawn curves. They turned out to represent wavefunctions whose main portion is outside the classical potential barrier; they are without physical significance, their position depending randomly on the size of the basis.

Therefore we calculated Fig. 1 in the following way: we used a basis ranging from 3s to 15s. Eigenvalues were calculated for $i_{max} = 50$ values of F, distributed according to Eq. (13). Beginning with $F_1 = 0$, the eigenvalues obtained for successive F_i were combined in a way to gie a smooth curve for each level. The routine stopped joining points when it could no longer find a unique continuation (corresponding to a quasidiscrete level) amongst neighbouring points.

The dashed curves in Fig. 1 show the field strength where, for each j_c, the drawing is stopped in Fig. 2 in order to ensure a quasidiscrete character and sufficient precision of all levels drawn, in spite of the smaller range of our jl coupled basis.

The Fine Structure of the Stark Levels and the Wavefunctions

In Fig. 2 we give the result of a calculation using the purely jl coupled basis. The range of the spectrum is the same as in Fig. 1. At the end of each curve we now print the absolute magnetic quantum number $|M|$ of the total angular momentum J, only in the case of $|M| = 0$ we add a "+" or "−" to characterize the symmetry of the corresponding wavefunction. One can see that including fine structure in the basis gives rise to a splitting of the rough structure into a large number of components.

The number of existing components in the fine structure of an $|m_1|$ (rough structure-) level is the number of different combinations of m_1, m_c and m_s that give a definite $M \geqslant 0$ ($M = m_1 + m_c + m_s$) under the condition $m_1 = \pm|m_1|$. The result is given in Table 1.

Some of the components cannot be resolved in Fig. 2 because they have almost exactly the same shift. Especially $|m_1| = 2$ and 3 show only two groups of fine structure components, and it can be supposed from the quantum numbers $|M|$ that the group with higher energy corresponds to $|m_c| = 1/2$ and the lower one to $|m_c| = 3/2$, which is confirmed by the structure of the wavefunction to about 99% purity when displayed in a j_c, n, 1, m_1, m_c, m_s–basis. Such wavefunctions were tabulated by Ziegelbecker,[7] examples are given in the paper of Ziegelbecker and Schnizer.[17]

The situation is somewhat similar for $|m_1| = 0$ and $|m_1| = 1$ levels, but usually the purity of $|m_c|$ is around 90% or even less. However, in the special case of M = 0+ one even finds almost pure symmetric or antisymmetric mixtures of $|m_c| = 3/2$ and $|m_c| = 1/2$.

We want to add that all levels $|m_1| \geqslant 3$ usually have only very small admixtures of other $|m_1|$. Therefore they will not, or only very weakly, be seen in the visible transitions of neon I due to the selection rules $\Delta m_1 = 0, \pm 1$ and because there are only 3p (3p') final states ($|m_1| \leqslant 1$).

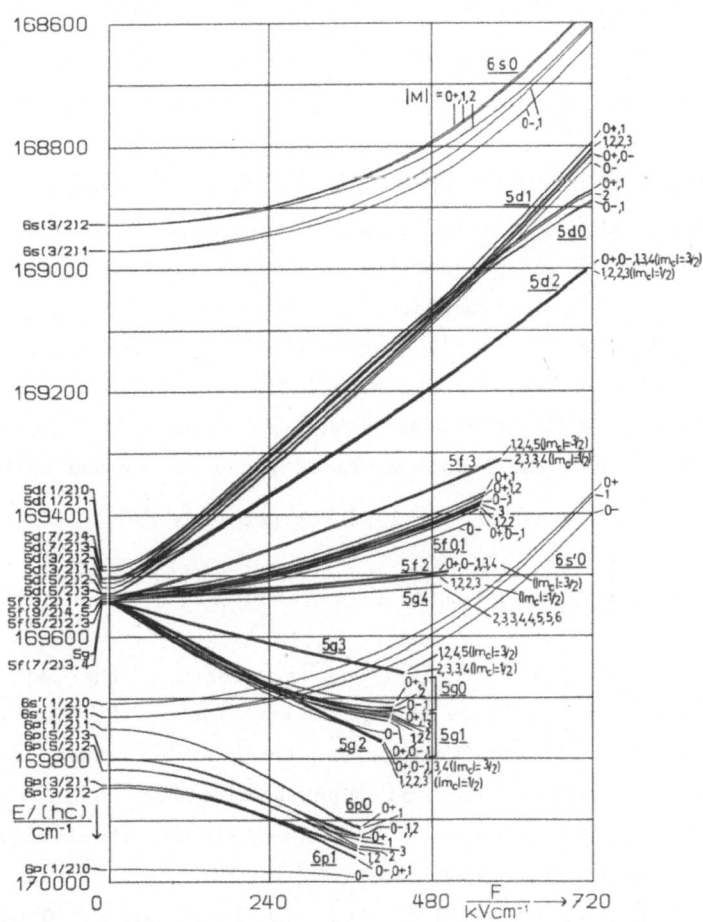

Fig. 2   The quasi–discrete Stark spectrum of the neon I levels 6s to 6p calculated using the assumption of exact jl coupling at zero field. The underlined designations refer to n, 1, $|m_1|$ (see Fig. 1), the numbers at the end indicate the absolute total magnetic quantum numbers $|M|$ contained in each curve plus reflection symmetry S in the case $|M| = 0$. In the Stark effect 0+ levels do not interact with 0– levels, in the same way as levels with different $|M|$ do not interact.

Table 1. The number of fine structure components belonging to a given rough structure component.

core level $\quad j_c$	3/2				1/2		
rough structure comp. $|m_l|$	0	1	2	≥3	0	1	≥2
total number of comps.	5	10	9	8	3	5	4
of them comps. with $|M|$=0+	1	2	1		1	0	
0−	1	2	1		1	1	.
1	2	3	2	1	1	2	1
2	1	2	2	2..		1	2..
3		1	2	2....			1....
4			1	2....			....
5				1....			..

A Nomenclature for the High Field Stark Effect of Noble Gases

Based on the rigorous symmetry properties of the perturbed wavefunctions and on the possibility to characterize them additionally by predominant values of intermediate magnetic quantum numbers (as described in the previous section: $|m_c|$) Ziegelbecker[7] developed a nomenclature to characterize single levels in the high field Stark effect at regions wher the $|m_l|$ groups are clearly separated. We present it here with slight modifications to meet the requirements of group theory and uniformity.

The newly suggested high field Stark notation is

$$n \; l^{j_c} \; |m_l| (\,(|m_c|, \; |M_K|; \cdot e) \; |M| \; S \; . \tag{14}$$

j_c (... "prime" if $j_c = 1/2$), n, l and $|m_l|$ refer to the rough structure of Stark effect. n and l are the generalization of the familiar quantum numbers at F = 0. $|M|$ and S are only quantum numbers that are conserved exactly in the Stark effect, as described earlier. In the case of $|M| = 0$ the reflection symmetry S is dropped because the levels S = 1 and S = −1 are degenerate. (We repeat that for $|M| = 0$ the reflection symmetry is given by $S = (-1)^{l-J}$ of the level at F = 0 and appears directly in the relative sign of coupling coefficients c_l similar to those of (9) if the wavefunction is expanded in an l_c, s_c, n, l, s, m_{l_c}, m_{s_c}, m_l, m_s-basis. In a basis where $\vec{l}_c$ and $\vec{s}_c$ are coupled to a $\vec{j}_c$ already, one has to take into account the reflection symmetry $(-1)^{l_c+s_c-j_c}$ of the basis function $|j_c \; m_c\rangle$).

If there are two or three levels with equal quantum numbers j_c, n, 1, $|m_1|$, $|M|$ and S (see Table 1), further quantum numbers are needed to achieve a unique labelling of these Stark levels. A sophisticated study of the wavefunctions of neon I at different electric field strengths showed that the best choice seems to be $|m_c|$ and $|M_K|$, which are therefore introduced into the Stark nomenclature in spite of the fact that they are often not well defined. Sometimes even an almost pure symmetric or antisymmetric mixture of different $|m_c|$ or $|M_K|$ occurs; in this case we suggest giving both values participating in the mixture, in the order of their magnitude if decideable, separated by the "mixing sign" by which the relative sign of the coupling coefficients belonging to equal m_1 (not: $|m_1|$) is understood (see examples below).

In order to avoid misunderstandings due to not well defined quantum numbers in the brackets in (14), we introduced "e." It stands for "+" or "−" and shall indicate the energetical position if the exact wavefunction is not known and if the quantum numbers outside the brackets would designate two different levels. "+," "o," or "−" will be used in the single case when they designate three levels (cf. Table 1).

Examples: according to Fig. 2, the 5d, $|m_1|$ = 0 level with the lowest energy (F ≥ 600 kV/cm) is named 5d0(1/2, 1/2)0+. The lower 5d, $|m_1|$ = 1, $|M|$ = 0+ level is named 5d1(1/2 + 3/2, 1/2;−)0+ according to its wavefunction data[17] at 600 kV/cm. The higher one is called 5d1(3/2 − 1/2, 1/2;+)0+ (the $|m_c|$ − mixture is practically symmetric here, so that the choice of their order must be an arbitrary one). The three 5d$|m_1|$ = 1, $|M|$ = 1 levels are named (in ther energetical order) 5d1(3/2 + 1/2, 1/2;−)1, 5d1(1/2, 3/2;o)1 and 5d1(1/2 − 3/2, 1/2;+)1 according to the wavefunction data at 600 kV/cm (see Ziegelbecker[7]). Here again the three states are almost pure.

The advantage of this nomenclature can be seen most clearly if we consider levels that have undergone anticrossings with other levels of equal $|M|$ and S but different j_c (such levels cross each other in our calculation, but in reality there is no rigorous jl coupling, causing a smaller or wider energy gap between them even if they seem to cross): theoretically it would be possible to designate a Stark level by the name of the level at F = 0 plus the absolute magnetic quantum number $|M|$. This is advantageous especially up to field strengths where no anticrossings occur to that level. However, when using this notation at field strengths beyond a region of anticrossings (e.g. 6s'−5g at about 250 kV/cm, see Fig. 2, or also 4d'−5p), levels of e.g. 6s'-displacement will have to be given the name of some 5g- or 5f- level because they originate from there. On the one hand, this would require exact knowledge of the arrangement of these quasi-degenerate levels that usually cannot be resolved experimentally, on the other hand, beyond an anticrossing this "usual" designation will contain no useful information about the level.

On the contrary, the newly suggested notation cannot be used around anticrossings and sometimes not at zero field, but it directly exhibits the character of the wavefunction outside these regions and will always call a level of 6s'-displacement a 6s'-level. Because it contains magnetic quantum numbers, it will be helpful for predicting intensities of trnasitions between higher excited levels at strong electric fields an thus can be checked experimentally. Eventual deviations between experimental data and those calculated by Ziegelbecker[7] would indicate the need of better wavefunctions in the theoretical calculations.

CONCLUDING REMARKS

It was shown in the chapters above that the main character of the Stark splitting of the neon atom can be quite well understood and systematized if one considers the rough structure (following from level positions and the matrix elements in a j_c, n, l, m_l basis) and the fine structure wavefunctions. For this purpose we used a relatively simple model, which is based on a one electron radial function (for the excited electron) and a two particle (electron, core) angular momentum wavefunction represented by a definite coupling scheme (jl coupling), which gives a good quantitative description of the neon atom in an electric field.[7] In spite of the fact, that rigorous jl coupling does not allow the description of anticrossings between levels assigned to different angular momenta of the core, the model allowed to identify most of the experimentally observed components, to label many of them using the new nomenclature,[9] and also to find new components on the original exposures.

It is likely that, using improved wavefunctions such as those derived from multichannel quantum defect theory (MQDT) or self consistent field methods, a detailed computation of a few "uncertain" assignments and of the anticrossing regions (e.g. 6s'-5g and 5p-4d') will be possible.

ACKNOWLEDGEMENTS

We would like to thank Prof. H. Jäger, Institut für Experimentalphysik, and Prof. B. Schnizer, Institut für Theoretische Physik of the Technische Universität Graz, for their continuous interest and encouragement and for many helpful discussions. This work was supported by the Austrian Fonds zur Förderung der Wissenschaftlichen Forschung under contract No. P6242P.

REFERENCES

1. H. Bethe, Quantenmechanik der Ein- und Zweielektronenprobleme, in Handbuch der Physik XXIV/1, eds. H. Geiger and K. Scheel (Springer, Berlin, 1933).

2. N. Ryde, Atoms in Electric Fields (Almqvist and Wiksell, Stockholm, 1976).

3. D. A. Harmin, "Electric Field Effects in Rydberg Atoms," Comments At. Mol. Phys. 15, 281 (1985).

4. T. Bergeman, "Numeric Calculations of the Stark Effect in Neutral Hydrogen," in H-Workshop-Session held in Los Alamos, Nov. 1981.

5. L. Windholz, "Stark Effect of Ar I - Lines," Phys. Scripta 21, 67 (1980).

6. H. Jäger and L. Windholz, "Stark Effect of Ne I - Lines (I)," Phys. Scripta 29, 344 (1984).

7. R. Ch. Ziegelbecker, "Berechnung des Starkeffekts von Edelgasatomen unter Annahme reiner jl-Kopplung am Beispiel des Ne I," thesis, Technische Universität Graz (1986).

8. R. D. Cowan, The Theory of Atomic Structure and Spectra, eds. D. H. Sharp and L. M. Simmons, Jr. (University of California Press, Berkeley-Los Angeles-London, 1981).

9. R. Ch. Ziegelbecker, H. Jäger and L. Windholz, "Stark Effect of Ne I - Lines (II)," Phys. Scripta (to be published).

10. Ch. E. Moore, "Atomic Energy Levels I," Circular of the National Bureau of Standards 467 (U.S. Government Printing Office, Washington, D.C., 1949).

11. G. Racah, "On a New Type of Vector Coupling in Complex Spectra," Phys. Rev. 61, 537 (1942).

12. Ch. Froese-Fischer, "The Hartree-Fock Method for Atoms" (Wiley, New York, 1977).

13. S. Feneuille, M. Klapisch, E. Koenig and S. Liberman, "Determination Theorique des Probabilities de Transition $2p^53p - 2p^53s$ dans le Spectre du Neon I," Physica <u>48</u>, 571 (1970).

14. W. Gordon, "Zur Berechnung der Matrizen beim Wasserstoffatom," Ann. Phys. <u>2</u> (5), 1031 (1929).

15. B. L. Van der Waerden, "Group Theory and Quantum Mechanics," eds. B. Eckmann, J. K. Moser and J. K. Van der Waerden (Springer, Heidelberg-New York, 1977).

16. A. R. Edmonds, "Angular Momentum in Quantum Mechanics," eds. E. Wigner and R. Hofstadter (Princeton University Press, Princeton, 1957).

17. R. Ch. Ziegelbecker and B. Schnizer, "Calculation of the Stark Effect of Neon I Using jl Coupled Wavefunctions," to appear in Z. Phys. D (1987).

AN ATOMIC INTERFEROMETER

H. C. Bryant

Physics and Astronomy Department
The University of New Mexico
Albuquerque, NM 87131

Recently-reported ripples in the H^- photodetachment cross section near threshold in an external electric field are interpreted as interference fringes in a dual beam interferometer. The role of the photon's coherence length in the resolution of the ripples enters in a natural way.

When the photodetachment cross section of H^- in an external electric field is measured in π polarization near threshold, [1] the result is a ripple-like modulation on the zero-field cross section. Two formal derivations of this effect have been given[1,2] as well as an heuristic one.[1] In this paper I wish to pursue the analogy between the mechanism involved and that of a Michelson interferometer, one of whose arms is very short, illustrated by Fig. 1.

The atom plays the role of both the beam splitter and the short arm with mirror M2. The photodetachment process converts the incident laser photon into an electron wave acket with the same temporal coherence as that of the photon. Since the laser light is π polarized, the ejected electron tends to be directed parallel or antiparallel to the external DC electric field, and its wave function, of course, can be imagined to go both ways. The electron which finds its potential energy rising as it leaves the atom is eventually "reflected" by the electric field and at a time τ after its ejection finds itself passing the parent atom again, heading toward the detector. There are, therefore, two pathways along which the electron can arrive at the detector, and, if the electron's coherence length is long enough, the amplitudes for these pathways can interfere, giving rise to the observed fringes. By varying the electric

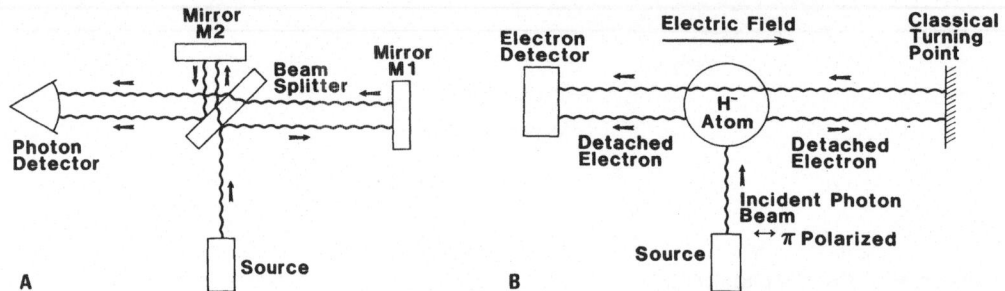

Fig. 1A. A Michelson interferometer with one very short arm. Mirror M1 may be varied in distance from the beam splitter.

Fig. 1B. At atomic interferometer. The H⁻ ion functions as both the beam splitter and the short arm. Moreover, at the atom the light wave is converted into an electron wave. The electric field serves as the mirror M1 whose effective distance is altered by varying the field strength.

field, the delay time τ can be varied, for a given photon energy. Changing the electric field for the atomic interferometer is equivalent to changing the distance to mirror M1 in the Michelson case.

In what follows we pursue the analogy analytically.

Let the amplitudes for the pathway via mirror M1 (arm 1) and mirror M2 (arm 2) be f_1 and f_2, respectively, and the r.m.s. angular frequency spread of the incident photon be s.

Then, considering the simple case first,

$$f_2(t) = A \int_{-\infty}^{\infty} \exp\left\{-(\omega - \omega_0)^2/2s^2 - i\omega t\right\}d\omega. \tag{1}$$

We take $\hbar\omega_0(=E)$ to be the central energy of the electron as it is ejected (E = photon's energy – electron affinity of neutral atom). Integration yields[4]

$$f_2(t) = \sqrt{2\pi} \, As \exp\left\{-t^2 s^2/2 - i\omega_0 t\right\}. \tag{2}$$

The amplitude of the other branch is more complicated. For an electron of energy E moving against an electrical potential energy eFx, we have

$$\frac{1}{2} mv^2 + eFx = E, \tag{3}$$

where x lies along the field direction with origin at the atom.

Thus

$$\tau = 2 \int_0^L \frac{dx}{v} \ , \tag{4}$$

where

$$L = E/eF, \tag{5}$$

so that

$$\tau = \sqrt{8mE}/eF \ . \tag{6}$$

We can then write analogously to Eq. 2:

$$f_1(t) = \sqrt{2\pi} \ As \ \exp\left\{-(t - \tau)^2 s^2/2 - i(\omega_0 t - \alpha)\right\} \tag{7}$$

The phase shift α is given by

$$\alpha = \phi_\tau + \Delta \ , \tag{8}$$

where ϕ_τ is due to the presence of the electrostatic potential V, and Δ is the delay at reflection due to penetration of the barrier. The potential V affects the phase of the wave just as does the vector potential in the Aharanov-Bohm effect[5]:

$$\phi_\tau = \int_0^\tau Vdt/\hbar \ , \tag{9}$$

where $V = eFx$. So

$$\phi_\tau = \frac{2eF}{\hbar} \int_0^L \frac{xdx}{v} \ , \tag{10}$$

where

$$v = \left\{\frac{2}{m} (E - eFx)\right\}^{1/2}. \tag{11}$$

We find

$$\phi_\tau = \frac{2}{3} \frac{E}{\hbar} \sqrt{\frac{8mE}{eF}} = \frac{2}{3} \omega_0 \tau \ . \tag{12}$$

Thus

$$f_1(t) = \sqrt{2\pi}\; As\; \exp\{-[t - \tau]^2 s^2/2 - i\omega_0(t - \tfrac{2}{3}) + i\Delta\}. \tag{13}$$

Now, the complete amplitude for an electron from the photodetachment process to arrive at the detector at time t is

$$f(t) = f_1(t) + f_2(t). \tag{14}$$

The cross section α is

$$\sigma = \int\limits_{-\infty}^{\infty} dt\; f^*(t)\; f(t). \tag{15}$$

We find

$$\int\limits_{-\infty}^{\infty} dt\; f_1^*(t)f_1(t) = \int\limits_{-\infty}^{\infty} f_2^*(t)f_2(t)dt = 2\pi^{3/2}\; A^2 s \tag{16}$$

and

$$\int\limits_{-\infty}^{\infty} dt\; f_1^*(t)f_2(t) = 2\pi A^2 s^2 \int\limits_{-\infty}^{\infty} \exp\{-[t^2 + (t - \tau)^2]s^2/2$$

$$- i\,\tfrac{2}{3}\,\omega_0\tau - i\Delta\}dt. \tag{17}$$

So,

$$\int\limits_{-\infty}^{\infty} f_1^*(t)f_2(t)dt = 2\pi^{3/2}\; A^2 s\; \exp[-\tau^2 s^2/4 - i(\tfrac{2}{3}\,\omega_0\tau + \Delta)]. \tag{18}$$

Thus

$$\sigma = 4\pi^{3/2}A^2 s[1 + e^{-\tau^2 s^2/4}\; \cos(\tfrac{2}{3}\,\omega_0\tau + \Delta)]. \tag{19}$$

Now we make two ad hoc modifications.

1. Replace $4\pi^{3/2}A^2 s$ by $\sigma_0 E^{3/2}/(E + E_B)^3$ to model the field when $F \to 0$ $(\tau \to \infty)$. E_B is the binding energy.

2. Insert an efficiency η in front of the second term to account for the fact that the two beams 1 and 2 do not completely overlap. Thus $0 < \eta < 1$.

Thus

$$\sigma = \sigma_0 E^{3/2} [1 + \eta e^{-s^2 \tau^2/4} \cos(\phi)]/(E + E_B)^3 , \qquad (20)$$

where

$$\phi = 2/3 \ \omega_0 \tau + \Delta . \qquad (21)$$

In order to have destructive interference (dark fringes, dips) in the cross section we must have

$$\phi = (2n - 1), \ \text{where} \ n = 1,2 \ ..., \qquad (22)$$

so

$$\frac{2}{3} \ \omega_0 \tau = (2n - 1)\pi - \Delta = \frac{4E}{3\hbar} \frac{\sqrt{2mE}}{eF} . \qquad (23)$$

Solving for E we find

$$E = (\frac{\hbar^2}{2m})^{1/3} (eF)^{2/3} \{\frac{3}{4}[(2n - 1)\pi - \Delta]\}^{2/3}. \qquad (24)$$

Table 1

m	a_m'	$[3\pi(4n-3)/8]^{2/3}$	Δ
1	1.01879	1.11546	1.770497
2	3.24820	3.26163	1.619248
3	4.82010	4.82632	1.598103
4	6.16331	6.16713	1.589771
5	7.37218	7.37485	1.585328
6	8.48849	8.49051	1.582569
7	9.53545	9.53705	1.580689
8	10.52766	10.52897	1.579327
9	11.47506	11.47616	1.578294
10	12.38479	12.38573	1.577484

Comparing with the simple theory using the Airy function,[1] where we found,

$$E = -(\frac{\hbar^2}{2m})^{1/3}(eF)^{2/3}a'_n .$$ (25)

we can identify

$$-a'_n = [\frac{3}{4}(2n - 1) - \frac{3}{4}\Delta]^{2/3} .$$ (26)

It can be shown[6] that for large n, asymptotically,

$$-a'_n \rightarrow [3\pi(4n - 3)/8]^{2/3} .$$ (27)

From this we can identify

$$\Delta = \pi/2 .$$ (28)

Table 1 displays how Δ would have to vary with n to give perfect agreement. Figure 2 plots $\Delta - \pi/2$ versus n. Since our approach is that of physical optics rather than quantum mechanics, we are pleased with the degree of agreement.

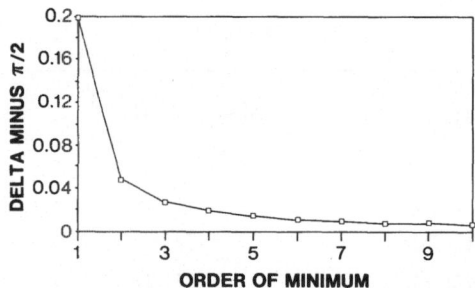

Fig. 2. The value of Δ required to fit Eq. 26 minus its asymptotic value, $\pi/2$, versus the order of the minimum. This figure illustrates the rapid convergence of Δ.

Numerical modeling can be done making a reasonable assumption about η. We expect that the divergence of the reflected wave would increase with τ, so that

$$\eta = T/\tau, \ \tau > T,$$ (29)

where T is a constant not too different from the time for the detached electron to go one Bohr radius. For convenience, a more ad hoc function is proposed:

$$\eta = \frac{1}{1 + \tau/T} , \tag{30}$$

where T is a characteristic time given by a characteristic distance ρ divided by the velocity of the electron in the vicinity of the atom v_0, where

$$v_0 = \sqrt{2E/m} . \tag{31}$$

We note that η, for $\tau \gg T$, is proportional to F/E and that for the nth fringe, $E_n \sim F^{2/3}$. Therefore η for the nth fringe goes like $F^{1/3}$. This seems to be in accord with Rau and Wong.[3] Figures 3 through 6 display photodetachment spectra obtained using this model (Eq. 20) with parameters adjusted to simulate the data. We believe these results will be useful in designing future experiments. These results fail very near threshold since no tunneling is built into the model.

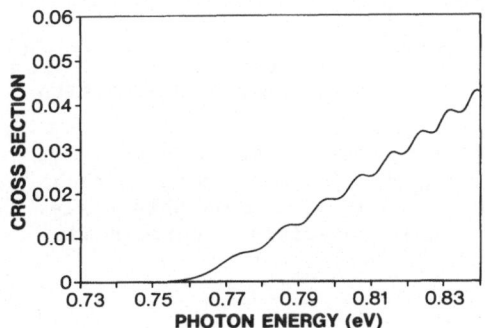

Fig. 3. Photodetachment cross section (Eq. 20) for s = 0 and ρ = 35 Å at a field of 34,472 V/cm (6.7×10^{-6} a.u.).

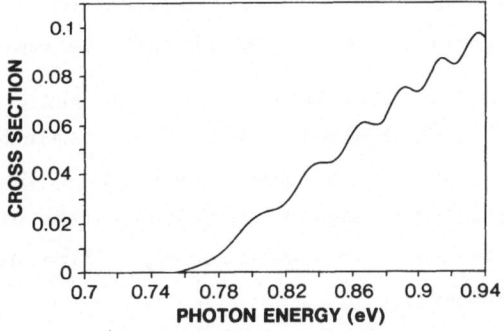

Fig. 4. Photodetachment cross section (Eq. 20) for s = 0 and ρ = 20 Å at a field of 144,065 V/cm (2.8×10^{-5} a.u.).

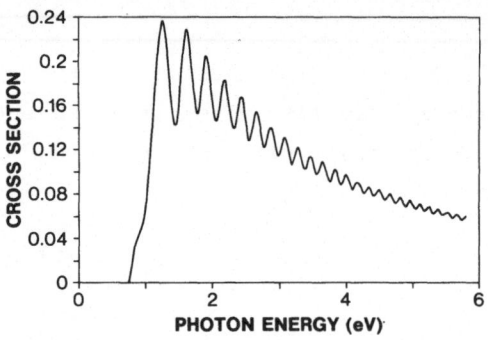

Fig. 5. Photodetachment cross section (Eq. 20) for s = 0 and ρ = 20 Å at a field of 5 MV/cm (9.7×10^{-4} a.u.).

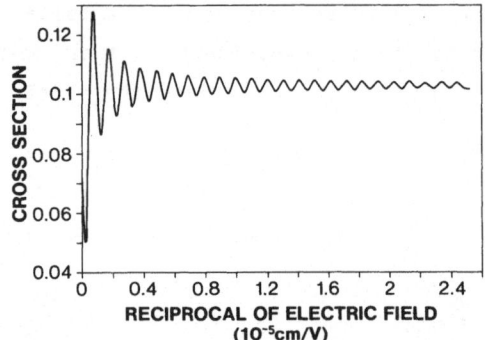

Fig. 6. Photodetachment cross section versus 1/F (cm/V) at a photon energy 0.9542 eV (200 meV above threshold). Note that the ripples are equally spaced on this scale. s = 0 and ρ = 20 Å. 2.5 on the abscissa corresponds to 40 kV/cm and $\tau = 7.5 \times 10^{-13}$ sec.

Post script. After this paper was presented, the author became aware of the seminal paper of I. I. Fabrikant[7] entitled "Interference effects in photodetachment and photoionization of atoms in a homogeneous electric field," in which it is shown, in a rigorous way, that the oscillatory energy dependences of cross sections of the type treated here are due to interference between classical trajectories.

The author has benefited from comments and suggestions from many of his colleagues. Discussions with W. P. Reinhardt, A.R.P. Rau, R. A. Reeder, V. Yuan, C. R. Quick and J. E. Stewart have been particularly helpful. This work is supported by the Division of Chemical Sciences, Office of Basic Energy Sciences, Office of Energy Research, U.S. Department of Energy.

REFERENCES

1. H. C. Bryant, A. Mohagheghi, J. E. Stewart, J. B. Donahue, C. R. Quick, R. A. Reeder, V. Yuan, C. R. Hummer, W. W. Smith, Stanley Cohn, W. P. Reinhardt, L. Overman. Phys. Rev. Lett. $\underline{58}$, 2412 (1987).

2. W. P. Reinhardt, in Atomic Excitation and Recombination in External Fields, edited by M. H. Nayfeh and C. W. Clark (Gordon & Breach, New York, 1985) p. 85.

3. A.R.P. Rau, private communication; A.R.P. Rau and Hin-Yiu Wong, Phys. Rev. A, to be published.

4. In this paper we make repeated use of the relation

$$\int_{-\infty}^{\infty} \exp\{-P^2 x^2 \pm qx\}dx = \frac{\sqrt{\pi}}{P} \exp \left(\frac{q^2}{4P^2}\right)$$

   from I. S. Gradshteyn and I. M. Ryzhik, Table of Integrals, Series and Products, (Academic Press, New York, 1980) p. 307.

5. T. T. Wu and C. N. Yang, Phys. Rev. D $\underline{12}$, 3845 (1975).

6. M. Abramowitz and I. A. Stegun (eds.), Handbook of Math Functions, NBS (1964).

7. I. I. Fabrikant, Zh. Eksp. Teor. Fiz. $\underline{79}$, 2070 (1980) (Sov. Phys. JETP $\underline{52}$, (6) (1980)).

EFFECT OF POLARIZABILITY OF ALKALI DIMERS IN ELECTRON AND POSITRON SCATTERING

N. Bhattacharya and D. N. Tripathi

Department of Physics
Banaras Hindu University
Varanasi 221005, India

ABSTRACT

The effect of the unusual polarizability of the alkali dimers (Li_2 and Na_2) have been examined regarding the electron and positron scattering by them. Rigorous close coupling calculations have been made to determine various cross sections and it has been reported that the polarization potential gives rise to a Ramsauer Townsand Minima and shape resonance in the total scattering cross sections. Contribution of the polarization part of the interaction potential in yielding the overall scattering cross sections (elastic and inelastic) is enormously large.

INTRODUCTION

The polarizability of the alkali dimers have been measured[1,2] as well as calculated[3] and are found to be very high. These dimers are very interesting as simple scattering targets because of their H_2 like character. The strongly attractive potential of alkali dimers and its dominancy over the static potential (due to quadrupole moment) is expcted to exhibit some interesting features such as Ramsauer Townsand Minima and the shape resonance.[4,5] This idea that the alkali molecules should have a large scattering cross section due to its large polarization has been the point of discussion of many physicists in the "Round Table Discussion on Future Directions in Positron-gas Research" during the "International Conference on Positron Scattering and Annihilation of Gases," July 1981 at York University. However, studies, experimental and theoretical, on electron (positron) scattering by alkali dimers have not yet been done in this context. This motivated the authors to take up the investigation of finding the effect of the polarization potential on the scattering cross sections of the projectiles due to alkali molecules. In the present note we report the results of our close coupling calculations for the scattering of the electrons and positrons by the Li_2 and Na_2 molecules in their ground electronic ($^1\Sigma_g^+$) and vibrational state in the energy range of 0.5 eV – 50 eV.

INTERACTION POTENTIAL

The interaction potential contains the static (due to quadrupole moment) and the induced part (due to dipole polarization) in it. The total interaction potential for the electron (positron) – molecule is written as:

$$V(r,\theta) = [(-\alpha_0/2R^4)] + [(-\alpha_2/2R^4) - (Q/R^3)]P_2(\cos\theta) - \text{for electron,}$$

and

$$V(r,\theta) = [(-\alpha_0)/2R^4] + [(-\alpha_2/2R^4) + (Q/R^3)]P_2(\cos\theta) - \text{for positron.}$$

r is the positron vector of the electron (positron) with respect to the center of mass of the molecule and θ is the angle between r and internuclear axis, $P_2(\cos\theta)$ is the Legendre Polynomial. The contents used are given in the table below.

Molecule	Sym. comp. of polarizability α_0(a.u.)	Asym. Comp. of polarizability α_2(a.u.)	Quadrupole moment Q(a.u.)	Rotational constant B(a.u.)
Li_2	202.8[a]	51.9[a]	9.0[b]	0.0000415
Na_2	297.4[c]	210.8[c]	21.3[c]	0.00000070

a – Ref. 3, b – Ref. 11, c – Ref. 12

RESULTS AND DISCUSSION

The electron and positron scattering cross sections have been calculated by solving the coupled Schrödinger equation[6] in the integral equation algorithm of Sams and Kouri[7] with total angular momentum representation.[8] To avoid singularities at the origin a hard core radius of about 3-4 a_0 has been taken which satisfies the convergence criterion. All the results reported here are fully converged.

The elastic scattering cross sections for the electrons (positron) scatterig by Li_2 and Na_2 are shown in Figs. (1-4) in which the only

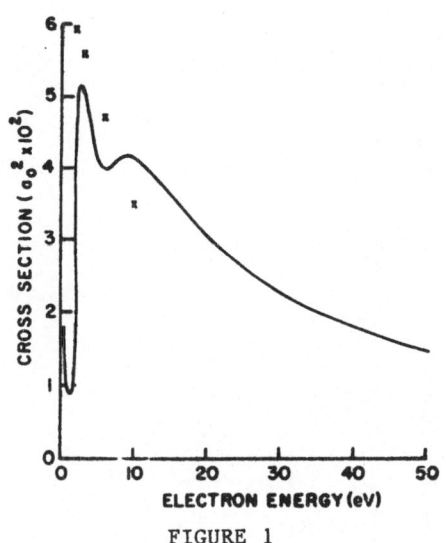

FIGURE 1

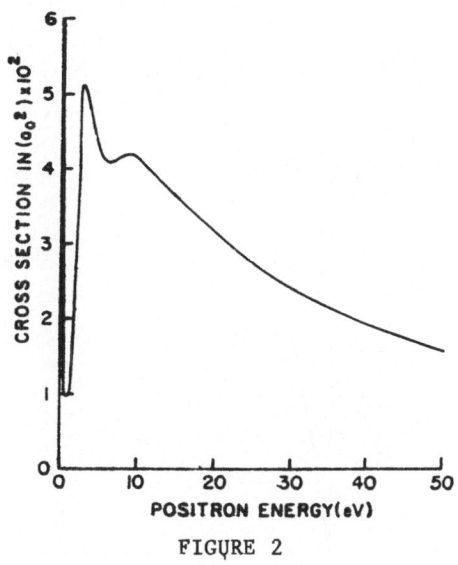

FIGURE 2

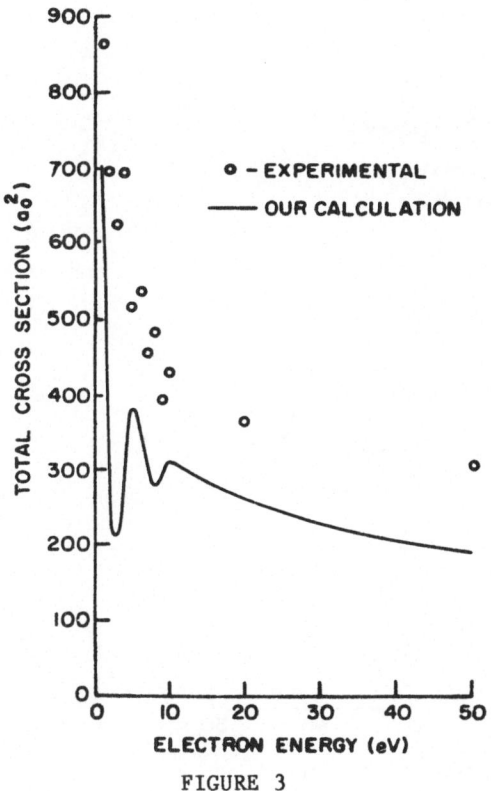

FIGURE 3

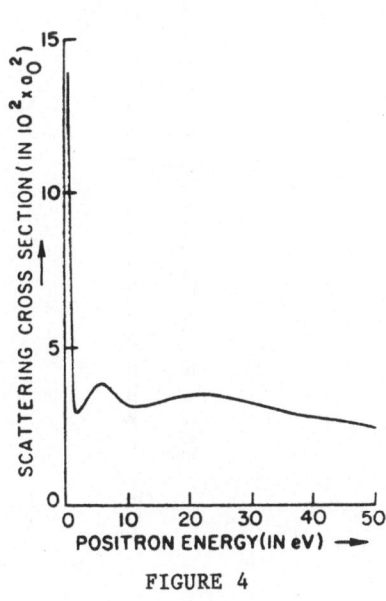

FIGURE 4

189

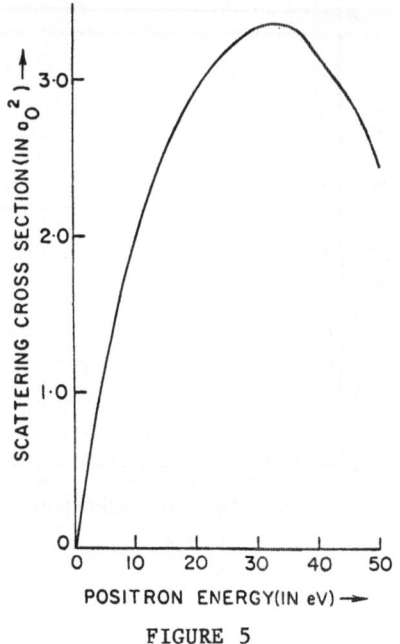

FIGURE 5

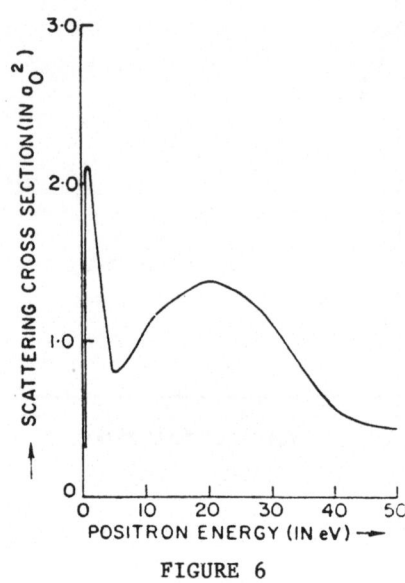

FIGURE 6

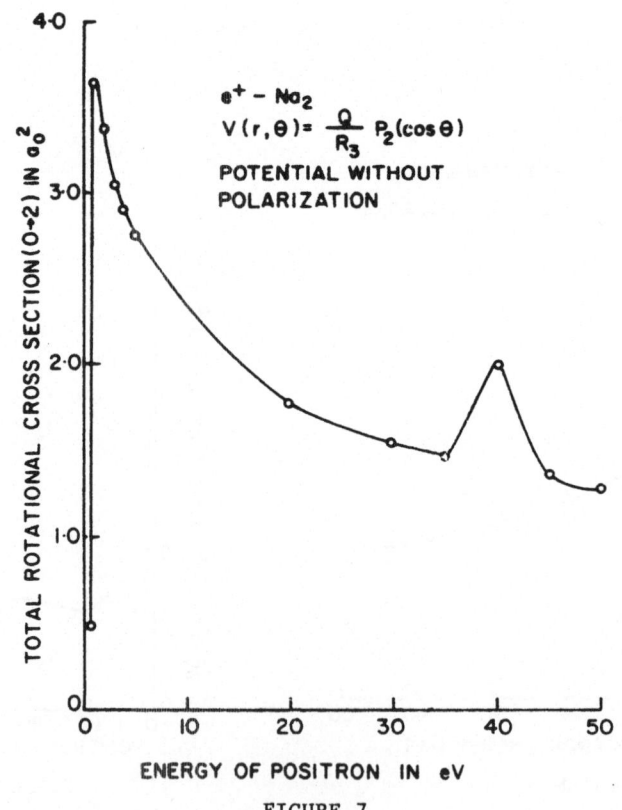

$$e^+ - Na_2$$
$$V(r,\theta) = \frac{Q}{R_3} P_2(\cos\theta)$$
POTENTIAL WITHOUT
POLARIZATION

FIGURE 7

available results for the absolute cross sections for electron scattering due to Miller and Kasdan[9,10] are also displayed. It shows an excellent agreement with our results. Our calculation depicts some structure viz; a minima in the lower energy side (1 eV for Li_2 and 3 eV for Na_2) followed by a sharp peak. This, based on the method of partial cross section analysis, is predicted to be a shape resonance. The minima appearing in the cross section curve is proposed to be the Ramsauer Townsand Minima. Again, the reasoning put forth is based on the partial wave analysis and it is caused by the strong attractive potential created by the large polarizability of Li_2 and Na_2 molecules.

The calculations have been made for e^+-Na_2 scattering by excluding the polarization potential component from the total interaction. In Fig. 5 the potential considered contains only the static interaction. It does not show any structure, either shape resonance or Ramsauer Townsand Minima. A comparison is made between Fig. 6, which shows the rotational excitation cross section for $j = 0 -- 2$ transition from e^+ - Na_2 scattering with polarization, and Fig. 7, for the same transition without polarization. The difference in the shape of the rotational excitation for the $j = 0 -- 2$ transition arising due to the noninclusion of the polarization potential is insignificant. Further, it can be seen that the magnitude of the elastic cross section increases by a factor more than 100 due to the inclusion of the polarization contribution of the potential. Thus, the present calculation confirms that the enormously large scattering cross sections of the electron (positron) scattering by the alkali molecules arise only due to their large polarizability.

ACKNOWLEDGEMENT

The authors are grateful to Professor D. K. Rai and Professor S. N. Thakur for their encouragement during the course of this work.

REFERENCES

1. R. W. Molof, T. M. Miller, H. L. Schwartz, B. Bederson and J. T. Park, J. Chem. Phys. 64, 1816 (1974).

2. R. W. Molof, H. L. Schwartz, T. M. Miller and B. Bederson, Phys. Rev. 10, 113 (1974).

3. D. A. Dixon, R. A. Eades and D. G. Truhlar, J. Phys. B 12, 2714 (1979).

4. N. Bhattacharya and D. N. Tripathi, IV ICPEAC, Brighton (U.K.).

5. N. Bhattacharya and D. N. Tripathi, Ind. J. Pure Appl. Phys. (in press).

6. A. K. Pandey, D. N. Tripathi and R. S. Singh, Ind. J. Pure Appl. Phys. 18, 408 (1980).

7. W. N. Sams and D. J. Kouri, J. Chem. Phys. 51, 4809, 4815 (1969).

8. A. M. Arthurs and A. Dalgarno, Proc. Roy. Soc. (London) A 256, 540 (1960).

9. T. M. Miller and A. Kasdan, J. Chem. Phys. 59, 3913 (1973).

10. T. M. Miller, A. Kasdan and B. Bederson, Phys. Rev. A 25, 1777 (1982).

11. Bo. Jonson and Bjorn, J. Chem. Phys. 74, 4566 (1981).

12. W. Muller, FB Chemie, der Universität Kaiserslautern, W. Germany, private communication.

HIGHLY EXCITED BARIUM RYDBERG ATOMS IN EXTERNAL

ELECTRIC, MAGNETIC, AND COMBINED FIELDS

H. Rinneberg and J. Neukammer

Institut Berlin Physikalisch-Technischen, Bundesanstalt
Abbestraße 2-12, 1000 Berlin 10, FRG

and

M. Kohl, A. König, K. Vietzke, H. Hieronymus, and
H.-J. Grabka

Institut für Atom- und Festkörperphysik, Freie Universität
Berlin, Arnimallee 14, 1000 Berlin 33, FRG

We have observed Rydberg states of barium with principal quantum numbers up to n = 520 in zero external fields. At lower principal quantum numbers we have studied barium Rydberg atoms in external electric, magnetic, and parallel fields. Because ofthe effective suppression of stray electric fields we have observed quasi-Landau resonances in external magnetic fields in the order of tens of Gauss. The energy dependence of the spacing between neighboring quasi-Landau resonances was found to be in excellent agreement with calculations involving classical trajectories. In addition the Stark effect of the n = 46, |m| = 0, 1, 2 diamagnetic manifolds has been investigated. The recorded spectra clearly exhibit the influence ofthe non-hydrogenic part of the Coulomb potential of the ionic core. The observed line positions and intensities have been accounted for by diagonalizing the corresponding Hamiltonian in a truncated spherical basis set.

I. RYDBERG STATES WITH HIGH PRINCIPAL QUANTUM NUMBERS IN ZERO EXTERNAL FIELDS

About seventy-five years ago, when Niels Bohr published his theory of the atom,[1] he discussed the properties of states with high principal quantum numbers. At that time the principal series of lithium and sodium had been followed up to principal quantum numbers n = 42 and n = 59, respectively. The behaviour of Rydberg states (n = 25) of alkali atoms in external electric and magnetic fields was first studied by Segrè and co-workers[2,3] around 1935. More recently, the narrow bandwidths and high output powers of today's dye lasers have been exploited by several groups to excite Rydberg states of rubidium,[4] cesium,[5] strontium,[6] and barium[7] with principal quantum numbers ranging between n = 150 and n = 300. In this contribution we show spectra of Rydberg states of barium with principal quantum numbers up to n = 520. Such highly excited atoms are extremely sensitive toward extenal fields. In particular, stray electric fields which cause Stark mixing present the main obstacle for observing Rydberg states with even higher principal quantum numbers. In outer space, however, Rydberg states with principal quantum numbers up to n = 733 have been detected by means of radioastronomy.[8]

We used two tunable cw dye lasers (Fig. 1) with a bandwidth of about 1 MHz to excite 6sns $^1S_0$ and 6snd $^1D_2$ Rydberg states of barium. Starting from the atomic ground state, the Rydberg states were reached via the 6s6p $^1P_1$ intermediate level. Both laser beams intersected a well-collimated (500:1) atomic beam of barium perpendicularly. The Rydberg atoms were ionized in collisions with barium atoms in the ground state or in the 6s6p $^1P_1$ intermediate level. The barium ions produced were detected by means of a channeltron followed by conventional counting

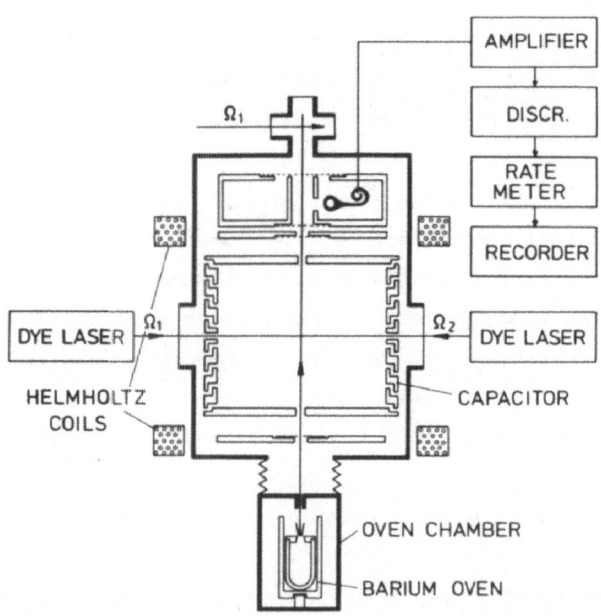

Figure 1.   Experimental set-up used to record Rydberg states with high principal quantum numbers. The Helmholtz coils shown in the figure produced magnetic fields up to 120 Gauss.

electronics. In order to reduce stray electric fields the interaction volume was carefully shielded by a plate capacitor. Nine concentric rings between both capacitor plates served to correct for fringe fields. All inner surfaces of the capacitor were covered with graphite. Components of stray electric fields directed along the atomic beam axis were compensated applying a corresponding voltage to the capacitor plates. Great care was taken to direct the atomic beam along the symmetry axis of the capacitor in order to minimize perpendicular components of spurious electric fields. The atomic beam was aligned by translating and tilting of the entire oven chamber. The orientation of the beam was monitored by observing te fluorescence of the resonance transition in a seperate compartment on top of the main scattering chamber. The oven and scattering chambers (Fig. 1) were pumped separately and connected by a diaphragm to allow for differential pumping. Typically, the pressure in the main scattering chamber was about $5 \cdot 10^{-10}$ mbar. In order to suppress the Stark effect caused by the motion of the barium Rydberg atoms in the earth's magnetic field, it was compensated using three mutually perpendicular pairs of Helmholtz coils.

The spectrum shown in Fig. 2 was recorded by scanning the frequency of the second laser while the first laser was stabilized to the resonance transition $6s^2 \, ^1S_0 \rightarrow 6s6p \, ^1P_1$. The strong signals correspond to $6snd \, ^1D_2$ states, which can be followed up to n = 500 in Fig. 2. At principal quantum numbers below n = 360 $6s(n + 2)s \, ^1S_o$ Rydberg states can be discerned as small signals halfway between the 6snd and $6s(n + 1)d \, ^1D_2$ resonances. With increasing principal quantum number, the amplitudes of the recorded signals are seen to decrease rapidly. The signal at n = 500 is smaller that that at n = 441 by a factor of about 4.5, whereas a ratio of 1.5 is expected from the n^{-3} scaling law for the absorption cross section. This discrepancy is explained by the presence of stray electric fields which cause Stark mixing of te $6snd \, ^1D_2$ with neighboring Rydberg states. This results in a line broadening as well as a distribution of oscillator strength. From Stark spectra taken aat lower principal quantum numbers we infer that signals of $6snd \, ^1D_2$ Rydberg states disappear, when $3(F/F_0)n^5$ equals unity ($F_0 = 5.14 \, 10^9$ V/cm). It follows from this relation that stray electric fields were reduced to about 50 μV/cm in our present set-up. It is to be expected that Rydberg states with considerably higher principal quantum numbers can be observed in future. Such highly excited atoms allow to measure extremely small electric fields and to probe their spatial distribution.

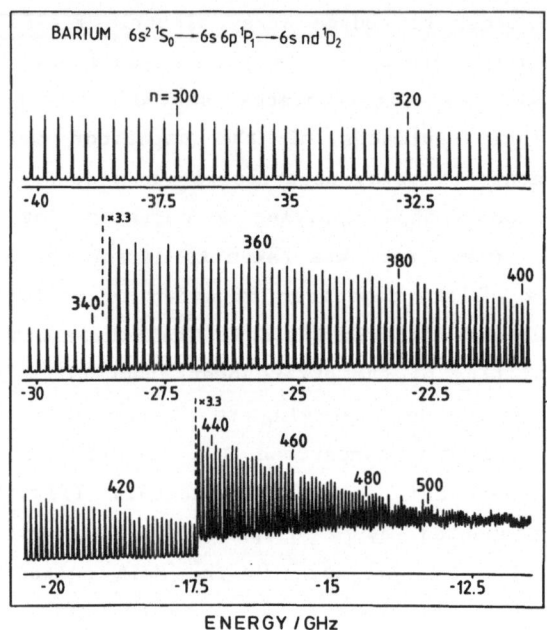

Figure 2. Spectrum of 6snd $^1D_2$ (289 ≤ n ≤ 500) Rydberg states of barium. Below n = 360 6sns $^1S_0$ states are observed as small signals halfway between corresponding 6snd $^1D_2$ resonances.

II. BARIUM RYDBERG ATOMS IN EXTERNAL FIELDS

II.1 Low Field Quasi-Landau Resonances

Quasi-Landau resonances which were discovered in barium by Garton and Tomkins in 1969,[9] have gained renewed interest because of experimental[10,11] and theoretical[12-15] advances within the last two years. In particular, Holle et al.[10] and Main et al.[11] observed modulations in the absorption cross section of hydrogen which were explained by various three-dimensional classical orbits of the electron in the combined Coulomb and diamagnetic potential.[14,15] On the other hand, the Garton-Tomkins type quasi-Landau resonances correspond to a periodic motion of the elctron in the plane perpendicular to the external field.[16-18] Up to now, quasi-Landau resonances have been studied experimentally in external magnetic fields in the order of tens of kGauss. The amplitudes of the modulations observed in Sr and Ba by Lu et al.[19] strongly depended on the magnetic field strength. No quasi-Landau resonances could be detected below 10 kGauss. However, we have recorded quasi-Landau resonances at magnetic field strengths as low as 60 Gauss. In contrast to the results obtained by Lu et al.[19] we found the amplitudes of the Garton-Tomkins type resonances (m = 0) to be independent of the magnetic field strength between 60 Gauss and 7 kGauss. The upper part of Fig. 3 shows a typical recording of m = 0 quasi-Landau resonances taken at B = 120 Gauss in the vicinity of the zero field ionizatio limit (E = 0 GHz). The polarization vectors of both linearly polarized laser beams were oriented parallel to the magnetic field. In Fig. 3 the separation between neighboring quasi-Landau resonances amounts to $\Delta\nu$ = 502(10 MHz corresponding to ΔE = 1.5 $h\nu_c$, ν_c being the cyclotron frequency. For comparison we show m = -1 quasi-Landau resonances excited using $\delta-\pi$ polarized laser beams. The spacing $\Delta\nu$ = 212(5) MHz between neighboring resonances corresponds to ΔE = 0.64 $h\nu_c$. These resonances are caused by a three-dimensional periodic motion of the outer electron.[11,15]

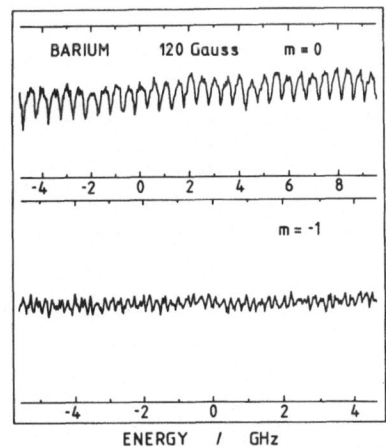

Figure 3. Low field quasi-Landau resonances. The count rate (vertical axis) ranges from 4000 to 10,000 counts/s indicated by the lower and upper horizontal lines, respectively. The time constant was set to 1 second.

The Fourier transforms of the spectra taken at 120 Gauss are shown in Fig. 4. The time is measured in units of the cyclotron period T_c = 2.98 · 10^{-9} sec. Prominent signals appear at the time of the first recurrence of the orbiting electron at T = 0.67 T_c, and T = 1.56 T_c for the m = 0 and m = -1 resonances, respectively. In addition, the second recurrence of the electron can be clearly discerned in the upper trace of Fig. 4.

We have measured the energy dependence of the spacing between neighoring quasi-Landau resonances in an external magnetic field of 736 Gauss. The field was produced by an electromagnet incorporated into an atomic beam apparatus (Fig. 5). Because of the reduced size of the scattering chamber, compared to the one shown in Fig. 1, stray electric fields could be suppressed down to about 2 mV/cm. This allowed to observe Rydberg states in zero field up to principal quantum numbers n = 250 only. The spectrum shown in Fig. 6 extends from the n-mixing to the quasi-Landau region. In zero external field, the binding energy of $E = -13$ cm^{-1} corresponds to the principal quantum number n = 92. At 736 Gauss the overall splitting of the n = 92 diamagnetic manifold ($\Delta E = 0.5$ cm^{-1}) is about twice as large as the separation $\Delta E_{n,n-1} = 2R/n^3$ between neighboring Rydberg states in zero field. In upper trace,

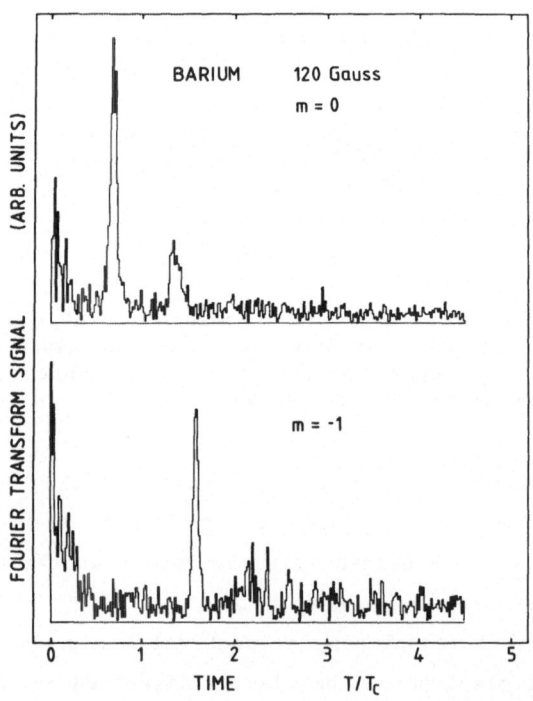

Figure 4.   Fourier transforms of the spectra shown in Fig. 3.   The time scale is given in units of the cyclotron period.

individual diamagnetic components are resolved. Their envelope shows a periodic modulation, which develops into the quasi-Landau resonances with increasing excitation energy. The lower six traces of Fig. 6 show about 180 quasi-Landau resonances. At the highest energy shown in Fig. 6, the separation between neighboring resonances amount to 1.16 $h\nu_c$.

In Fig. 7 the separations between neighboring quasi-Landau resonances, taken as the spacings between the recorded minima (Fig. 6), have been plotted versus the excitation energy. Theoretical values which are represented by the solid line, were calculated by evaluating the corresponding phase integrals within a WKB-treatment.[16-18] Alternatively, the separations between neighboring m = 0 quasi-Landau resonances were obtained by computing the orbital frequency of classical trajectories in the combined Coulomb and diagmagnetic potentials as a function of the excitatio energy.[20] As can be seen from Fig. 7, the experimental frequency separations which vary between 3.5 ν_c and 1.16 ν_c are in good agreement with the theoretical predictions. This demonstrates that prominent aspects of a quantum mechanical system can be accounted for by solving Newton's equation of motion in the semiclassical region.

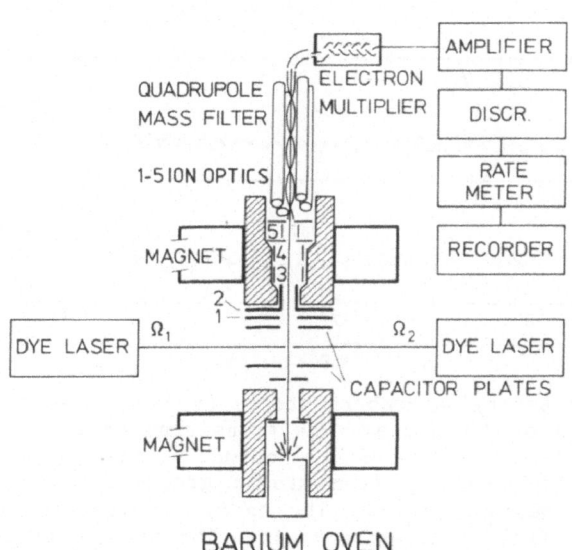

Figure 5. Experimental set-up used to study Ba Rydberg states in the magnetic fields up to 7 kGauss.

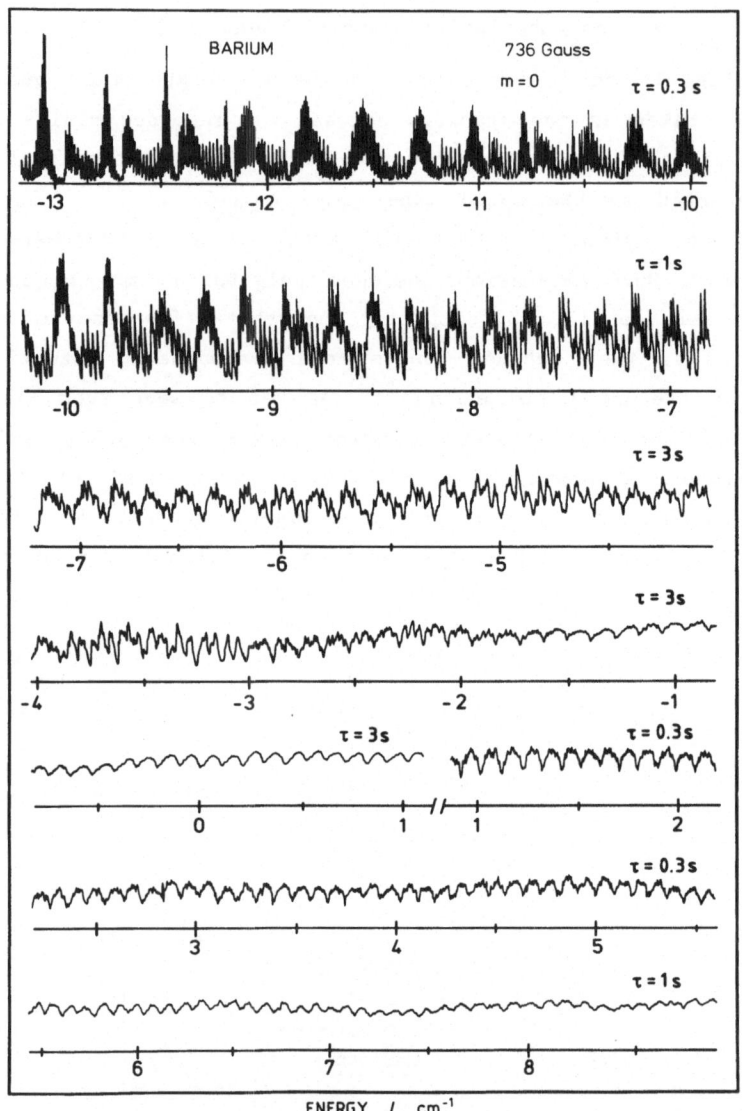

Figure 6.   Even parity Ba Rydberg states (m = 0) in an external magnetic
field of 736 Gauss.   The transition from the n-mixing to the
quasi-Landau region is shown.   The Rydberg states were
excited starting from atomic ground state via the 6s6p $^1P_1$
intermediate level.   The energy is measured relative to the
zero field ionization limit.   The time constants are given in
the figure.

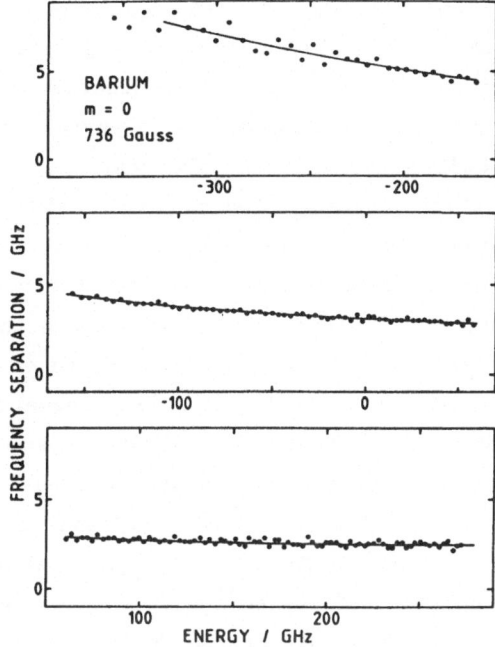

Figure 7. Separations between neighboring quasi-Landau resonances as a function of the excitation energy. The energy is measured with respect to the zero field threshold. The solid line was obtained calculating the phase integral given in Ref. 18.

II.2 ℓ-Mixing in Parallel Electric and Magnetic Fields

Contrary to the quasi-Landau resonances discussed in the previous section in terms of a hydrogenic model, Stark and diamagnetic manifolds of barium clearly exhibit the influence of the non-hydrogenic part of the Coulomb potential of the ionic core at field strengths, sufficiently weak for n to remain a good quantum number (ℓ-mixing region). The presence of core penetrating low angular momentum states, i.e. states with large quantum defects, results in avoided crossings and a redistribution of the oscillator strength. Previously Rydberg states of alkalis in external electric[21-25] and magnetic[23,26-28] fields have been studied extensively, including the influence of the atomic core on the spectra. The Stark effect of the n = 30 diamagnetic manifold of lithium in parallel fields was reported by Cacciani et al.[29] while the effect of crossed fields ($\vec{F} \perp \vec{B}$) on Rydberg states of cesium was studied by Gay and Delande.[28]

As particular example we discuss the Stark effect of the n = 46 |m| = 0, 1, 2 diamagnetic manifolds of barium. Electric fields up to 8 V/cm were applied while the magnetic field was kept constant at B = 3043 Gauss. Starting from the $6s^2$ $^1S_0$ ground state, the Rydberg states were excited via the 6s6p $^1P_1$ intermediate level using the

experimental set-up shown in Fig. 5. The spectrum of the n = 46, m = 0
diamagnetic manifold in zero electric field is displayed in Fig. 8, with
the diamagnetic components labelled by the quantum number k.[30] We have
observed all diamagnetic omponents with even parity (even k), twenty-one
of which are included in the figure. The remaining two components with
k = 44 and k = 0 which converge towards the 6s50s $^1S_0$ and 6s49d $^1D_2$
states in zero magnetic filed, respectively, lie outside te frequency
range covered in Fig. 8 because ofthe large quantum defects of these low
angular momentum states. The spectrum clearly shows the existence of two
classes of diamagnetic components, called vibrator (42 > k > 32) and
rotator (30 > k > 2) states. In first order, these states exhibit the
term sequence of an anharmonic vibrator and rigid rotator,
respectively.[30,31] The vibrator and rotator states are associated with
classical orbits which differ in their orientation relative to teh
external magnetic field. In the ℓ-mixing region the quantity
$\Lambda = 4A^2 - 5A_z^2$ is a constant of motion[31] rather than the Runge-Lenz vector
$\vec{A}$ is oriented along the major axis of the Kepler ellipse and proportional
to its numberical eccentricity. For vibrator states $\vec{A}$ moves on a twofold
hyperboloid ($\Lambda < 0$) and hence is oriented predominantely parallel or
antiparallel to the magnetic field (Fig. 9). The two surfaces correspond
to the twofold degeneracy of odd parity and even parity vibrator states
in hydrogen. In barium, however, this degeneracy is lifted due to the
presence of core penetrating low angular momentum states. Rotator states
are characterized by the motion of the Runge-Lenz vector on the one-fold
hyperboloid $\Lambda > 0$. In hydrogen the sequence of even parity and odd

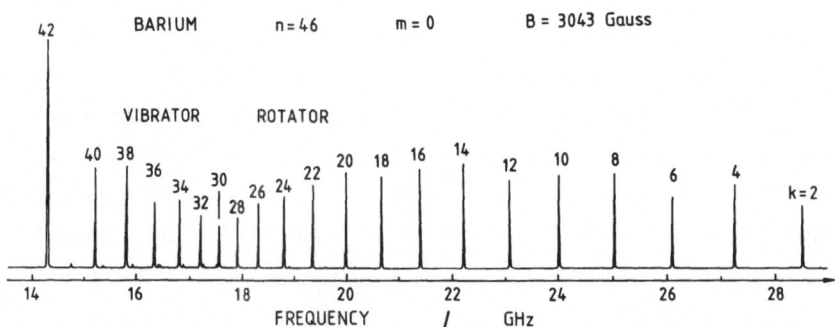

Figure 8. Spectrum of the even parity, n = 46, m = 0 diamagnetic
manifold of barium. The frequency is measured with respect
to the 6s50s $^1S_0$ state in zero field.

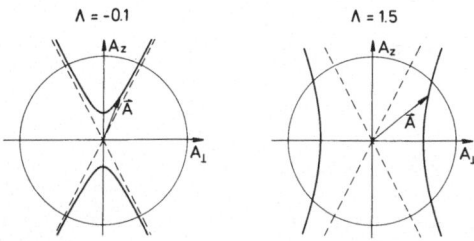

Figure 9. Orientation of the Runge–Lenz vector $\vec{A}$. The tip of the vector moves on a twofold ($\Lambda < 0$) and onefold ($\Lambda > 0$) hyperboloid, associated with vibrator and rotator states, respectively.[26,28]

parity rotator states resembles that of a rigid rotator. In barium, however, two rotator-like term sequences are needed to describe even parity and odd parity rotator states, respectively, which are shifted relative to each other. It follows from the discussion above the vibrator states have a non-vanishing electric dipole moment contrary to rotator states. Therefore in parallel electric and magnetic fields the doubly degenerate vibrator states exhibit a linear Stark splitting whereas rotator states undergo a quadratic Stark effect, provided the electric field strength is sufficiently small. In order to estimate the relative influence of the external electric and magnetic field, the field-induced shift of the uppermost Stark component and that of the highest rotator state are compared by calculating the ratio

$$\beta = \frac{12}{5} \left(\frac{F}{F_0}\right)\left(\frac{B}{B_0}\right)^{-2} n^{-2} \tag{1}$$

with $B_0 = 2.35 \times 10^9$ Gauss. The overall energy dependence of rotator and vibrator states of barium with increasing electric field strength is shown in Fig. 10. The quantity β varies from $\beta = 0.0$ to $\beta = 1.0$ at $F = 8$ V/cm. Previously[29], the Stark effect of the $n = 30$, $m = 0$ diamagnetic manifold of lithium was studied within the range of $0 < \beta < 0.2$. By comparing Fig. 10a and Fig. 10b experimental data and calculated values are see to agree well. For electric field strengths below 300 mV/cm vibrator states have been omitted from Fig. 10 for clarity. These data are shown in Fig. 13 on an expanded scale. The linear dependence of the energy of the vibrator states on the electric field strength is evident from Fig. 10. Rotator states, on the other hand, exhibit a transition from the quadratic to the linear Stark effect

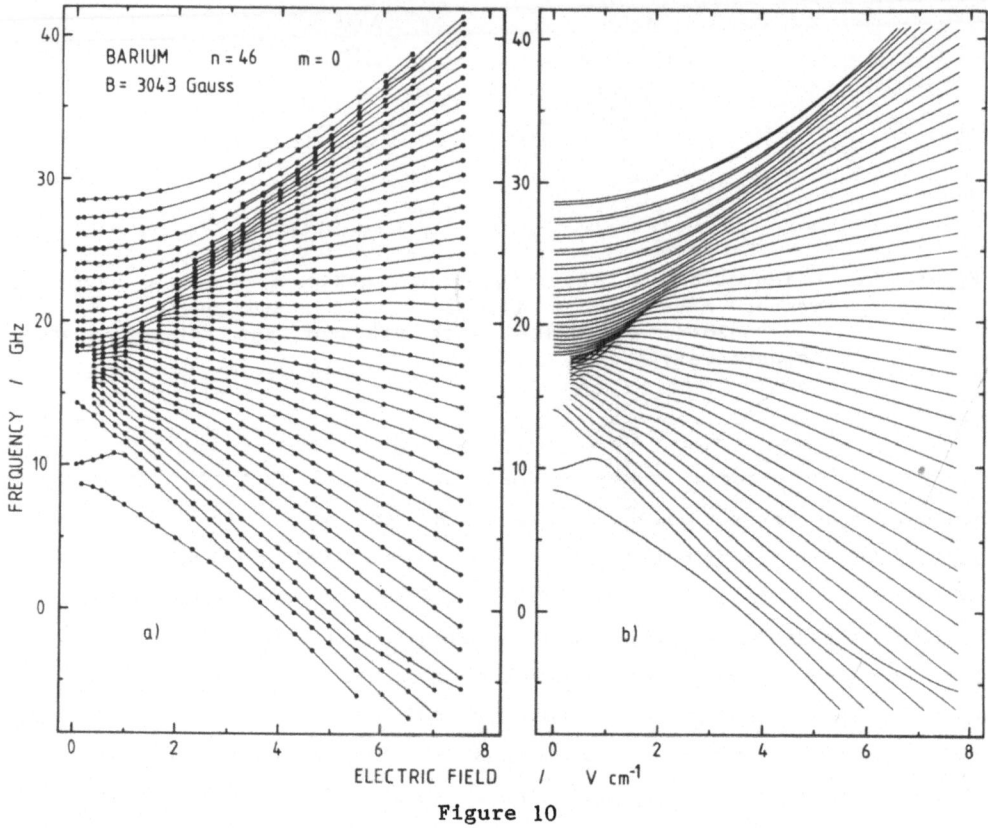

BARIUM   n = 46   m = 0
B = 3043 Gauss

FREQUENCY / GHz

ELECTRIC FIELD / V cm^{-1}

Figure 10

with increasing field strength.   Fig. 10b was obtained by diagonalizing the Hamiltonian

$$H = H_0 + \mu_B \ell_z B + ezF + \frac{e^2 B^2}{8 m_e c^2} r^2 \sin^2\theta \tag{2}$$

within a truncated spherical (m = 0) basis set.   The electric and magnetic fields have been taken along the quantizatio (z) axis. H_0 is the Hamiltonian of the barium atom in zero external fields.   Starting from 6s6p $^1P_1$ intermediate level, singlet states are excited predominantly in an optical transition.   Hence apart from spin-orbit effects included in H_0, the spins of the two valence electrons have not been taken into account in Eq. (2).   The following zero field wave functions $|6sn\ell \, ^1L_\ell$, m = 0$\rangle$, given in order of increasing energy have been used to set up the Hamiltonian: $|47d \, ^1D_2\rangle$, $|49s \, ^1S_0\rangle$, $|49p \, ^1P_1\rangle$, $|45g \, ^1G_4\rangle$, $|45f \, ^1F_3\rangle$, $|48d \, ^1D_2\rangle$, $|50s \, ^1S_0\rangle$, $|50p \, ^1P_1\rangle$, $|46g \, ^1G_4\rangle$, $|46f \, ^1F_3\rangle$, $|46\ell \, ^1L_\ell\rangle$ (5 < ℓ < 45), $|49d \, ^1D_2\rangle$, $|51s \, ^1S_0\rangle$, $|51p \, ^1P_1\rangle$, $|47g \, ^1G_4\rangle$, $|47f \, ^1F_3\rangle$, and $|50d \, ^1D_2\rangle$.   In order to account for the non-hydrogenic part in the Coulomb potential of the Ba$^+$ ionic core, experimental term values have been used as diagonal elements of H_0.   When calculating the matrix elements of the Stark and diamagnetic terms of Eq. (2), quantum defects were taken into account. This was done by using the expansions of the matrix elements $\langle 6sn\ell|r|6sn'\ell'\rangle$ and $\langle 6sn\ell|r^2|6sn'\ell'\rangle$ in terms of quantum defects δ_ℓ and $\delta_{\ell'}$, within a WKB approximation.[32,33] Quantum defects δ_ℓ were obtained from zero field spectra and found to be zero for $5 < \ell < 45$. We have included all states into Fig. 10b, evolving from the zero field levels $6s50s$ $^1S_0$, $6s50p$ $^1P_1$, $6s46g$ $^1G_4$, $6s46f$ $^1F_3$, and $6s46\ell$ $^1L_\ell$ ($5 < \ell < 45$). The latter states with zero quantum defect comprise the so-called incomplete hydrogenic manifold. With the origin of the energy scale of Fig. 10 chose to coincide with the energy of the $6s50s$ $^1S_0$ state in zero fields, the coponent originating from $6s49d$ $^1D_2$ state lies beyond 40 GHz and is not shown in the figure. Because of their small quantum defects, at B = 3043 Gauss the $6s46f$ $^1F_3$ and $6s46g$ $^1G_4$ state can no longer be distinguised from the incomplete hydrogenic manifold. On the contrary at low electric field strengths (F < 1 V/cm) two states are well seperated from the manifold, which can be traced back adiabatically to the $6s50s$ $^1S_0$ and $6s50p$ $^1P_1$ states in zero electric and magnetic fields. With increasing electric field stength, these components are seen (Fig. 10) to join the manifold through a series of avoided crossings.

Contrary to hydrogen, in zero electric field the barium rotator states appear as close-lying doublets (Fig. 10b) of even and odd parity. Since the rotator states are not mixed appreciably by the second order Stark effect, parity is approximately conserved up to electric field strengths, where the transition to the linear Stark effect occurs. In Fig. 11 we show a spectrum of the top of the rotator band at an external electric field of 851 mV/cm. Since the spectrum was obtained starting from the $6s6p$ $^1P_1$ intermediate level, the strong signals correspond to even parity components into the odd parity rotator states. In Fig. 11 even and odd parity rotator states are labelled by even and odd values of the quantum number k. For comparison we show a calculated spectrum in the lower trace of Fig. 11. The ratio of the transition amplitudes $\langle 6s6p\ ^1P_1|z|6sns\ ^1S_0\rangle$ and $\langle 6s6p\ ^1P_1|z|6snd\ ^1D_2\rangle$ used in the calculation was derived from experimental spectra taken in zero electric and magnetic field.

Contrary to rotator states, even and odd parity vibrator levels are strongly mixed by small external electric fields due to their large permanent electric dipole moments. As can be seen from Fig. 12a and Fig. 12b, an electric field of 36 mV/cm causes extensive mixing of odd ($43 > k > 31$) and even parity ($42 > k > 32$) vibrator states. Due to this Stark mixing all vibrator states contain even parity components (6s50s, 6s49d) in their wave function and can therefore be excited from the 6s6p $^1P_1$ intermediate level. In addition neighboring vibrator states strongly repel each other. This is illustrated in Fig. 13, where the

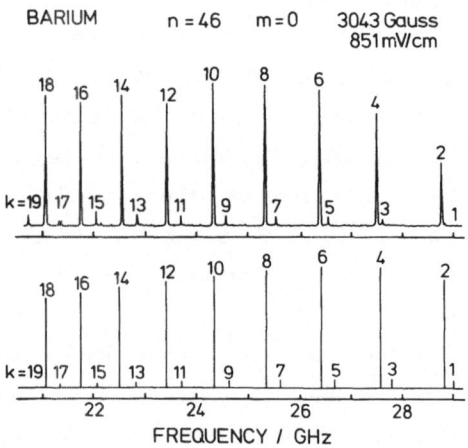

BARIUM   n = 46   m = 0   3043 Gauss
                                851 mV/cm

Figure 11.   Top of the n = 46 rotator band of barium in an electric field of 851 mV/cm.   Upper trace: experimental spectrum; lower trace: calculated spectrum.   Even (odd) k components have approximately even (odd) parity.   The origin of the frequency scale coincides with the 6s50s $^1S_0$ state in zero external fields.

energies of vibrator states have been plotted versus the electric field strength.   In Fig. 13a, experimental data are represented by dots which are connected by straight lines.   The lowest lying even parity rotator states are included for comparison.   The splitting between even and odd parity vibrator states in zero electric field is evident from Fig. 13. With increasing electric field strength the vibrator states undergo a series of avoided corssings.   For all vibrator states, the avoided crossings take place at the same electric field stengths.   The size of the avoided crossings does not change appreciably with electric field strength.   However, the avoided crossings are smaller at the top compared to the bottom of the vibrator band.   At an avoided crossing a rearrangement of the mixed wave functions of the two interacting states occurs.   This rearrangement results in the formation of two wave

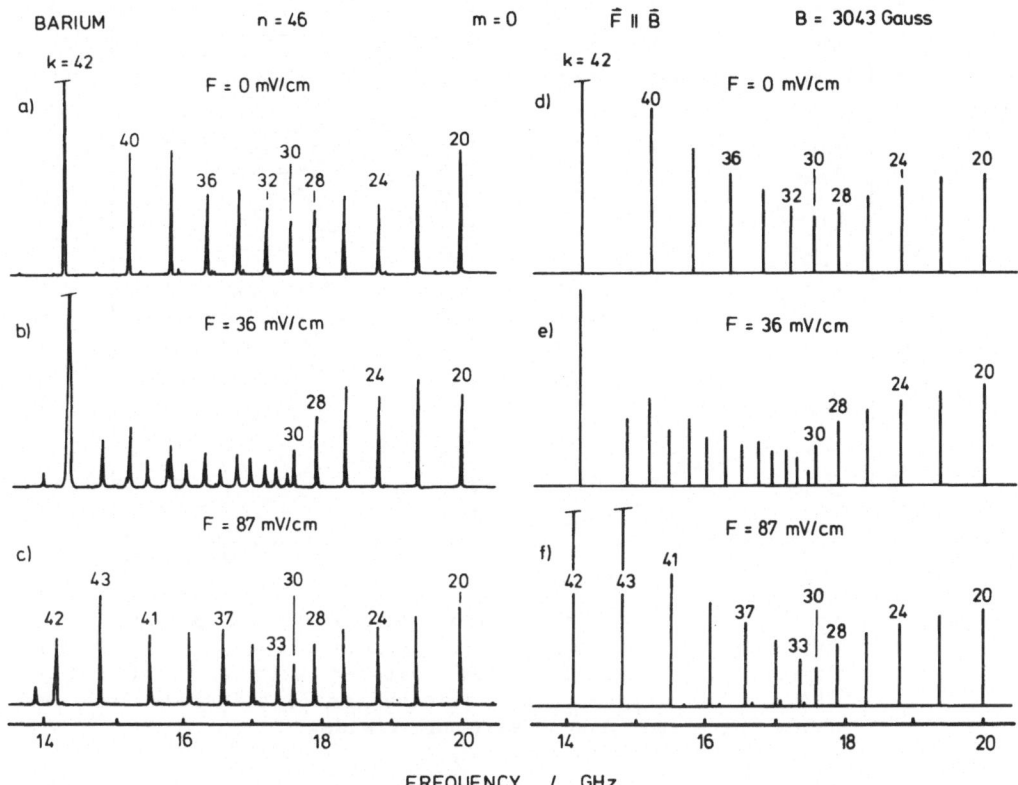

Figure 12. Mixing of even parity and odd parity vibrator states of the n = 46, m = 0 diamagnetic manifold of barium by small electric fields. Experimental spectra (a-c) are compared with calculated ones (d-f). The origin of the frequency scale coincides with the 6s50s $^1S_0$ state in zero external fields.

functions of definite parity, i.e. odd parity and even parity, respectively. Hence only one of these states can be reached from the intermediate level by an electric dipole transition. This explains the simplicity of the spectrum recorded at 87 mV/cm and shown in Fig. 12c. Contrary to the spectrum at zero electric field, in Fig. 12c only vibrator components with odd k quantum numbers have been observed. It should be noted that a particular signal (Fig. 12c) has been labelled by the quantum number k of that vibrator state in zero electric field, from which it evolves adiabatically. Theoretical intensities (Fig. 12d-f) as well as energies (Fig. 13b) calculated as a function of external electric field strength agree well with the experimental data. The k = 42 component (Fig. 12a-c and Fig. 13a) can be traced back adiabatically to the 6s46g $^1G_4$ state in zero field. There exists an avoided crossing between the $^1G_4$ (k = 42) state and a triplet state at a field strength of

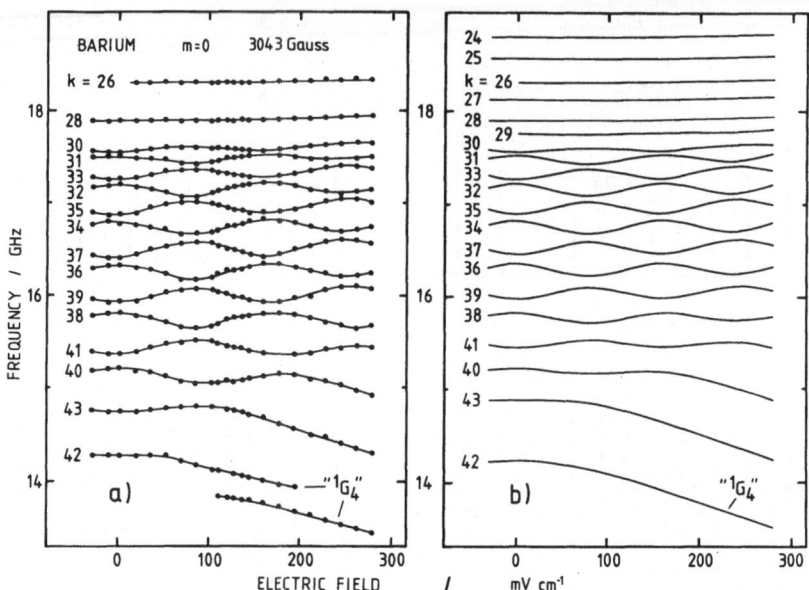

Figure 13. Stark effect of the vibrator states of the n = 46, m = 0 diamagnetic manifold of barium (see Fig. 10a,b). Experimental data (13a) are compared with calculated ones (13b). The origin of the vertical scale coincides with the $6s50s\ ^1S_0$ state in zero external fields.

about F = 150 mV/cm. Since triplet states have not been included in the calculations, this avoided crossing is not reproduced in Fig. 13b.

The interaction of the incomplete n = 46 hydrogenic manifold with core penetrating low angular momentum states determines the splitting between even and odd parity vibrator states as well as rotator states in zero electric field. Therefore these splittings depend on the magnetic quantum number m. As can be seen from Figs. 13, 14a, and 14c, the splitting between odd and even parity vibrator states is small for the m = 0 and m = −2 manifolds in zero electric field, whereas the odd parity m = −1 vibrator states lie approximately halfway between neighboring even parity m = −1 vibrator states. It should be noted that even (odd) parity vibrator states are labelled by even (odd) k quantum numbers for $|m|$ = 0, 2 whereas the reverse holds for $|m|$ = 1 components. As discussed above and illustrated in Fig. 13 for the m = 0 manifold, the size of avoided crossings at small electric field stengths is determined by the splitting between odd and even vibrator states in zero external field. The same applies to the m = −1 and m = −2 manifolds. The m = −2 manifold (Fig. 14e) shows a pattern of avoided crossing similar to the one observed for the m = 0 vibrator states (Fig. 13). In contrast the m = −1 vibrator states are not shifted appreciably by external electric

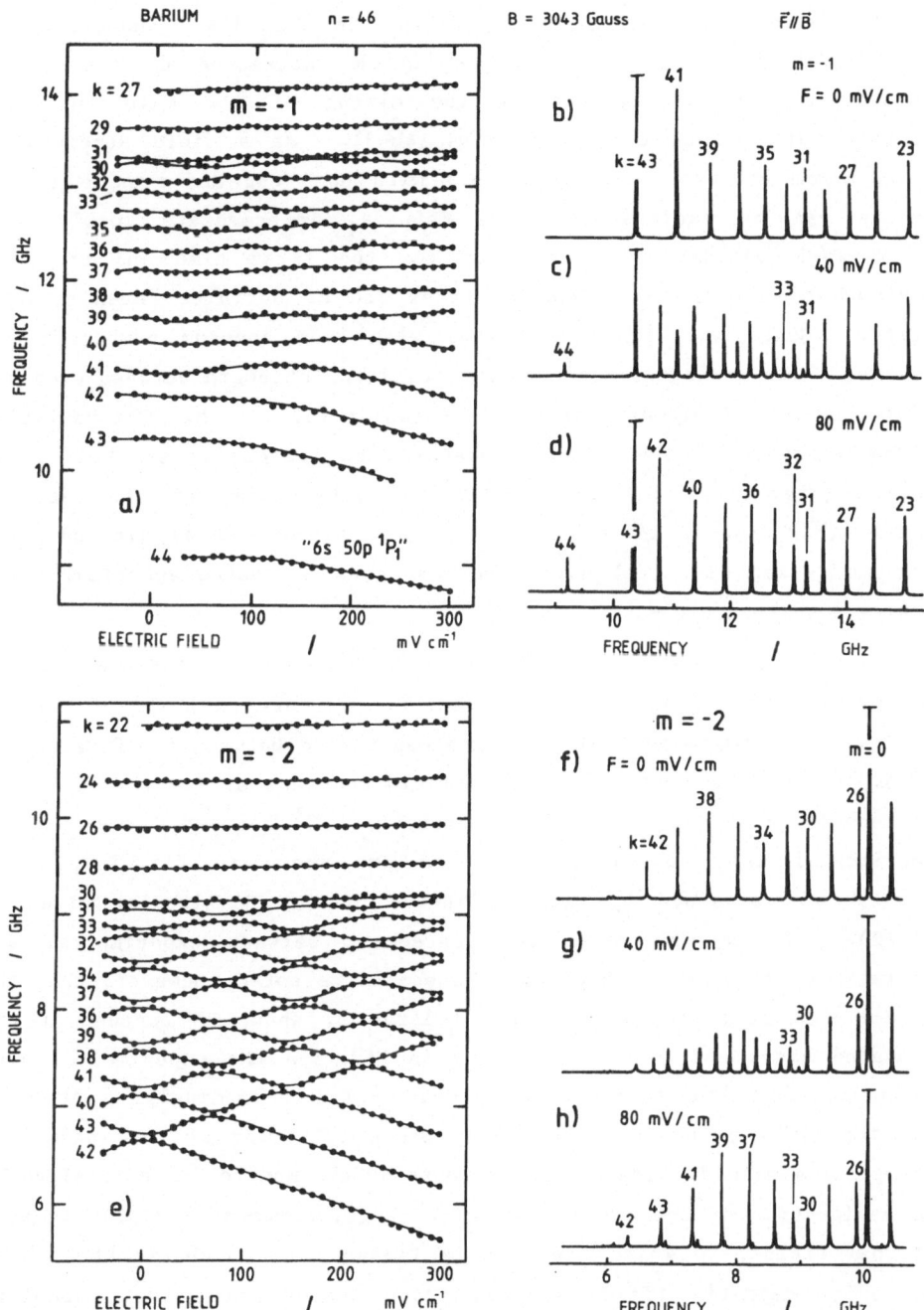

Figure 14. Stark effect of the n = 46, m = −1 and m = −2 vibrator
states. In Figs. 14a and 14e the energies of the m = −1 and m = −2
vibrator states have been plotted versus the electric field strength.
The spectra of the m = −1 and m = −2 vibrator states are shown for
selected values of the electric field strength. The spectra show the
mixing of even and odd parity vibrator states at F = 40 mV/cm as well as
the formation of vibrator states of definite parity (F = 80 mV/cm). The
origin of the frequency scale corresponds to the 6s50s $^1S_0$ Rydberg state
in zero external fields in each case.

fields smaller than 300 mV/cm, as shown in Fig. 14a. However, the intensities of the recorded signals which correspond to the m = −1 vibrator states clearly exhibit the mixing of the wave functions occurring at avoided crossings (Fig. 14b–d). At a field strength of F = 40 mV/cm vibrator states (m = −1) with both odd and even k quantum numbers can be excited from the $6s6p$ $^1P_1$ intermediate level. At F = 80 mV/cm avoided crossing occur for the first time and the wave functions of the m = −1 vibrator states are of definite (even or odd) parity. Hence only vibrator states with even quantum numbers k are observed. The analogous exchange of oscillator strength between even and odd parity m = −2 vibrator states is shown in Fig. 14f–h. The variation of the energies of the vibrator states with increasing electric field stength (Fig. 14a, 14e) as well as the intensities of the recorded signals (Fig. 14b–d, 14f–h) have been well reproduced by diagonalizing the Hamiltonian (Eq. (2)) using the analogous truncated spherical basis sets as described above.

In the last part of this section we discuss the influence of an external magnetic field on the n = 46, m = 0 Stark manifold of barium. In Fig. 15 the magnetic field strength was varied between B = 0 Gauss and B = 5.0 kGauss, keeping the electric field constant at F = 1 V/cm. This corresponds to a variation of the parameter β^{-1} between $0 < \beta^{-1} < 20$. Experimental data represented by dots are shown in Fig. 15a, while Fig. 15b was obtained by diagonalization of the Hamiltonian given in Eq. (2). As can be seen, good agreement between experimental and theoretical values were obtained. It should be noted, however, that the overall splitting was calculated too large by about 2% because of the truncated basis set used. Therefore in Fig. 15a the experimental data are comparted with theoretical values scaled accordingly. At zero magnetic field we observe the linear Stark splitting of the incomplete hydrogenic manifold. The components outside the manifold are labelled by the designation of the corresponding low angular momentum states in zero external fields, to which they can be traced back adiabatically. With increasing magnetic field strength these states enter the incomplete manifold via a series of avoided crossings. At B = 1.1 kGauss which corresonds to β = 1 a pronounced increase in the slope of the rotator-like components is observed. At higher field strengths, the magnetic field dominates and hence the components arrange themselves to form the rotator and vibrator bands.

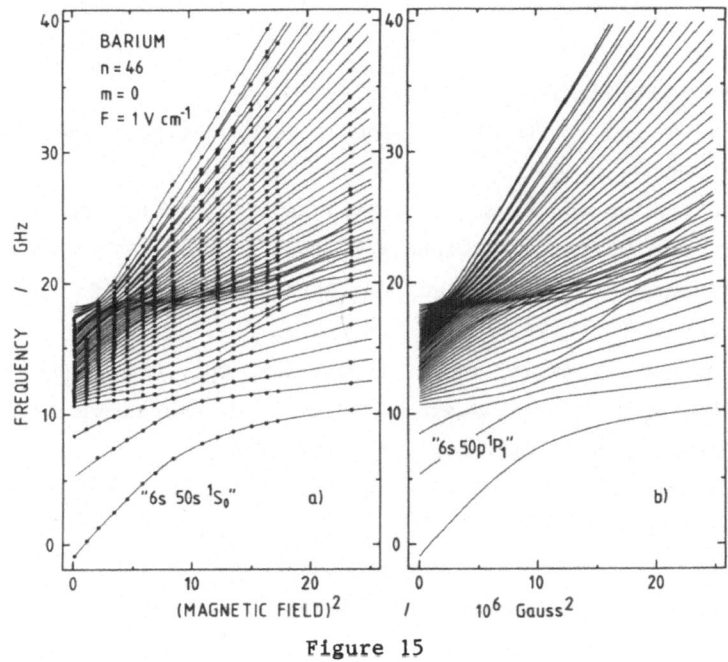

Figure 15

III. OUTLOOK

In the previous section we have discussed the influence of parallel electric and magnetic fields on Rydberg atoms in the ℓ-mixing region. The recorded spectra have been interpreted quantitatively by straightforward diagonalization of the corresponding Hamiltonian using a truncated spherical basis set. Such high resolution studies should be extended to perpendicular electric and magnetic fields, as well as to the n-mixing region. Of particular interest is a study of the Stark effect of quasi-Landau resonances. Last but not least we would like to point out one important application of high-n Rydberg atoms. Measurements of fundamental importance often require the absence of external electric fields. Because of their extreme snesitivity, Rydberg atoms may be used to determine the size of spurious electric fields and to probe their spatial distriution.

This work was supported by the Deutsche Forschungsgemeinschaft, Sonderforschungsbereich 161. We thank Mrs. Braun for patiently typing the manuscript.

REFERENCES

1. N. Bohr, Phil. Mag. 26, 1 (1913)

2. E. Amaldi and E. Segrè in: Zeeman Verhandelingen, Martinus Nijhoff (Haag 1935), p. 8.

3. F. A. Jenkins and E. Segrè , Phys. Rev. 55, 52 (1939).

4. B. P. Stoicheff and E. Winberger, Can. J. Phys. 57, 2143 (1979).

5. D. Delande, C. Chardonnet, F. Biraben, and J. C. Gay, J. Phys. (Paris) 43 (C2), 97 (1982).

6. R. Beigang, W. Makat, and A. Timmermann, Opt. Comm. 49, 253 (1984).

7. H. Rinneberg, J. Neukammer, G. Jönsson, H. Hieronymus, A. König, and K. Vietzke, Phys. Rev. Lett. 55, 382 (1985).

8. A. A. Konovalenko, Pis'ma v Astron. Zh. (Sov. Astron. Lett. (USA)) 10, 846 (1984).

9. W.R.S. Garton and F. S. Tomkins, Astrophys. J. 158, 839 (1969).

10. A. Holle, G. Wiebusch, J. Main, B. Hager, H. Rottke, and K. H. Welge, 2789 (1986).

12. W. P. Reinhardt in: Atomic Excitation and Recombination in External Fields, M. H. Nayfeh an C. W. Clark, eds., Gordon and Breach Science Publ. (New York 19850, p. 85.

13. M. A. Al-Laithy, P. F. O'Mahony, and K. T. Taylor, J. Phys. B19, L773 (1986).

14. M. L. Du and J. B. Delos, Phys. Rev. Lett. 58, 1731 (1987).

15. D. Wintgen and H. Friedrich, Phys. Rev. A36, 131 (1987).

16. A. R. Edmonds, J. Phys. (Paris) 31, (C4), 71 (1970).

17. A. F. Starace, J. Phys. B6, 585 (1973).

18. J.A.C. Gallas and R. F. O'Connell, J. Phys, B15, L309 (1982).

19. K. T. Lu, F. S. Tomkins, and W.R.S. Garton, Proc. Roy. Soc. Lond. A362, 421 (1978).

20. J. Neukammer, H. Rinneberg, K. Vietzke, A. König, H. Hieronymus, M. Kohl, H.-J. Grabka, and G. Wunner, to be published.

21. M. L. Zimmerman, M. G. Littman, M. M. Kash, and D. Kleppner, Phys. Rev.A20, 2251 (1979).

22. E. Luc-Koenig, S. Liberman, and J. Pinard, Phys. Rev. A20, 519 (1979).

23. D. Kleppner, M. G. Littman, and M. L. Zimmerman in: Rydberg States of Atoms and Molecules, R. F. Stebbings and F. B. Dunning, eds., Cambridge Univ. Press (Cambridge, 1983), p. 73.

24. D. A. Harmin, Phys. Rev. A30, 2413 (1984).

25. D. A. Harmin in: Atomic Excitation and Recombination in External Fields, M. H. Nayfeh and C. W. Clark, eds., Gordon and Breach Science Publ. (New York 1985), p. 39.

26. P. Cacciani, E. Luc-Koenig, J. Pinard, C. Thomas, and S. Liberman, Phys. Rev. Lett. 56, 1124 (1986).

27. P. F. O'Mahony and K. T. Taylor, Phys.Rev.Lett. 57, 2931 (1986).

28. J. C. Gay and D. Delande in: Atomic Excitation and Recombination in External Fields, M. H. Nayfeh and C. W. Clark, eds., Gordon and Breach Science Publ. (New York 1985), p. 131.

29. P. Cacciani, E. Luc-Koenig, J. Pinard, C. Thomas, and S. Liberman, Phys. Rev. Lett. 56, 1467 (1986).

30. D. R. Herrick, Phys. Rev. A26, 323 (1982).

31. E. A. Solov'ev, JETP Lett. 34, 265 (1981).

32. V. A. Davydkin and A. Yu. Makarenko, Opt. Spectrosc. (USSR), 53, 327 (1982).

33. V. A. Davydkin and B. A. Zon, Opt. Spectrosc. (USSR), 51, 13 (1981).

EXPERIMENTAL INVESTIGATION AND ANALYSIS OF RYDBERG ATOMS

IN THE PRESENCE OF PARALLEL MAGNETIC AND ELECTRIC FIELDS

J. Pinard, P. Cacciani, S. Liberman, E. Luc-Koenig,
and C. Thomas[+]

Laboratoire Aimé Cotton*
Centre National de la Recherche Scientifique
Campus d'Orsay
91405 Orsay, Cedex, France

ABSTRACT

A complete quasi-hydrogenic diamagnetic manifold is experimentally investigated as well as the evolution of the structure obtained by adding a small electric field parallel to the magnetic field. The experiment has been performed on lithium atoms excited in a one step process in $M = 0$ odd Rydberg states. The pure diamagnetic structure has revealed the existence of two classes of states which have been previously predicted by the analysis of an approximate constant of the motion $\Lambda = 4A^2 - 5A_z^2$ ($\vec{A}$ is the Runge-Lenz vector) and the semi-classicaly theory has proved to be a useful tool to interpret entirely the structure. In the presence of an electric field, the two classes of states evolve differently, three classes are now observed which are to be related to the existence and the properties of the approximate constant of the motion:

$$\Lambda_\beta = 4A^2 - 5A_z^2 + 10A_z\beta \quad (\beta = 12F/5n^2B^2 \text{ in a.u.}).$$

The complete analysis of these structures using a semi-classical quantization of Λ_β has been also performed and new properties of the corresponding action integral have been pointed out; it concerns a

---

[+]Present address: Bureau International des Poids et Mesures, Pavillon de Breteuil, 92310 Sevres, France.
*Associated with the University Paris-Sud.

peculiar analogy existing between the diamagnetic spectrum in the presence of an electric field and the diamagnetic spectrum obtained for n-manifold and different M values.

INTRODUCTION

During the past decade a continuously growing interest has been found in the experimental and theoretical studies of the behavior of atoms, in particular hydrogen atoms, perturbed by a strong magnetic field[1] (i.e. magnetic field such that the magnetic force is comparable with or greater than the Coulomb force). This interest is due to the difficulty to understand numerous peculiar features observed in the energy spectra through the corresponding, and although very simple, nonrelativistic Hamiltonian. This Hamiltonian can be reduced to:

$$H_\gamma = 1/2 \ p^2 - 1/r + 1/8 \ \gamma^2(x^2 + y^2) = H_o + H_D$$
where
$$\gamma = B/B_c \text{ and } B_c = 2.35 \ 10^5 T \ . \tag{1}$$

The paramagnetic term $H_p = 1/2 \ \gamma L_z$ has been omitted (as L_z is a constant of the motion for the problem it only introduces a global shift of the energy and thus it can be disregarded). The term of interest is the so-called "diamagnetic Hamiltonian" H_D 1/8 $\gamma^2(x^2 + y^2)$ = 1/8 $\gamma^2\rho^2$ in which the total parity is conserved, but whose main effect is to break the symmetry of the Coulomb problem H_o and thus to reduce the number of independent constants of the motion from 3 to 2: E and L_z. The problem is then a two-dimensional one but a nonseparable one.

However, if a pure quantum treatment of the problem is almost impossible, the study of the classical electron trajectories using classical mechanics has been found to be very fruitful to explain a number of features occuring in the atomic spectra in the vicinity of the ionization limit.[2] Such a method has even permitted to demonstrate the existence of chaotic motion for the electron;[3] this peculiar behavior of the atom in a magnetic field has given rise recently to original developments.

Independently of the studies for which H_D cannot be considered as a perturbation, numerous experimental and theoretical investigations have been done in the so-called "inter-l mixing" region when the diamagnetic energy ΔE_D is weaker than the energy spacing between two consecutive n-manifolds of the zero field Hamiltonian. The condition for such a situation can easily be expressed by $5\gamma^2 n^7/16 \ll 1$. In this low field case the theory of classical secular pertubations has permitted to find a complete set of 3 constants of motion E_0, L_z and

$$\Lambda = 4A^2 - 5A_z^2 \tag{2}$$

(where $\vec{A} = \vec{L} \wedge \vec{p} + \vec{r}/r$ is the Runge-Lenz vector) L_z is an exact constant of motion whereas E_0 and Λ are approximate ones which are conserved to within an accuracy of a B^4 term.[4] The quantization of E_0, L_z and Λ can be performed in a semi-classical Bohr-Sommerfeld theory and, it gives the values $E_0 = -1/2 \ n^2 \ L_z = M$ and Λ_k. In these conditions the total energy of the states of a given n-manifold is

$$E_k = E_0(n) + 1/16 \ \gamma^2 n^2 \ (n^2 + n^2 \Lambda_k + M^2) \tag{3}$$

Simultaneously with this classical study, a pure quantum theory of the low field case has been developed[4,5] showing first that the Hamiltonian is separable, and secondly that there is a coexistence of two different symmetries for the states belonging to the same n-manifold.

The aim of this paper is primarily to report on an experimental study of the structure of a complete diamagnetic n-manifold. The spectra have been recorded for M = 0 odd Rydberg states of the lithium atom.[6] It has revealed the presence of two different well separated classes of states which are analyzed using the properties of the constant of motion Λ. Secondly we will present a study of the modification of the diamagnetic structure obtained by adding a small electric field parallel to the magnetic field:[7] in these conditions, M is still a good quantum number, however the parity π_t is no longer conserved. This experiment will permit observation of three types of evolution for the states depending on their original class and which will be easily linked to the existence of the symmetry properties mentioned above. A theoretical analysis of the atom in the presence of electric and magnetic fields has been done using a new constant of motion Λ_β and the same quantization method as that used in the study of the pure diamagnetic multiplet. Finally we will point out another peculiar feature found by studying and comparing the two integrals of motion leading to the Λ values at M = 0 on one hand and to the Λ_β values at M = 0 on the other hand.

THE EXPERIMENT

The experimental set-up is schematized on Fig. 1. The atoms of an atomic beam of lithium are subjected at right angle to the interaction with the light of a UV laser source tunable over a wide range in the region of λ = 230 nm. This source allows us to excite the lithium atoms directly from the ground state to Rydberg states in a one step process. In the interaction region a magnetic field as well as an electric field can be applied, both of them being directed along the atomic beam axis such that motional electric field effects are avoided. The magnetic field strength up to 6 teslas is created by a pair of superconducting coils in the Helmholtz position. Atoms in highly excited states are detected using the field ionization technique: the produced electrons are accelerated and detected by a surface barrier diode insensitive to the magnetic field.

The lithium has rather low quantum defect of the p states δ_p = 0.053, and consequently the odd diamagnetic spectrum obtained in π polarization (M = 0) using a direct excitation scheme (π_t = −1) can be considered as hydrogenic. However non−hydrogenic effects appear as soon as the wave functions of the states involve the s states for which δ_s = 0.35. This will be the case in the even M = 0 states of lithium appearing when an electric field is applied.

THE STRUCTURE OF THE DIAMAGNETIC MANIFOLD IN ODD RYDBERG STATES OF LITHIUM

A typical recording of the diamagnetic manifold n = 30, M = 0 (π_t = −1) of lithium is shown in Fig. 2. The strength of the magnetic field B = 1.65T is weak enough and corresponds to the "inter−1 mixing"

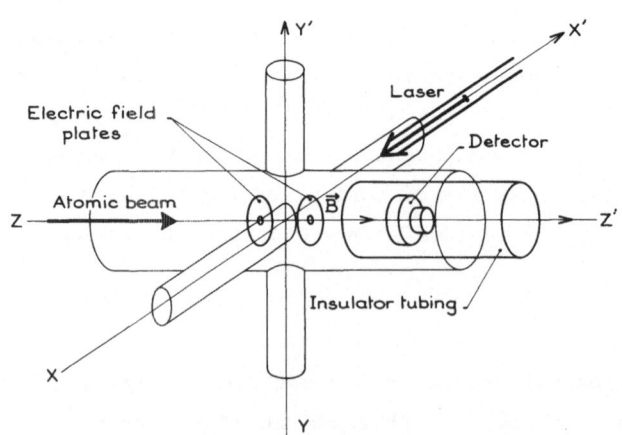

Fig. 1 The experimental set−up.

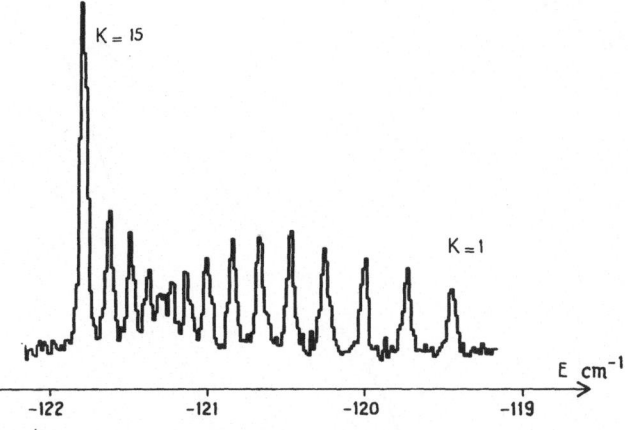

Fig. 2 Recording of the n = 30, M = 0 odd diamagnetic manifold of lithium in the present of a magnetic field B = 1.65T.

regime, but it is sufficient to resolve all the components (15) of the manifold. One must notice that, using the direct, π polarized excitation scheme, the oscillator strength is approximately uniformly spread over the whole structure such that every component can be observed (this will not be the case in σ excitation). A detailed analysis of the spectrum reveals a peculiar repartition of the different components in the energy range. This situation can be better visualized on a diagram (Fig. 3) representing the energy difference between consecutive components versus the K numbers (K is an integer labelling the components in the order of the decreasing energy). We observe, in the vicinity of K = 10 a singularity separating two different behaviors.

This phenomenon can be understood in a pure classical way by an analysis of the approximate constant of motion $\Lambda = 4A^2 - 5A_z^2 = A^2(4 - 5\cos^2\theta)$ where θ is the angle $(\vec{B}, \vec{A})$ and $A^2 \leq 1 - M^2/n^2$ is the modulus of the Runge-Lenz vector which is egal to the eccentricity of the Kepler ellipse describing the rapid motion of the electron. If M = 0 the Λ values are restricted to the interval $[-1, +4]$, then two cases can be defined, represented in Fig. 4.

(a) $\Lambda < 0$: The extremity of $\vec{A}$ moves slowly (due to the weak diamagnetic perturbation) on a portion of a two-fold hyperboloid limited by the sphere $|\vec{A}| = 1$, the presence of the two sheets symmetrical with respect to the z = 0 plane gives rise to a degenerance of states with oppositve parity. In this case, the motion of the electron is stretched along the $\vec{B}$ direction and thus the mean value $\langle \rho^2 \rangle$ is small; these states

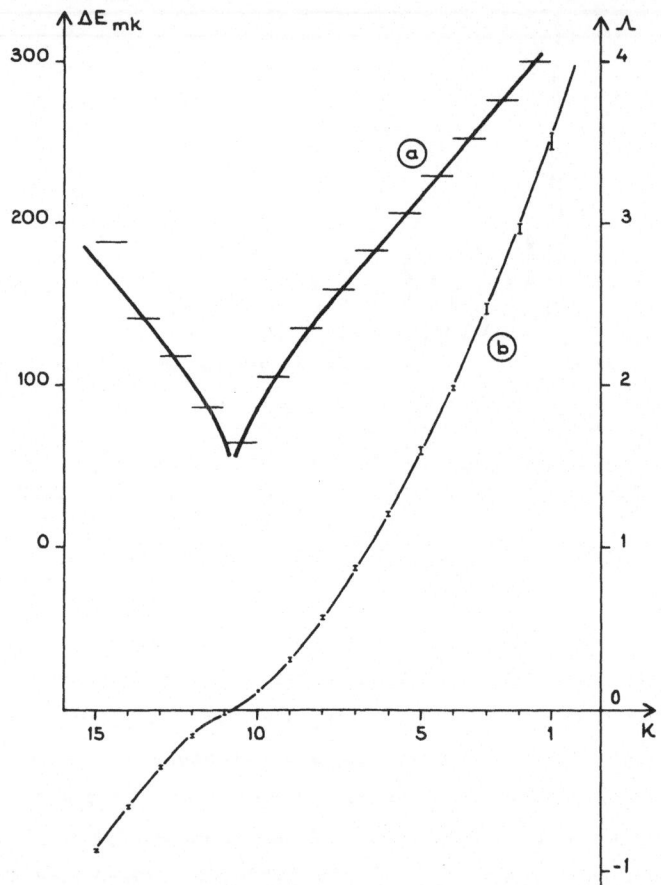

Fig. 3   Analysis of the diamagnetic structure shown in Fig. 2.
(a) Experimental spacing of the K-components versus the K-values.
(b) Λ values deduced from the experiment and calculated using semi-classical quantization theory.

correspond to the lower components of the manifold. The $\vec{A}$ vector can be parallel to the field and its motion presents a librational symmetry.

(b) $\Lambda > 0$: The extremity of $\vec{A}$ moves on a portion of a one-fold hyperboloid. In this case the motion of the electron is mainly localized near the plane $z = 0$ so that $\langle \rho^2 \rangle$ is large; these states, which are nondegenerate, correspond to the higher components of the manifold. The $\vec{A}$ vector is never parallel to the field and the corresponding motion is of a rotional symmetry.

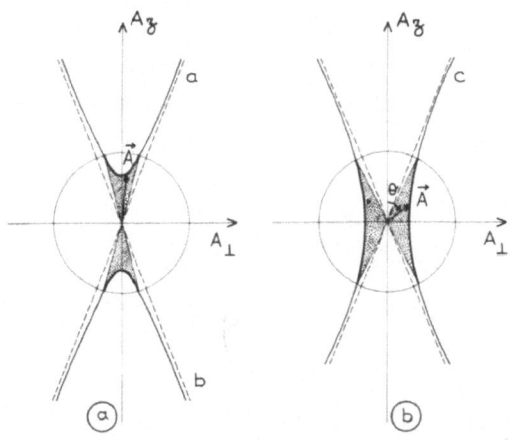

Fig. 4   Variation of the Runge–Lenz vector $\vec{A}$ under the influence of a $\vec{B}$
field in the z direction (M = 0 $\gamma^2 n^7 \ll 1$).   The extremity of $\vec{A}$
moves on a hyperboloid Λ = const. and is bounded by a sphere
$|\vec{A}| < 1$.   The domain of variation depends on the sign of Λ:
a) $\Lambda < 0$ b) $\Lambda > 0$.

Consequently the diamagnetic states of a given manifold split up
into two classes of different symmetry depending on the sign of Λ, the $\vec{A}$
motion is localized in two different domains of space limited by the
double cone $\cos^2\theta_o$ = 4/5 associated with the value Λ = 0.   Let us
emphasize that for M = 0 states the existence of a librational or a
rotational type of symmetry is tightly connected to the possibility or
the impossibility for the $\vec{A}$ vector to be parallel to $\vec{B}$.   This definition
of the symmetry of the states remains meaningful when an electric field
is applied in the $\vec{B}$ direction; in this new situation as discussed below,
the symmetry cannot be defined by the sheet of the hyperboloid swept by
the $\vec{A}$ vector.

The values of Λ_k corresponding to each of the components of the
manifold can be deduced from the experimental spectra using the formula
(2).   These values have been reported on Fig. 3, they show that (even for
a not purely hydrogenic spectrum) the singularity observed in the energy
spacing distribution is to be connected with the value Λ = 0.   This
confirms the existence of two, well separated, classes of states in the
diamagnetic manifold.

The quantized values Λ_k have also been calculated from the semiclassical quantization method used in Refs. 4 and 8.   Using as a
generalized angular coordinate the angle θ, we can deduce the

corresponding generalized conjugate momentum which is $L_\perp(\theta)$, the component of the instantaneous $\vec{L}$ vector on the direction perpendicular to the plane $(\vec{B}, \vec{A})$. Versus the different constants of motion and the angle θ, the expression of $L_\perp(\theta)$ is:

$$L_\perp(\theta) = [n^2(1 + \frac{\Lambda}{1 - 5\sin^2\theta}) - \frac{M^2}{\sin^2\theta}]^{1/2} .\qquad (4)$$

In the phase plane $(\theta, L_\perp(\theta))$ the Bohr Sommerfeld quantization rules give:

$$\int_{\theta_1}^{\theta_2} L_\perp(\theta) \, d\theta = \pi(k + 1/2)$$

where θ_1 and θ_2 are the roots of the equation $L_\perp(\theta) = 0$.

In Fig. 5, we have represented the different evolution curves in the phase plane $(\theta, L_\perp(\theta))$ for all the calculated quantized values of Λ corresponding to the diamagnetic manifold $n = 29$ $M = 0$. Here also we can observe that the phase plane is separated in three nonoverlapping regions. The evolution curves, characteristic of the electron motion,

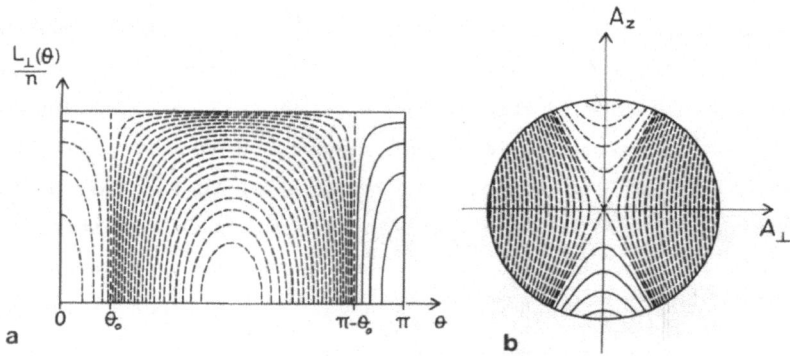

Fig. 5   (a) Left side: Trajectories of the operating point in the phase plane θ, $L_\perp(\theta)$. The different curves correspond to calculated quantized values of Λ_k.
(b) Right side: Corresponding motion of the extremity of the $\vec{A}$ vector.

being entirely in one of the three regions depending on the sign of Λ, and for $\Lambda < 0$ on the sign of A_z. For a chosen $\Lambda < 0$ value, there exist two distinct $(\theta, L_\perp(\theta))$ curves symmetrical with respect to $\theta = \pi/2$. One can also notice that, depending on the sign of Λ the integration has to be performed on different domains of θ. This gives rise to a description of the diamagnetic manifold using 2 sets of quantum numbers. The corresponding values Λ_k have been also reported on Fig. 3. The agreement between experimental and theoretical values is rather good. This shows that the semi-classical model is able to give a good representation of the diamagnetic spectrum.

Let us remark that the singularity observed at $\Lambda = 0$ as well in the experimental spectra as in the theoretical studies leads to the presence of a "separatrix" in the phase space and it is tightly linked to the possibilty of chaotic motion.[3]

EXPERIMENTAL STRUCTURE OF A LITHIUM M = 0 MANIFOLD IN THE PRESENCE OF PARALLEL ELECTRIC AND MAGNETIC FIELDS

A second experiment has been done by adding an electric field $\vec{F}$ in the same direction as the $\vec{B}$ field.   In these conditions M remains an exact quantum number but the parity π_t is no longer a good one; consequently n components are expected to appear in the (n, M = 0) manifold. On the other hand it is easy to imagine that, according to the $\vec{F}$ field strength two different types of evolution must appear for the two classes of states of the diamagnetic manifold. The degeneracy of the lowest states will be removed and as these states are stretched along the $\vec{F}$ direction, they are very sensitive to the electric field and should exhibit a linear Stark effect. On the contrary the highest nondegenerate states of the manifold have weak z component and thus are weakly sensitive to the electric field: they should present a quadratic Stark effect. One should notice that in this region the even parity components should appear in the energy spectrum equally spaced between the odd components.

A series of experimental recordings of the n = 30 M = 0 diamagnetic manifold of lithium at B = 2.33T and for increasing electric field is presented in Fig. 6. As expected an actually different behavior is observed in the lower ($\Lambda < 0$) and in the upper ($\Lambda > 0$) parts of the spectrum. Important and complex modifications appear in the low energy range: the energy of the lowest components decreases rapidly, almost linear with F; simultaneously additional lines are observed at inter-mediate energy. The upper part of the manifold is weakly perturbed and, surprisingly the parity mixing induced by F does not lead to the appearance of additional lines. In fact this behavior is typical of the nonhydrogenic character of the even $\pi_t = 1$ M = 0 diamagnetic manifold of the lithium atom where the large quantum defect δ_s induces a complete modification of the structure; in the lower part of the manifold odd and even states alternate, while pairs of quasi-degenerate odd and even states are present in the upper part.[9]

In order to study the evolution of the structure versus the electric field strength we have reported on a diagram (Fig. 7a) the energies deduced from our experimental spectra. The results have been compared to theoretical calculations obtained by diagonalization of the total Hamiltonian of the Li atom. The two results agree perfectly. The

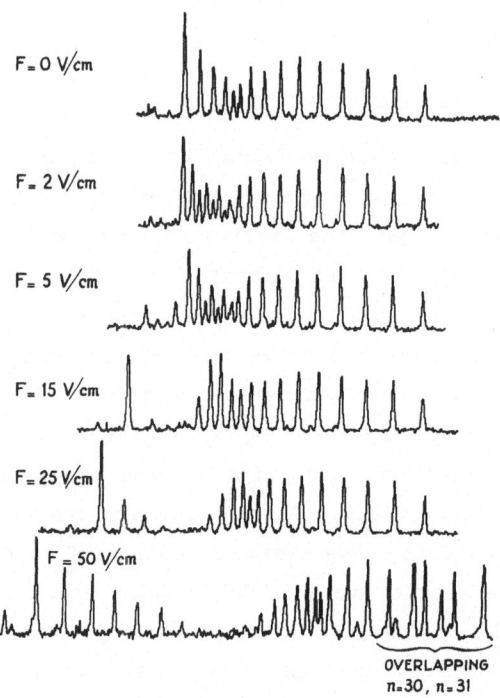

Fig. 6 Experimental recordings of n = 30 M = 0 manifold of Li in the presence of parallel F and B fields. B = 2.33T and F is increasing. For F = 50V/cm a relative important overlapping between n = 30 and n = 31 manifolds is observed.

corresponding map calculated for hydrogen is shown in Fig. 7b. It shows that, for both H and Li atoms, the global features are very similar especially for what concerns the existence and the properties of the three classes of states (these classes will be defined in the next paragraph by analyzing the hydrogen spectrum) as well as the existence of a peculiar well defined frontier between the class labelled III and the other two. However the nonhydrogenic character of the M = 0 even states of lithium induces some significant differences: the states I and II exhibit large anticrossings which do not appear in the hydrogen spectrum, and the upper states of the manifold present the previously mentioned quasi-degeneracy.

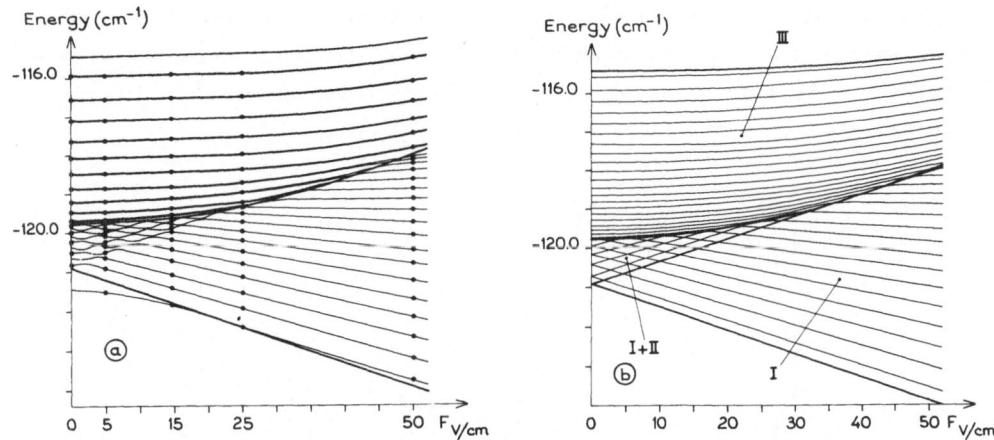

Fig. 7 Maps representing the evolution with increasing F of the structure of the n = 30 M = 0 B = 2.33T manifold.
(a) Lithium: full curves result from calculations by diagonalization. The points represent experimental results. One can observe that for F = 50V/cm no inter-n-mixing effect can be detected.
(b) Hydrogen: map calculated in the same conditions. The global features are identical; the differences are due to the core effects on the s states in lithium.

INTERPRETATION OF THE EVOLUTION OF THE STARK STRUCTURE OF THE DIAMAGNETIC MULTIPLET OF HYDROGEN IN A SEMI-CLASSICAL THEORY

The Hamiltonian for the hydrogen atom in the presence of parallel $\vec{B}$ and $\vec{F}$ fields in the following where H_γ is given in Eq. 1

$$H_\beta = H_\gamma - fz \tag{5}$$

In the low fields limit, the first order perturbation theory within a fixed n-manifold is valid and the operator relation[10] $\vec{r} = +3/2n^2 \vec{A}$ can be used. From the expression f the diamagnetic energy (Eq. 3) the total $(\vec{B} + \vec{F})$ perturbed energy shift ΔE_β is[11]

$$\Delta E_\beta = \gamma^2 \, (_n^2 n^2 + n^2 \Lambda + M^2)/16 + 3n^2 fA_z/2$$

which can be reduced to a form similar to Eq. 3:

$$\Delta E_\beta = \gamma^2 n^2 (n^2 + n^2 \Lambda_\beta + M^2)/16 \tag{6}$$

where

$$\Lambda_\beta = 4A^2 - 5(A_z - \beta)^2 + 5\beta^2 \tag{7}$$

In atomic units $\beta = 12fn^2/5n^4\gamma^2$, ($f = F/F_c$, $F_c = 5.14 \ 10^9 V/cm$).

The β parameter measures the relative strength of the linear Stark effect with respect to the diamagnetic interaction. Λ_β commutes with H_o therefore Λ_β is a constant of motion valid to first order within terms in B^2 and F.

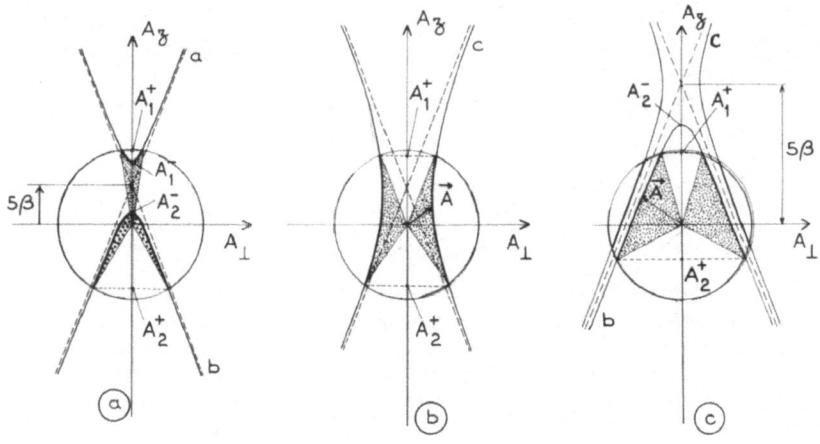

Fig. 8   Variations of the Runge–Lenz vector $\vec{A}$ under the influence of $\vec{B}$ and F fields parallel to the z direction ($M = 0$ $\gamma^2 n^7 \ll 1$ $fn^5 \ll 1$). The hyperboloid is defined by Λ_β = cont (Eq. 7).
(a,b) β < 1/5 three classes of states exist associated with the three sheets a, b and c of the hyperboloid limited by the double cone.
(c) β > 1/5 the two sheets b and c presented here intersect the sphere along two circle: all the corresponding states belong to the same class. The double cone is no longer a frontier between the classes.

In an analysis identical to that performed for the pure diamagnetic multiplet one can observe that, for a given energy state, the extremity of $\vec{A}$ moves on the hyperboloid of revolution:

$$4A^2 - (A_z - 5_\beta)^2 = \Lambda_\beta - 25\beta^2 \tag{8}$$

This equation is of the same type as Eq. 2, except that, now, the hyperboloid is not centered at the origin but is shifted on the z-axis by an amount of 5β. Nevertheless the evolution of A remains limited by the sphere $|\vec{A}| = 1$.

Figure 8 shows typical secular variations for the $\vec{A}$ vector, that can be connected with the different behaviors of the diamagnetic states of hydrogen under the influence of an increasing electric field (Fig. 7b). Two different situations are to be considered according to the value of β with respect to the limiting value β = 1/5:

(a) $\beta < 1/5$: The top of the double cone is located inside the sphere. Therefore the extremity of $\vec{A}$ can sweep the three types of sheets a, b, or c (Figs. 3a and 3b). The three types of sheets correspond to different spatial localizations and symmetries for the $\vec{A}$ motion, defining three classes of states; sheets a and b correspond to librational motion located respectively near $\theta = 0$ and π meanwhile sheet c describes a rotational motion. The asymptotic cone $\Lambda_\beta = 25\beta^2$ determines a frontier – the so-called separatrix or barrier – between librational and rotational states. From Eq. 7 one can obtain the Λ_β extension for each class, the limiting values corresponding either to the sphere $|\vec{A}| = 1$ respectively at the points $\cos\theta = +1$ and -1 and along the circle $\cos\theta = \beta$. As a result

$$
\beta < 1/5 \quad
\begin{array}{llll}
\text{class I:} & \text{sheet b} & -10\beta - 1 \leq \Lambda_\beta \leq 25\beta^2 & \\
& & & \text{libration} \\
\text{class II:} & \text{sheet a} & +10\beta - 1 \leq \Lambda_\beta \leq 25\beta^2 & \\
& & & \\
\text{class III:} & \text{sheet c} & 25\beta^2 \leq \Lambda_\beta \leq 5\beta^2 + 4 & \text{rotation}
\end{array}
$$

As β increases, the tangent cone is shifted in the z-direction, the number of states in class II increases whereas the number decreases in class I; this class disappears completely for $\beta = 1/5$. The Λ_β extension and the number of states in class III vary slowly (quadratically) with β .

(b) $\beta > 1/5$: There remain only two classes corresponding to either a librational (class I) or a rotational (class III) symmetry (Fig. 8a sheet b and Fig. 8c). Let us emphasize that, for rotational states, the extremity of $\vec{A}$ can sweep a sheet of either type b or c. Consequently for $\beta > 1/5$ the double cone is no longer a frontier. The transition between librational and rotational states appears when the sheet b becomes internally tangent to the sphere $|\vec{A}| = 1$ at the point $\theta = 0$: now the frontier corresponds to $\Lambda_\beta = 10\beta - 1$ and is called a "quasi-barrier." From Eq. 7 one deduces the Λ_β extension for each class as above:

$$
\beta > 1/5 \quad
\begin{array}{llll}
\text{class I:} & \text{sheet b} & -10\beta - 1 \leq \Lambda_\beta \leq +10\beta - 1 & \text{libration} \\
\text{class II:} & \text{sheet b or c} & 10\beta - 1 \leq \Lambda_\beta \leq 4 + 5\beta^2 & \text{rotation}
\end{array}
$$

As β increases, the number of states in class III decreases, and the class disappears at $\beta = 1$. For $\beta > 1$, only librational states exist, and with increasing β the structure of the n-manifold becomes very similar to the linear Stark one (A_z = cont).

The Λ_β limiting values for all the classes are drawn on Fig. 9; they match very well with the energy diagram of Fig. 7b. In this diagram librational and vibrational states do not overlap; in particular for $\beta < 1/5$ the class II states are transformed into class I states without crossing the frontier $\Lambda_\beta = 25\beta^2$ which acts as a "barrier."[11]

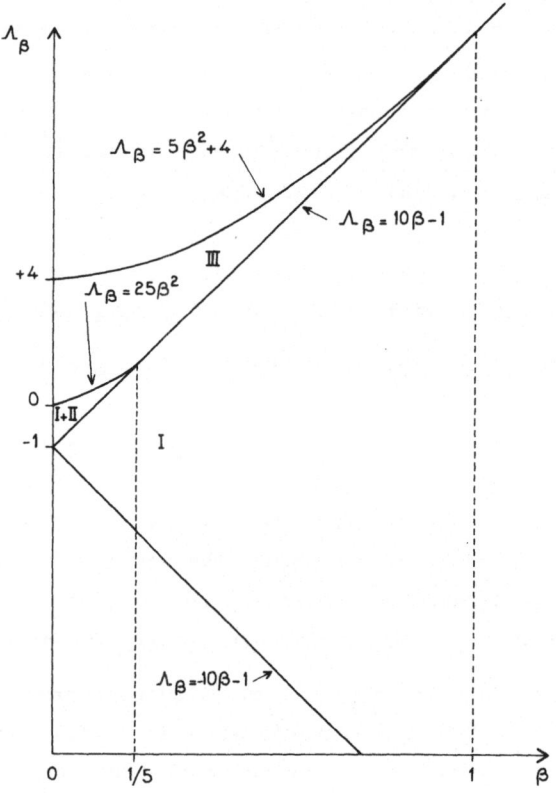

Fig. 9 Curves representing the limiting Λ_β values for the different three classes as a function of β.

The detailed analysis of the Stark structure of the diamagnetic multiplet requires now the quantization of the Λ_β values. This can be done using the same variable θ and its conjugate moment $L_\perp(\theta)$ as those used for the pure diamagnetic structure.*

However $L_\perp(\theta)$ is not such a simple function of θ. In the case $M = 0$, $L_\perp(\theta)$ (Eq. 4) becomes equal to:

$$L_\perp(\theta) = n\sqrt{1 - A^2} = L(\theta) \tag{9}$$

---

*Another semi-classical quantization has been recently performed using as angle and action variables respectively the argument of the perihelion and the modulus L of the angular momentum.[12]

where A is the solution of the equation (deduced from Eq. 7)

$$(4 - 5\cos^2\theta)A^2 + 10\beta\cos\theta\ A - \Lambda_\beta = 0 \qquad (10)$$

Figure 10a represents trajectories of the operating point in the phase plane $(\theta, L_\perp(\theta))$ for different values of Λ_β and $\beta = 0.16$. These curves correspond to the hyperboloids shown on Fig. 10b. As can be seen $L_\perp(\theta)$ may be a multivalued function of θ, the action integral to be evaluated corresponds to the area S limited by the curve and the horizontal axis. For each of the three classes the quantized values of $\Lambda_{\beta k}$ are given by solving the equation

$$S(\beta, \Lambda_{\beta k}) = (k + 1/2)\ \frac{\pi}{n}\ . \qquad (11)$$

The results of the quantization are summarized in Fig. 11 where we can observe the evolution of the pure diamagnetic structure towards a quasi-Stark structure. The first part of the figure $(0 < \beta < 0.2)$ perfectly matches the map obtained by a pure diagonalization of the diamagnetic Hamiltonian for the hydrogen atom (Fig. 7b).

Analytical studies of the action integral[14] permit also to demonstrate that in hydrogen, for $\beta < 1/5$, states of the classes I and II cross by pairs, all the crossings occuring at the same β_p value (in lithium due to the core effects these crossings are changed into anticrossings). They show that these "collective crossings" are equidistant in β, their abscissae being equal to $p/n\sqrt{5}$ (p integer).

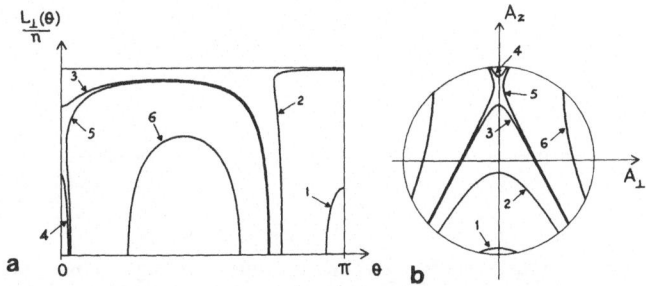

Fig. 10 (a) Left side: curves representing the evolution of the point in the phase plane $(\theta, L_\perp(\theta))$ for $\beta = 0.16$ and different values Λ_β. The curves may be multivalued function of θ.
(b) Right side: corresponding motion of the extremity of the $\vec{A}$ vector.

In this paragraph we want to point out some analogies found between the calculated structure for a pure diamagnetic multiplet $(\beta = 0)$ for

nonvanishing M values and the Stark structure of the diamagnetic manifold
with M = 0 such as it has been studied above. Let us recall that, for
β = 0, the quantized values of $\Lambda_{\beta k}$ for the constant of motion Λ are given
by Eq. 4:

$$I(\Lambda, \frac{M}{n}) = \int [1 - \frac{\Lambda_{Mk}}{4 - 5\cos^2\theta} - \frac{M^2}{n^2\sin^2\theta}]^{1/2} d\theta = (k + 1/2) \frac{\pi}{n}$$

and that analysis of this integral shows that the librational states
(Λ < 0) no longer exist[13] when $|M| > n/\sqrt{5}$. Such a situation also exists
in the presence of parallel fields (β ≠ 0 M = 0): the librational states
of the class I disappear for β > 1/5. The two conditions are equivalent
if we consider the transformation β → $|M|/n\sqrt{5}$.

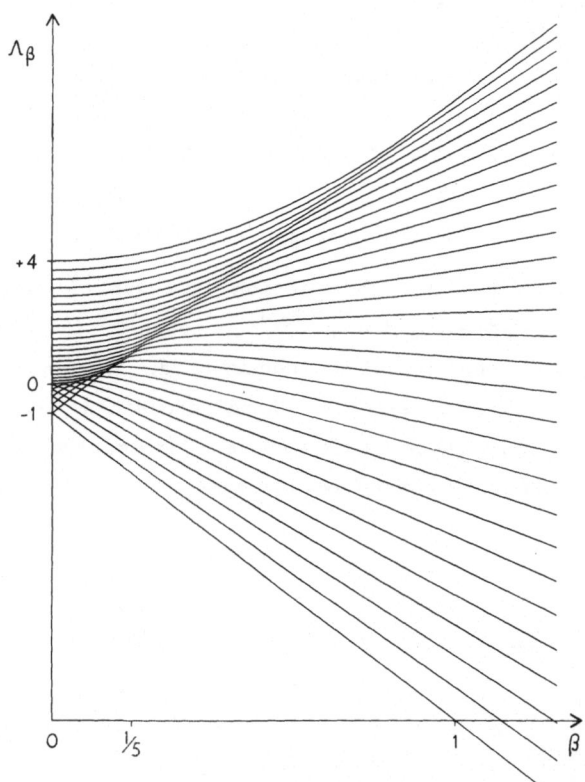

Fig. 11 Map resulting from calculations using the semi-classical
        quantization method. In the range β < 1/5 this map is quite
        similar to the one obtained by diagonalization (Fig. 7b).

Furthermore for integer M, these β values coincide with the
abscissae β_p of the "collective crossings" of the librational states and
the number of states belonging to class II is then identical to the
corresponding number of degenerate states Λ < 0 for the diamagnetic
multiplet (n, M). In fact, more than these few comparisons, one can make

a complete analogy between the two integrals using the transformation $\beta_M \rightarrow |M|/n\sqrt{5}$ an it can be shown that:

$$\Lambda_{Mk} = \Lambda_{\beta_M,k} - 25\beta_M^2 \tag{12}$$

or equivalently: $\Lambda_{Mk} + 5M^2/n^2 = \Lambda_{\beta_M,k}.$

This identity is well seen on Fig. 12 where we have plotted simultaneously the $\Lambda_{\beta k}$ values and the calculated Λ_{Mk} values shifted of $5M^2/n^2$. A corresponding relation may be found for the two energy spectra disregarding the paramagnetic term

$$\Delta E_D(M,k) = \Delta E_\beta(\beta_M,k) - \gamma^2 n^2 M^2/4 \tag{13}$$

It shows that surprisingly, the diamagnetic spectrum at $M \neq 0$ is identical with the Stark spectrum of the $M = 0$ diamagnetic multiplet for an electric field such as $\beta_M = |M|/n\sqrt{5}$. Including the paramagnetic term the two spectra are shifted by an amount of:

$\gamma M/2 - \gamma^2 M^2/4.$

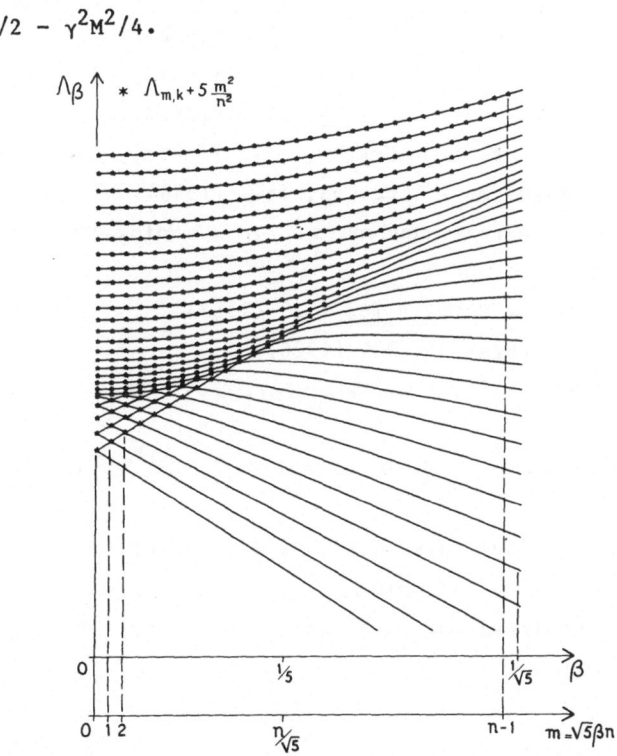

Fig. 12 Comparison of the Λ_β values (full line) associated with the $n = 30$ $M = 0$ manifold with increasing β and the Λ_M values for the pure $n = 30$ diamagnetic manifold ($\beta = 0$) for different M values. To facilitate the comparison the $\Lambda_{Mk} + 5M^2/n^2$ values, represented by *, are reported using the horizontal scale $M = \sqrt{5}\beta n$.

CONCLUSION

We have completely analyzed the evolution of the diamagnetic structure under the influence of an electric field (parallel to the magnetic field). The semi-classical treatment is able to reproduce all the peculiar phenomena observed in the spectrum and to classify unambiguously all the states. Careful analysis of the action integrals has permitted us to point out new peculiar analogies between the diamagnetic spectrum ($M \neq 0$) and the Stark perturbed diamagnetic $M = 0$ spectrum; these similitudes correspond to one most striking feature of the diamagnetic problem, which has to be studied more thoroughly.

REFERENCES

1. J.-C. Gay and D. Delande, Comments At. Mol. Phys. 13, 275 (1983).

2. J. Main, G. Wiesbusch, A. Holle and K. H. Welge, Phys. Rev. Lett. 57, 2789 (1986).

3. D. Delande and J.-C. Gay, Comments At. Mol. Phys. 19, 35 (1986).

4. E. A. Solov'ev, Sov. Phys. J.E.T.P. 55, 1017 (1982).

5. D. R. Herrick, Phys. Rev. A 26, 323 (1982).

6. P. Cacciani, E. Luc-Koenig, J. Pinard, C. Thomas and S. Liberman, Phys. Rev. Lett. 56, 1124 (1986).

7. P. Cacciani, E. Luc-Koenig, J. Pinard, C. Thomas and S. Liberman, Phys. Rev. Lett. 56, 1467 (1986).

8. J. B. Delos, S. K. Knudson and D. W. Noid, Phys. Rev. A 28, 7 (1983).

9. P. A. Braun, J. Phys. B 16, 4323 (1983).

10. W. Pauli, Z. Phys. 36, 336 (1926).

11. P. A. Braun and E. A. Solev'ev, Sov. Phys. J.E.T.P. 59, 38 (1984).

12. R. L. Waterland, J. B. Delos and M. L. Du, Phys. Rev. A 35, 5064 (1987).

13. P. A. Braun, Sov. Phys. J.E.T.P. 57, 492 (1983).

14. P. Cacciani, S. Liberman, E. Luc-Koenig, J. Pinard and C. Thomas, submitted for publication in J. Phys. B.

NEGATIVE ION PHOTODETACHMENT IN A STATIC EXTERNAL FIELD

Chris H. Greene

Department of Physics and Astronomy
Louisiana State University
Baton Rouge, Louisiana 70803

INTRODUCTION

The photodetachment of negative ions in an external magnetic field is a prototype for nonseparable quantum mechanical systems currently being studied in a variety of contexts and in this volume in particular. The motion of an escaping photoelectron begins with spherical symmetry at small distances, while at large distances the diamagnetic term in the Hamiltonian permits escape of the photoelectron only along the field axis. In this outer region, therefore, the electron wavefunction must have the cylindrical form expected for a Landau-level-type "quantized continuum." This problem represents one of the simplest possible types of nonseparability since the Schroedinger equation is locally separable in either spherical or cylindrical coordinates at essentially each point in space. This simplification will be seen below to permit an analytical solution through a "local frame transformation" of the type used by Harmin[1] and Fano[2] to solve the nonhydrogenic Stark effect problem. In that problem the required transformation is from spherical to parabolic coordinates, while in the present photodetachment problem a spherical-to-cylindrical transformation is necessary.

Aside from general interest as a nonseparable quantum mechanical system, the photodetachment of negative ions sheds light on threshold laws as well, in particular their dependence on the dimensionality of the relevant configuration space. Early studies by Pavlov[3] and by Larson and coworkers[4] pointed out that a uniform magnetic field in effect changes the final three-dimensional electron continuum state into a one-dimensional continuum state since the electron can only escape to infinity in a direction parallel to the magnetic field. As a consequence, the photodetachment cross section at each Landau threshold energy E_{th} was predicted in Refs. 3 and 4 to <u>diverge</u> in proportion to $(E - E_{th})^{-1/2}$, just because the (box-normalized) density of states is modified from its three-dimensional form d^3p/dE (giving the Wigner threshold law $\sigma \propto (E - E_{th})^{1/2}$) to a one-dimensional form dp/dE. This point was later contested and clarified by Clark,[5] who showed that <u>any infinitesimal electron-atom interaction changes the divergent threshold behavior back into the usual Wigner law</u>. Strictly speaking this result does not conflict with the prediction of Refs. 3 and 4, since these treatments completely neglect the electron-atom interaction. It is a curious result, nonetheless, that any nonzero scattering length, regardless of its size, can produce such a dramatic alteration of the threshold law. On the other hand it is reminiscent of another well known result of one-dimensional quantum mechanics: the fact that a <u>free</u> electron in one dimension has no bound states, but if an <u>infinitesimal</u> attractive (one-dimensional) potential is present then the system is guaranteed to have at least one bound state, regardless of the strength of this potential.[6]

In this paper I will describe and amplify the physical content and implications of some recently completed theoretical work on negative ion photodetachment in an eternal magnetic[7] or electric[8,9] field. A considerable body of experimental work on these problems already exists, for a magnetic field by Larson's group[4,10] and for an electric field by Bryant et al.[11] The main achievement of recent frame transformation studies[7-9] has been to show that the photodetachment cross section in the presence of an external field normally factorizes into a product of the zero-field cross section multiplied by a density of states factor. The latter depends on both the short-range electron-atom scattering length (or reaction matrix for open-shell atoms) and on a field-dependent modulation factor characteristic of the frame transformation from spherical to cylindrical coordinates. This analytical factorization, as in the corresponding Rydberg Stark treatment of Refs. 1 and 2, bypasses the more difficult direct numerical solution of the final-state Schroedinger equation over a large region of space.

ANALYTICAL FORMULATION

A careful discussion of the required theoretical elements can be found in Refs. 1 and 7. Here I will highlight the major results and formulas. The starting point for the spherical-to-cylindrical frame transformations needed to describe photodetachment in a magnetic or electric field is the known connection between zero-field solutions of the free-particle Schroedinger equation in these two coordinate systems. The familiar spherical solutions involve Legendre functions and spherical Bessel functions:

$$f_{\ell m}(\vec{r}) = \frac{e^{im\phi}}{(2\pi)^{1/2}} N_{\ell m} P_{\ell m}(\cos\theta)(2k/\pi)^{1/2} j_{\ell}(kr). \tag{1}$$

The cylindrical solutions involve trigonometric functions and Bessel functions:

$$\psi_{qm}(\vec{r}) = \frac{e^{im\phi}}{(2\pi)^{1/2}} J_m[(k^2 - q^2)^{1/2}\rho](\pi q)^{-1/2}\cos[qz + (\Pi_z - 1)\pi/2], \tag{2}$$

in which the reflection parity under the operation $z \rightarrow -z$, Π_z, is a constant of the motion. The cylindrical separation constant q is just the z-component of the electron wavevector. The two sets of solutions (1) and (2) at a given electronic energy $\varepsilon = \frac{1}{2} k^2$ are related through a real, orthogonal transformation:

$$\psi_{qm}(\vec{r}) = \sum_{\ell} U_{q\ell}^{(B=0)} f_{\ell m}(\vec{r}). \tag{3}$$

The zero field frame transformation matrix is proportional to a Legendre function,

$$U_{q\ell}^{(B=0)} = \left[\frac{2\ell + 1}{kq}\right]^{1/2}\left[\frac{(\ell - m)!}{(\ell + m)!}\right]^{1/2} P_{\ell m}(q/k)(-1)^{[(\ell-m-1)/2]}. \tag{4}$$

In the presence of an external magnetic field the Schroedinger equation at small distances is almost exactly the same as for $B = 0$, but asymptotically the motion is profoundly affected, in particular giving the Landau-level solutions whose motion across the magnetic field is now quantized:

$$\psi_{nm}^{B}(\vec{r}) = \frac{e^{im\phi}}{(2\pi)^{1/2}} N_{nm} e^{-1/2\alpha\rho^2}(\alpha\rho^2)^{|m|/2} L_n^{(|m|)}(\alpha\rho^2)$$

$$(\pi q_n)^{-1/2}\cos[q_n z + (\Pi_z - 1)\pi/2]. \tag{5}$$

Here the subscript n relates to the number of modes of the wavefunction in ρ, and the free electron energy eigenvalues take the well-known form

$$\varepsilon = \alpha(2n + m + |m| + 1) + 1/2 \, q_n^2, \tag{6}$$

the term in α representing the amount of energy associated with motion in (ρ, ϕ) and the term $\frac{1}{2} q_n^2$ representing the energy associated with motion along the z axis (field axis).

Following arguments given by Fano,[2] the spherical B = 0 solutions and the cylindrical B ≠ 0 solutions obey virtually the __same__ Schroedinger equation at small distances. (Recall that typically $\alpha < 10^{-5}$.) Accordingly the two sets of solutions must be related at small radii by a linear transformation, namely

$$f_{\ell m}(\vec{r}) \sim \sum_{n=0}^{n_{max}} (U^{-1})_{\ell n} \psi_{nm}^B(\vec{r}).$$ (7)

Similarly a spherical solution irregular at r = 0 connects to an irregular cylindrical solution whose oscillations along the z axis lag ψ^B by 90°:

$$g_{\ell m}(\vec{r}) \sim \sum_n (U^T)_{\ell n} \chi_{nm}^B(\vec{r}).$$ (8)

The field-dependent frame transformation matrix U is Eq. (4) aside from a normalization constant,

$$U_{n\ell} = U_{q_n \ell}^{(B=0)} \left[\frac{2\alpha n!}{(n + |m|)!}\right]^{1/2} \left[\frac{2n + m + |m| + 1}{2}\right]^{|m|/2}.$$ (9)

Equations (7)-(9) allow us to specify in detail how a spherical continuum wavefunction near the nucleus evolves into a cylindrical channel expansion at large distances. In essence, these expressions show how the spherical wavenfunction projects its amplitude among the full set of energetically accessible Landau functions. We will need this projection for one reason only - to calculate the final state normalization integral which will simply renormalize the B = 0 photodetachment cross section.

The B = 0 continuum state of the negative ion is most efficiently described using quantum defect theory. For simplicity consider a single channel example in which a low energy photoelectron escapes with a single allowed value of the orbital angular momentum ℓ and projection m. Then its asymptotic, energy-normalized wavefunction is characterized by a phaseshift δ_ℓ:

$$\psi_\ell^{B=0} \xrightarrow[r \to \infty]{} \phi_A(\omega)[f_{\ell m}(\vec{r}) \cos\delta_\ell - g_{\ell m}(\vec{r})\sin\delta_\ell],$$ (10)

with ϕ_A the wavefunction of the (spinless) neutral atomic ground state. At low energies the phaseshift δ_ℓ is proportional to $k^{\ell+1/2}$, as is the whole wavefunction (10) itself near r = 0. The effect of a nonzero magnetic field can now be seen from Eqs. (7) and (8) to modify the final state (10) into the form

$$\Psi_\ell^B \xrightarrow[r \to \infty]{} \phi_A(\omega) \sum_{n=0}^{n_{max}} [\cos\delta_\ell (U^{-1})_{\ell n} \psi_{nm}^B(\vec{r}) - \sin\delta_\ell (U^T)_{\ell n} \chi_{nm}^B(\vec{r})]. \tag{11}$$

These $B \neq 0$ solutions Ψ_ℓ^B are no longer orthonormal, however, which requires us to introduce an orthonormalization matrix, termed a density matrix in Refs. 1, 2:

$$(D^{-1})_{\ell\ell'} \delta(E - E') \to \langle \Psi_\ell^B(E) | \Psi_{\ell'}^B(E') \rangle$$

$$= \delta(E - E')[\cos\delta_\ell \ H_{\ell\ell'}^{-1} \cos\delta_{\ell'} + \sin\delta_\ell \ H_{\ell\ell'} \sin\delta_{\ell'}], \tag{12}$$

where a modulation factor has been defined to be

$$H_{\ell\ell'} = \sum_{n=0}^{n_{max}} U_{n\ell} U_{n\ell'}. \tag{13}$$

Note that all of these quantities in (12) and (13) depend on m and on the electron energy ε. The index $n_{max}(\varepsilon)$ is the quantum number of the highest energetically accessible Landau channel and is readily identifiable from Eq. (6).

Knowledge of the density of states matrix $D_{\ell\ell'}$ permits finally the calculation of the total cross section for photodetachment using light of frequency ω:

$$\sigma^B \propto \omega \sum_{\ell\ell'} d_\ell D_{\ell\ell'} d_{\ell'},$$

with

$$d_\ell \equiv \langle \Psi_\ell^{B=0} | \hat{\epsilon} \cdot \vec{r} | \Psi_{gnd} \rangle . \tag{14}$$

Near threshold a single partial wave dominates (either s or p depending on the final state parity), and a strict factorization is obtained,

$$\sigma^B = \sigma^{B=0} D_{\ell\ell'}. \tag{15}$$

The modulation factor $H_{\ell\ell'}$ turns out to be important in its own right since near threshold $\delta_\ell \sim 0$ and $D_{\ell\ell'} \sim H_{\ell\ell'}$. It is shown for s and p waves in Fig. 1 for an easily accessible magnetic field strength. An explicit expression is especially trivial for the important case $\ell = 0$,

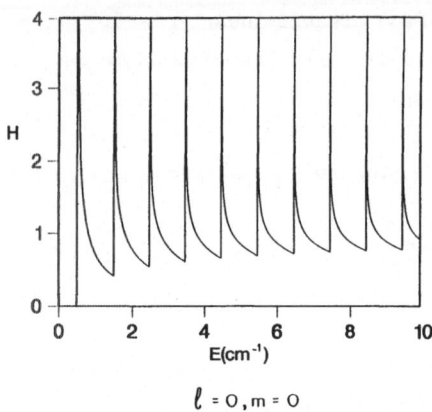

$\ell = 0, m = 0$

Fig. 1. Dimensionless s-wave modulating factor H is shown as a function of the photoelectron energy for a 1.07 T magnetic field.

$$H_{00} = \sum_{n=0}^{n_{max}} (2\alpha/kq_n).$$ (16)

Note the divergence of H_{00} at each Landau threshold, where $q_n \to 0$. This divergence is responsible for the series of asymmetric spikes in ,
Fig. 1. Note also that the p-wave modulation factor vaies drastically with m, as in fact the Landau threshold spikes are totally absent for $\ell = 1$, $m = 0$. This is physically reasonable since no ($\ell = 1$, $m = 0$) photoelectrons are ejected initially orthogonal to the field axis $[P_{10}(0)]$. An immediate consequence is that when p-waves dominate at threshold, the Landau oscillations will be prominent for σ-polarization but virtually absent for π-polarization of the incident light.

THRESHOLD LAW ANALYSIS

As discussed in the Introduction, very different forms for the threshold law have been derived by Larson et al.[4] and by Clark.[5] Eqs. (12), (15) and (16) now permit a derivation of each special case and its range of validity. In fact writing $\tan\delta_{\ell = 0} \sim ka$ at low photoelectron energy, taking the electron-atom scattering length to be a, the detachment cross section has the structure (assuming $\ell = 0$),

$$\sigma^B \underset{k \to 0}{\to} 2\alpha\bar{\sigma} \; (\sum_n \frac{1}{q_n})/[1 + 4\alpha^2 a^2 (\sum_n \frac{1}{q_n})^2].$$ (17)

A "reduced" zero-field cross section has been introduced, $\bar{\sigma} \equiv \sigma^{B=0}/k$, which is essentially a constant in the energy range considered here. Immediately above a new (n^{th}) Landau threshold (where $q_n \to 0$) the cross section reduces simply to

$$\sigma^B \to 2\alpha\bar{\sigma}\,\frac{q_n}{q_n^2 + 4\alpha^2 a^2} + \text{constant.} \tag{18}$$

This expression confirms Clark's analysis,[5] in that a Wigner threshold behavior $(E - E_{th})^{1/2}$ is generally obtained. In the special case $a = 0$ characteristic of <u>no electron–atom interaction</u>, however the divergent $(E - E_{th})^{1/2}$ threshold law of Ref. 4 is recovered. The cross section rises rapidly from its background value to a maximum value $\sigma^B_{max} \sim \sigma^{B=0}/2\delta_0 \sim \bar{\sigma}/2a$ which turns out to be nearly independent of energy in this range, and surprisingly, independent of the magnetic field strength. The energy at which this maximum occurs is $2\alpha^2 a^2$ above the threshold energy E_{th}, indicating that the Wigner threshold law is correct only for an extremely small energy interval of, for instance, $\Delta\epsilon \approx 10^{-5} \text{cm}^{-1}$ for S^- in a 1 Tesla field. Accordingly the visual appearance of the cross section is dominated by an apparent $(E - E_{th})^{-1/2}$ behavior which is ultimately truncated prior to its divergence. This is evident from the calculated S^- photodetachment cross section in Fig. 2. Owing to the multichannel nature of the zero-field detachment problem, there are additional complications such as a Zeeman fine structure to the cross section. A detailed treatment of these effects is described in Ref. 7.

We see from Eq. (18) and the following discussion that the theoretical treatment of Ref. 4 should adequately describe the photodetachment cross section in the near-threshold region where scattering phase shifts are constrained to be very small. This will not be true, though, in the vicinity of a low energy shape or Feshbach resonance since continuum phaseshift then rises rapidly by π radians. An example of the effect of a shape resonance on the Landau oscillations is

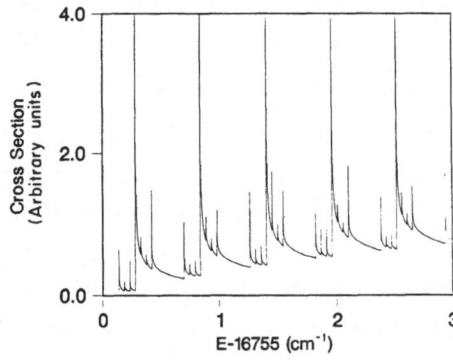

Fig. 2. Photodetachment cross section for S^- ground state ions in a 0.6 T magnetic field, with the polarization along the field.

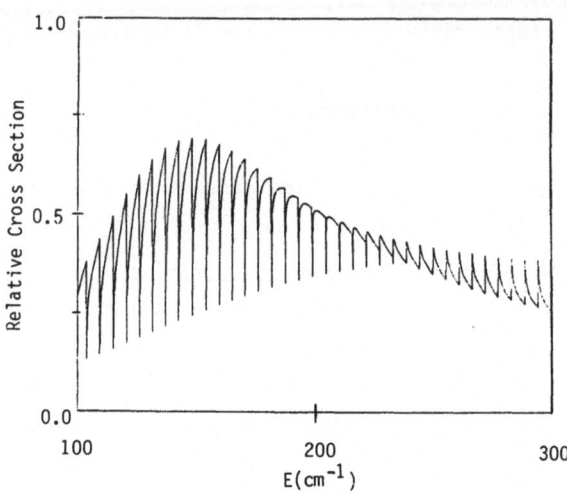

Fig. 3. Effect of a zero-field resonance on the Landau threshold behaviour, for σ^+ polarization in a 6 T magnetic field. In zero-field the resonance peak lies near 150 cm^{-1}.

shown for a low energy p-wave shape resonance in Fig. 3. The position and width of this resonance are appropriate for the $n = 2$ $^1p^0$ shape resonance in H$^-$. At the magnetic field used in Fig. 3, B = 6T, the Landau-level spacing is much smaller than the resonance width. As you move across the shape resonance the near-threshold profile changes distinctly from a threshold enhancement to a threshold window-type structure. Of course the escaping photoelectron does not have a pure $\ell = 1$ wavefunction since three channels (2sϵp, 2pϵs, 2pϵd) contribute, but the qualitative effect of the shape resonance on Landau threshold structures may be correctly described nevertheless.

THE STARK EFFECT IN NEGATIVE ION PHOTODETACHMENT

The frame transformation for negative ion photodetachment in an electric field is treated in very similar fasion.[8,9] The final state energy-normalized (regular) cylindrical solution in the presence of an electric field F (in a.u.) is now in place of Eq. (5);

$$\psi^F_{qm}(\vec{r}) = \frac{e^{im\phi}}{(2m)^{1/2}} J_m[(k^2-q^2)^{1/2}\rho](4/F)^{1/6} Ai[(2F)^{1/3}(z-q^2/2F)], \quad (19)$$

in which Ai denotes an Airy function and q is the continuous z-component of the electron wavevector. Treating only the case in which the photoelectron escapes initially with $\ell = 0$, the Stark frame transformation matrix is

$$U_{q0} = \pi^{1/2}(4/F)^{1/6}Ai[-q^2/(2F)^{2/3}]|k|^{-1/2}. \tag{20}$$

The s-wave modulation factor to be used now in place of Eq. (16) involves an integral over the continuous matrix index q,

$$H_{00}(\epsilon) = \int_{-\infty}^{\epsilon} d(1/2 \ q^2)(U_{q0})^2. \tag{21}$$

This modulation factor, as it does in the magnetic case, isolates the whole effect of the electric field. Equations (12), (14) and (15) remain correct for the Stark effect after replacing (16) by (21). The modulation factor is shown for a typical electric field strength in Fig. 4. It diverges at the detachment threshold as $|\epsilon|^{-1/2}$, exactly counterbalancing the zero-field cross section which vanishes in proportion to $|\epsilon|^{1/2}$, thereby leaving a finite cross section at threshold when the electric field is nonzero. In fact the cross section remains appreciable just <u>below</u> the detachment threshold, reflecting the ability

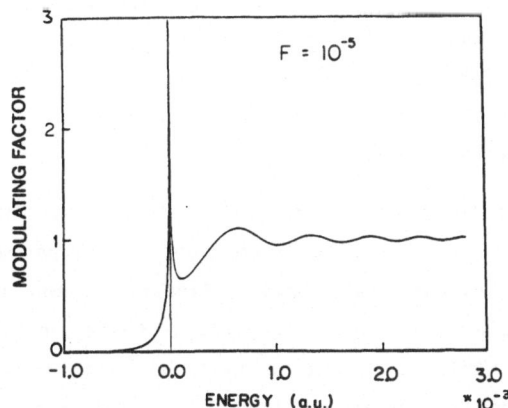

Fig. 4. Modulating factor appropriate for s-wave photodetachment in an electric field $F = 10^{-5}$ a.u. (From Ref. 8).

of the photoelectron to tunnel through the Stark barrier even at negative values of the photoelectron energy.

The oscillations in Fig. 4 can be visualized as resulting from an interference of two photoelectron escape pathways. One pathway is the direct one in which a photoelectron heads downfield after being liberated by the incident photon. On the second pathway the photoelectron initially heads upfield, eventually running into the electric field barrier and being reflected downfield. The constructive and destructive interference between these two pathways generates the maxima and minima of Fig. 4. In fact this simple argument can be made sufficiently quantitative to predict the positions of the oscillation maxima, giving a formula of Fabrikant,[12]

$$\varepsilon_n = (0.067Fn)^{2/3} \text{cm}^{-1}, \tag{22}$$

where F is the electric field in V/cm. An application of the preceding treatment has been developed in Ref. 9 for the H⁻ photodetachment experiment performed in an electric field by Bryant et al.[11] The frame transofrmation analysis shows good agreement with the observed field-dependent cross section. A time-dependent analysis by Reinhardt and Overman along the lines of Ref. 13 also provides an excellent description of the experimental results.

The Bryant et al.[11] experiment is apparently not the first to observe electric field-induced oscillations in negative ion photodetachment. In fact, Frey et al. presented experimental evidence in Fig. 7 of Ref. 14 for oscillations within a few tenths of a wavenumber above the $5p_{1/2}$ threshold in Rb⁻ photodetachment. However, these small oscillations were not interpreted as an electric-field effect in Ref. 14. They are shown in Ref. 15 to derive from a uniform field of 1.2 V/cm, which is very close to the ~ 1 V/cm retarding field used in Ref. 14 to distinguish fast and slow electrons.

REMAINING DIFFICULTIES OF THE THEORY

The quantum-mechanical formulation discussed in this treatment appears to correctly describe numerous features of negative ion photodetachment in an external field. Further extension to an arbitrary combination of electric and magnetic fields now appears to be straightforward. This treament cannot be considered as a complete solution to the problem, however, because in the case of magnetic field detachment, all effects associated with closed Landau channels have been neglected. Effects associated with closed zero-field channels are incorporated to the extent that they only affect the photoelectron phase shifts in the open channels. This neglect of the closed Landau channels means in essence that the theory is applicable only when the zero-field cross section shows no structures whose width is comparable to or less than the cyclotron frequency $\omega = 2\alpha$. For a very weakly bound, narrow Feshbach resonance, this approximation is rather certain to break down. In this case Feshbach resonances may become associated with specific Landau thresholds, as pointed out by Clark.[5] The extension of local frame transformation methods to deal with this more complicated situation would be an interesting topic for subsequent investigations. Similar closed-channel effects cause difficulties for the electric field formulation, both for the present negative ion problem and apparently at sufficiently negative energies in Harmin's treatment[1] of the Rydberg Stark effect. Channels which are only weakly closed appear to be described satisfactorily in the electric field photodetachment problem, as evidenced by the comparisons made in Ref. 9 with the experimental results for H^-.

ACKNOWLEDGEMENT

This work was supported in part by the National Science Foundation, and in part by the Alfred P. Sloan Foundation.

REFERENCES

1. D. A. Harmin, Phys. Rev. A 26, 2656 (1982).

2. U. Fano, Phys. Rev. A 24, 619 (1981).

3. G. G. Pavlov, Optika i Spectroscopiya 33, 1006 (1972).

4. W.A.M. Blumberg, W. M. Itano and D. J. Larson, Phys. Rev. A 19, 139 (1979).

5. C. W. Clark, Phys. Rev. A 28, 83 (1983).

6. S. Gasiorowicz, Quantum Physics, (Wiley, 1974), p. 297.

7. C. H. Greene, Phys. Rev. A (in press, 1987).

8. Hin-Yiu Wong, A.R.P. Rau and C. H. Greene, Phys. Rev. A (submitted).

9. A.R.P. Rau and Hin-Yiu Wong, Phys. Rev. A (submitted).

10. W.A.M. Blumberg, R. M. Jopson and D. J. Larson, Phys. Rev. Lett. 40, 1320 (1978).

11. H. C. Bryant, A. Mohagheghi, J. E. Stewart, J. B. Donahue, C. R. Quick, R. A. Reeder, V. Yuan, C. R. Hummer, W. W. Smith, S. Cohen, W. P. Reinhardt and L. Overman, Phys. Rev. Lett. 58, 2412 (1987).

12. I. I. Fabrikant, Sov. Phys. JETP 52, 1045 (1980).

13. W. P. Reinhardt, J. Phys. B 16, L635 (1983).

14. P. Frey, M. Lawen, F. Breyer, H. Klar and H. Hotop, Z. Phys. A 304, 155 (1982).

15. C. H. Greene and N. Rouze, to be published.

WAVEFUNCTION LOCALIZATION FOR THE HYDROGEN ATOM IN PARALLEL EXTERNAL ELECTRIC AND MAGNETIC FIELDS

S. Bivona,* P. F. O'Mahony** and K. T. Taylor**

*Istituto di Fisica Facolta di Ingegneria
Parco d'Orleans, 90128 Palermo, Italy

**Department of Mathematics
Royal Holloway and Bedford New College
Egham, TW20 0EX, Surrey, United Kingdom

ABSTRACT

A quantum mechanical study of a simple two dimensional problem identifies lines in coordinate space about which the probability densities are localized. These lines in general do not correspond to extrema in the system's static potential energy surface. The subject of this study has been the hydrogen atom in externally applied parallel electric and magnetic fields. Probability density plots of eigenstates, in which the electron experiences comparable influences from all three fields: Coulomb, electric and magnetic are each found to display a localization at a polar angle related to an invariant previously obtained in a classical study of the problem. The line about which localization occurs is found to vary with energy.

A general procedure to solve the Schrodinger equation for a particle moving in a multi-dimensional non-separable potential remains an important future goal in atomic and molecular physics. Given sufficient excitation energy the large number of degrees of freedom available to a particle moving in such a potential gives rise to a multiplicity of possible modes of excitation. The quantum spectrum therefore has a dense level structure and the classical motion of the particle is in general chaotic. Despite the complexity of the problem some general characteristics of the solutions to the Schrodinger equation have been learnt through the study of specific problems. In particular it has been shown that lines of potential extrema in the static potential energy surfaces coincide with regions of localization of the wavefunction and that these regions play a central role in calculating spectra and other physical observables.[1,2] For the systems studied however, localization has occurred about a direction in space that not only happened to be an axis of symmetry for the system, but was also a ridge line or extremum in the system's static potential energy. We present here a study of wavefunction localization in a two-dimensional system that does not have an obvious axis of symmetry and demonstrate that the concept of localization about lines of extrema of the static potential energy surface is a particular case of a more general phenomenon.

The electron in a hydrogen atom subjected to simultaneously imposed uniform electric and magnetic fields both directed along the z-axis provides an example of a two dimensional system with no axis of symmetry. The resulting static potential energy of the electron is, in atomic units

$$V(\rho,z) = \frac{-1}{\sqrt{\rho^2 + z^2}} + \frac{1}{2}\beta^2\rho^2 + Fz \qquad (1)$$

where β measures the magnetic field strength ($\beta = 10^{-5}$ corresponds to 4.7 tesla) and F the electric field strength (F = 4.86×10^{-9} corresponds to 25 V/cm). Since V does not depend on the azimuthal angle ϕ, motion in this coordinate can be separated off (and characterized by a fixed value of azimuthal quantum number m), so reducing the problem to that of finding eigenstates of a two dimensional Hamiltonian. Also lack of symmetry axis in V means no obvious symmetry axis in the Hamiltonian.

Eigenstates in the inter-ℓ and inter-n mixing regions of the spectrum have been obtained by diagonalizing the Hamiltonian within a basis set of Sturmian functions.[2] Attention has initially been directed to the inter-ℓ mixing region under conditions such that the electron experiences comparable influences from both external fields. By a classical approach Braun and Solioviev[3] have identified a parameter $\alpha = 3F/5n^2\beta^2$ (where n is the principal quantum number) that can be used to distinguish the three distinct regimes corresponding to either field dominant, or both comparable. Specifically $1/5 < \beta < 1/\sqrt{5}$ denotes the latter case. We have obtained energy levels and oscillator strengths for states in the inter-ℓ and inter-n mixing regimes at fixed field strengths

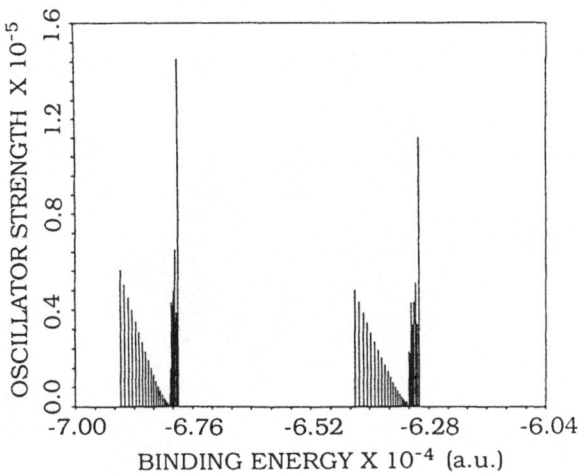

Fig. 1   The calculated oscillator strength as a function of the binding
energy (in a.u.) for photoabsorption from the ground state to
the n = 27 and n = 28 (m = 0), manifolds of a hydrogen atom in
parallel external magnetic and electric fields.

β = 3×10^{-6} and F = 4.896×10^{-9}.   Figure 1 displays the oscillator
strengths for the n = 27 and n = 28 manifolds, i.e. in ther inter-ℓ
mixing region of the spectrum.   The lower energy end of each manifold
shows behavior characteristic of a pure Stark spectrum, in that adjacent
energy levels are equally spaced and the oscillator strength decreases
monotonically with increasing energy.   The oscillator strengths at the
upper end of the manifold however are distributed in a more irregular
fashion.   The probability density in the ρ-z plane for the highest and
second highest energy levels in the n = 27 manifold are shown in
Fig. 2.   Figure 2(a) is a state clearly localized about some polar angle
θ_L.   There is however no ridge line in V, given by (1), at such an angle
for β = 3×10^{-6}, F = 4.86×10^{-9}.   Moreover we find that with fixed choice
of field strengths and hence fixed V, θ_L changes with n.   This is
demonstrated in Fig. 3 where we plot the corresponding states to those in
Fig. 2 but for the n = 32 manifold.   The angle θ_L has increased, with the
states now being localized closer to the z = 0 plane.   This demonstrates
the energy dependence of the lines of localization in coordinate space
and consequently shows that these lines will not in general coincide with
extremum lines of the static potential surface.

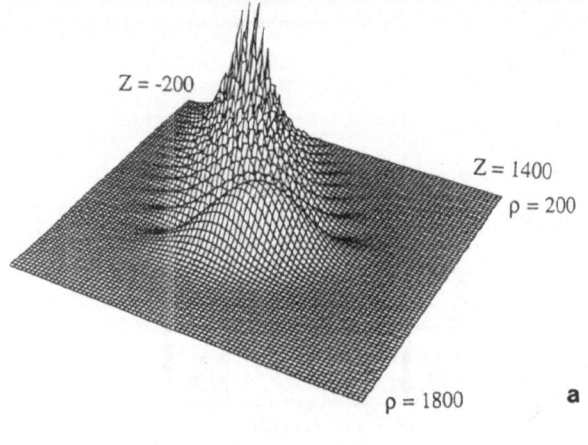

Z = -200

Z = 1400
ρ = 200

ρ = 1800   **a**

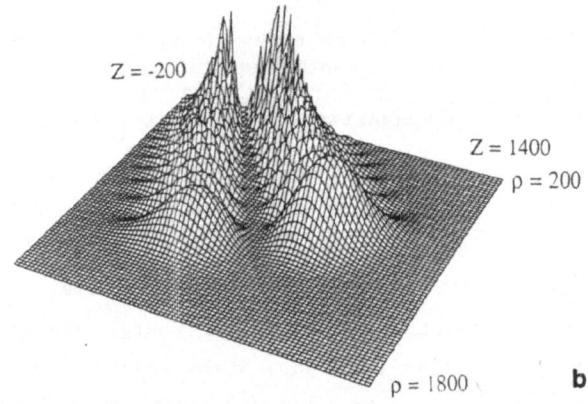

Z = -200

Z = 1400
ρ = 200

ρ = 1800   **b**

Fig. 2   Probability density as a function of ρ and z for states in the
n = 27 manifold.
(a) The state highest in energy.
(b) The second highest.

Braun and Solviev, using classical perturbation theory, have found
an adiabatic invariant, Λ, akin to that found in the purely magnetic case

$$\Lambda = 4A^2 - 5(A_z + \alpha)^2 + 5\alpha^2 \tag{2}$$

where $\underline{A}$ is the Runge–Lenz vector. We have in fact found that θ_L is given
by the maximum value of this invariant. This leads to the formula

$$\cos \theta_L = \alpha = \frac{3F}{5n^2\beta^2} . \tag{3}$$

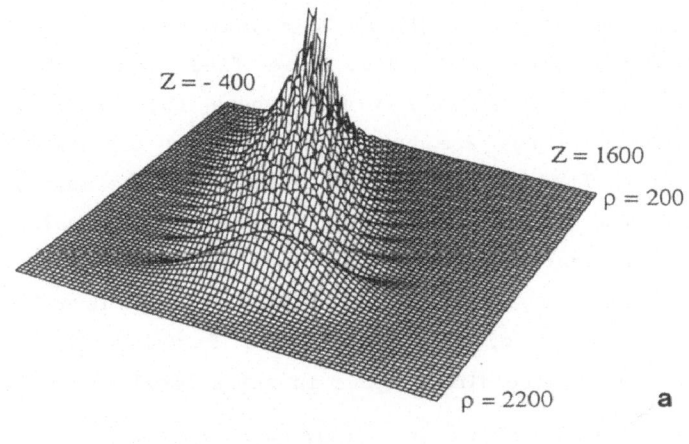

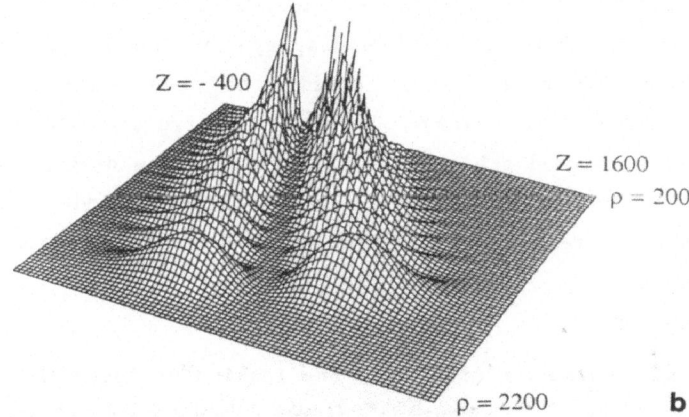

Fig. 3   Probability density as a function of ρ and z for states in the
         n = 32 manifold.
         (a) The state highest in energy.
         (b) The second highest.

This would give θ_L = 63.6° for the manifold featured in Fig. 2 and
θ_L = 71.6° for the n = 32 manifold in Fig. 3. These are the same angles
we get from the numerical calculations leading to Figs. 2 and 3. It may
be postulated that potential extrema lines identified by (3) would occur
when the problem is viewed in an appropriate coordinate system but these
potential lines would be strongly energy dependent.[4] In a given
n-manifold the system also exhibits a <u>local</u> separability, in that the
second and third highest lying lines have one and two nodes respectively
in a direction orthogonal to the line given by (3). Figures 2(b) and
3(b) illustrate this point for the second highest line in the n = 27 and
n = 32 manifolds.

On plotting the probability distributions of the <u>lowest</u> energy states of the manifolds in Fig. 1 we find that these are in fact localized along the negative z-axis where the invariant Λ is at a minimum. These states are hardly affected by the magnetic field and hence are Stark-like as mentioned above. Since the maximum and minimum of Λ correspond to the largest and smallest eigenvalues within the manifold it is clear that the eigenvectors of these eigenvalues will also indicate regions of localization.[5] However the invariant Λ has the additional advantage of giving a simple formula (3) identifying the specific regions of coordinate space in which localization occurs.

Finally, in accordance with (3), one can deduce that as n becomes larger, the angle of localization θ_L, approaches 90°. When F = 0 in (1), states localized about this line display the Quasi-Landau spacing of $3/2\, \hbar\omega$. Hence we predict this spacing to occur near the ionization threshold in this system also. In addition, frequencies characteristic of the shortest period orbits that pass through the origin should play a key role in the near threshold photoabsorption and photoionization spectru, as has already been observed for the purely magnetic case.[6]

ACKNOWLEDGEMENTS

S. B. was supported in part by the Italian Ministry of Education and the National Group of Structure of Matter. P. F. O'M. is supported by a grant from the Science and Engineering Research Council.

REFERENCES

1.   U. Fano and C. D. Lin, Atomic Physics 4 (Plenum, 1975) p. 47.

2.   C. W. Clark and K. T. Taylor, J. Phys. B 15, 1175 (1982).

3.   P. A. Braun and E. A. Soloviev, Sov. Phys., J.E.T.P. 59, 38 (1984).

4.   S. Watanabe and P. F. O'Mahony, J. Phys. B 20, 223 (1987); U. Fano, Proc. N. Bohr Centennial Conference, eds. J. Bang and J. De Boer (North-Holland, Amsterdam, 1985) p. 367.

5.   U. Fano, F. Robicheaux and A.R.P. Rau, Contributed paper XV I.C.P.E.A.C., Brighton, July 1987.

6.   M. A. Al-Laithy, P. F. O'Mahony and K. T. Taylor, J. Phys. B 19, L773 (1986); J. Main, G. Wiesbusch, A. Holle and K. H. Welge, Phys. Rev. Lett. 57, 2789 (1986).

SOME ASPECTS OF QUANTUM CHAOS

Marko Robnik

Max-Planck Institut für Kernphysik
6900 Heidelberg
West Germany

INTRODUCTION

This conference is vivid proof of the rapid propagation of ideas from the domain of theoretical quantum chaos into the field of atomic physics. The vast progress in the theoretical, computational and experimental works on the hydrogen atom in strong magnetic fields, as an example, demonstrates the transition that the general thinking in this field has undergone since the early beginnings.[1] The present status of this important problem, which may be considered a paradigm of quantum chaos in atomic physics is reported in many papers of these proceedings.[2]

Among the first attempts to study the statistical properties of atomic energy levels, such as the distribution of level spacings, was an early work by Rosenzweig and Porter[3] in 1960, which was still entirely in the spirit of random matrix theories, and went not sufficiently noticed although their way of thinking was pervading the well established statistical nuclear physics at the time. Their line of thought was continued in a new light when the quantum chaos began about two decades later.

The problem of quantum chaos may be stated as follows: are there any essential differences between the quantum systems which are classically integrable and those which are classically nonintegrable? The answer is positive at least as far as the stationary problem of energy spectra[4] and of the morphology of the eigenfunctions[5] is concerned. This lecture is not intended to give a review of quantum chaos, but rather to explain and to discuss two important aspects of quantum chaos. In the first place I discuss the role of antiunitary symmetries for the energy level statistics, and in the second part I discuss the morphology of eigenfunctions in the light of quantum integrability.[6]

PART I. ANTIUNITARY SYMMETRIES AND ENERGY LEVEL STATISTICS*

Introduction

One surprising development in the search for quantum chaos was the finding that the statistics of random matrix theories apply not only to the spectra of sufficiently complex systems with many coupled degrees of freedom, but also to relatively simple systems having a few degrees of freedom, provided that their classical dynamics is ergodic. A firm numerical evidence for this has been given for the first time by Bohigas, Giannoni and Schmit,[7] who studied the spectral fluctuations of the Sinair billiard which is a classically ergodic system. They have formulated the following conjecture supported by the numerical level spacings distribution and the spectral rigidity: spectral fluctuations of classically ergodic systems with time reversal symmetry obey the statistics of the Gaussian Orghogonal Ensemble (GOE) of random matrix theories, while in absence of time reversal invariance the statistics is that of the Gaussian Unitary Ensemble (GUE). (This was the stronger version of the conjecture, the weaker one requiring in addition the Kolmogorov entropy to be positive.)

The confidence in the conjecture put forward by Bohigas, Giannoni and Schmit has been strengthened by the numerical results of Seligman and Verbaarschot[26] and by Berry and Robnik,[5] who have demonstrated that in the absence of an antiunitary symmetry the spectral fluctuations are indeed consistent with the GUE statistics. There are also theoretical arguments in support of the conjecture, originally due to Pechukas,[15] and recently by Yukawa.[31] (The results of Pechukas and of Yukawa can be easily generalized to include the GUE case by allowing complex valued matrix elements v_{mn} (or f_{mn}) in Yukawa's notation.) For example, the exponent in the formula (9) of Yukawa's paper becomes 1 rather than 1/2, yielding the GUE level spacings distribution in the limit $N \to \infty$. The condition for the realness of $f_{mn}(t)$ is that there is an antiunitary A such that $AH(t) = H(t)A$ for all t, admitting a real representation of

---

*Part I is taken from M. Robnik, Lecture Notes in Physics 263, 120–130 (1986) (Proceedings of the Second International Conference on Quantum Chaos in Cuernavaca, Mexico, January 1986).

H(t), in which case the exponent is 1/2 and when N → ∞ one gets the GUE level spacings distribution.) Semiclassical theory can be a helpful tool in understanding the link between the ergodicity of classical dynamics and the statistics of spectral fluctuations of a given Hamiltonian H. The periodic orbit sum formula gives the semi-classicaly spectrum in terms of all classical periodic orbits.[1,2,9-13] It is a rather useless formula in calculating individual levels, but can be the decisive tool to express the statistics of spectral fluctuations in terms of the statistical laws governing the distribution of the classical closed orbits.[21] Berry[3] has calculated the semiclassical spectral rigidity which turns out to be the same as the leading term in the asymptotic expansion of the GOE or GUE spectral rigidity, depending on whether H has a discrete symmetry or not, no matter whether unitary or antiunitary!

We emphasize that we will deal only with systems whose classical dynamics is ergodic. For integrable systems one observes locally the Poisson statistics. For the generic systems in transition region between integrability and chaos one finds a smooth continuous transition of the spectral statistics.[4,14,19,25,28,29]

The Role of Antiunitary Symmetries

The significance of antiunitary symmetries is that if a Hamiltonian H has such a symmetry A, i.e. HA = AH, then there exists a large family of (complex) basis such that H has a real representation. (It is important that these so called A-adapted basis $\{\psi_\alpha\}$ defined by $A\psi_\alpha = \psi_\alpha$, are not related to the eigenbasis of the Hermitian operator H by an orthogonal transformation. So the realness is <u>not</u> a trivial one.) In fact an A-adapted basis can be constructed from arbitrary (complex) basis in Hilbert space if A satisfies certain conditions.

The time reversal T is a particular case of an antiunitary transformation. The general theory of such operators was given by Wigner,[30] Dyson,[8] and pp. 2-87 in Porter.[16] We review the essentials. In absence of spin, and using the coordinate representation, T can be represented simply by the complex conjugation operator K. This follows from the definition

$$TqT^{-1} = q, \quad TpT^{-1} = -p, \tag{1}$$

so that in the coordinate representation, $p = -\frac{i}{\hbar}\frac{\partial}{\partial q}$, we find T = K.

Now we discuss the antiunitary operators more generally. An operator A is <u>antilinear</u> if for any states ψ and ϕ and any complex numbers a and b

$$A(a\psi + b\phi) = a^*A\psi + b^*A\phi, \tag{2}$$

where * denotes complex conjugation. Since AK is a <u>linear</u> operator, it follows that every antilinear operator A can be written as a product of a linear operator L and the complex conjugation operator K, i.e.

$$A = LK. \tag{3}$$

Clearly $K^2 = 1$, and K is Hermitian, i.e. $K = K^+$. Thus $A^+ = KL^+$. For antilinear operators we cannot define the eigenvalues. An operator S is called a symmetry of H if it commutes with H, HS = SH, and if it preserves all transition probabilities, i.e. $|\langle\phi]\psi\rangle| = |\langle S\phi|S\psi\rangle|$ for any ϕ and ψ (S is an isometry). Thus, S preserves the physics of H.

Now there is a theorm by Wigner, discussed in Porter,[16] that every isometry S must satisfy $SS^+ = 1$ and either S is <u>linear</u> or <u>antilinear</u>, with no other possibility left. In the first case the symmetry S is termed <u>unitary</u> and in the latter case <u>antiunitary</u>. A unitary operator U preserves scalar products, i.e. $\langle\phi|\psi\rangle = \langle U\phi|U\psi\rangle$ for any ϕ and ψ. Any antiunitary A can be written, because of (3), as a product of some unitary operator U and K,

$$A = UK. \tag{4}$$

Clearly, $AA^+ = UKKU^+ = UU^+ = 1$. Therefore for any antiunitary A we have

$$\langle A\phi|A\psi\rangle = \langle UK\phi|UK\psi\rangle = \langle K\phi|K\psi\rangle = \langle\phi|\psi\rangle^* \tag{5}$$

for arbitrary ϕ and ψ.

Let H have an antiunitary symmetry A. In any basis $\{\psi_\alpha\}$ defined by

$$A\psi_\alpha = \psi_\alpha, \tag{6}$$

and called <u>A-adapted basis</u>, the matrix elements of H are real. The very simple argument is given in Porter,[16] pp. 2-87:

$$\langle\psi_\alpha|H\psi_\beta\rangle = \langle A\psi_\alpha|AH\psi_\beta\rangle^* = \langle A\psi_\alpha|HA\psi_\beta\rangle^* = \langle\psi_\alpha|H\psi_\beta\rangle^* \tag{7}$$

The question is whether an A-adapted basis exists for the given A. It seems that it does but I am able to prove it only for two classes of A.

The first class is defined by

$$A^n = 1, \text{ n integer.} \tag{8}$$

We generalize the arguments given by Porter.[16] Let $\{\phi_\alpha\}$ be any basis. Define the operator

$$R_+(A) = \sum_{\ell=1}^{n} A^\ell, \tag{9}$$

which has the property

$$AR_+ = R_+, \tag{10}$$

as a consequence of (8). Further define the first basis vector ψ_1 as follows

$$\psi_1 = R_+(A)a_1\phi_1, \quad A\psi_1 = \psi_1, \tag{11}$$

where a_1 is a complex number (chosen to achieve normalization). Next arrange that ϕ_2 and ψ_1 are orthogonal, i.e.

$$\langle\psi_1|\phi_2\rangle = 0. \tag{12}$$

Now define

$$\psi_2 = R_+(A)a_2\phi_2 \tag{13}$$

where a_2 is a complex constant. Obviously $A\psi_2 = \psi_2$, and even orthogonality is satisfied as can be easily verified:

$$\langle\psi_1|\psi_2\rangle = \langle\psi_1|\sum_{\ell=1}^{n} A^\ell a_2\phi_2\rangle = \langle\psi_1|a_2\phi_2\rangle = 0. \tag{14}$$

By appropriate choice of a_2 we can achieve $||\psi_2|| = 1$. By induction we can thus construct an orthogonal A-adapted basis $\{\psi_\alpha\}$ which is complete if $\{\phi_\alpha\}$ was complete.

The second class of antiunitary symmetries admitting an A-adapted basis consists of all A such that

$$A = UK, \quad U^n = 1, \quad n \text{ even.} \tag{15}$$

To construct the A–adapted basis we start from any __real__ orthonormal basis $\{\phi_\alpha\}$. (If ϕ_α are complex, then we decompose them in real and imaginary parts and consider their union which is made orthonormal.) We define the operators

$$R_+(U) = \sum_{\ell=1}^{n} U^\ell, \tag{16}$$

$$R_-(U) = \sum_{\ell=1}^{n} (-U)^\ell. \tag{17}$$

Clearly,

$$UR_+ = R_+, \quad UR_- = -R_-, \tag{18}$$

where the evenness of n is essential for the second equality. Now let us define

$$\phi_\alpha^+ = R_+(U)\phi_\alpha, \quad \phi_\alpha^- = R_-(U)\phi_\alpha, \tag{19}$$

so that ϕ_α^+ and ϕ_α^- are eigenvalues of U,

$$U\phi_\alpha^\pm = \pm\phi_\alpha^\pm. \tag{20}$$

They are orthogonal, or more generally, for any α and β

$$\langle \phi_\alpha^+ | \phi_\beta^- \rangle = \langle U\phi_\alpha^+ | U\phi_\beta^- \rangle = -\langle \phi_\alpha^+ | \phi_\beta^- \rangle = 0. \tag{21}$$

The complete A–adapted basis is constructed as follows

$$\psi_\alpha = \phi_\alpha^+ \pm i\phi_\alpha^-. \tag{22}$$

Indeed we verify

$$A\psi_\alpha = UK\psi_\alpha = U\phi_\alpha^+ \mp iU\phi_\alpha^- = \phi_\alpha^+ \pm i\phi_\alpha^- = \psi_\alpha. \tag{23}$$

The orthogonality

$$\langle \psi_\alpha | \psi_\beta \rangle = \begin{cases} 0 \text{ if } \alpha \neq \beta \\ \text{nonzero if } \alpha = \beta \end{cases} \tag{24}$$

is satisfied only for n = 2. This is easy to show. From (21) follows

$$\psi_\alpha | \psi_\beta \rangle = \langle \phi_\alpha^+ \pm i \ \phi_\alpha^- | \phi_\beta^+ \pm i \ \phi_\beta^- \rangle = \langle \phi_\alpha^+ | \phi_\beta^+ \rangle + \langle \phi_\alpha^- | \phi_\beta^- \rangle. \tag{25}$$

Now using (21) and $\langle U^\ell \phi_\alpha | U^\ell \phi_\beta \rangle = 0$ for any ℓ we get

$$\langle \phi_\alpha^+ | \phi_\beta^+ \rangle = \sum_{\substack{\ell,k\,=\,1 \\ \ell \neq k}}^{n} \langle U^\ell \phi_\alpha | U^k \phi_\beta \rangle = \sum_{\substack{\ell,k\,=\,1 \\ \ell < k}}^{n} \{ \langle U^{k-\ell} \phi_\alpha | \phi_\beta \rangle + \langle \phi_\alpha | U^{k-\ell} \phi_\beta \rangle \}. \tag{26}$$

On the other hand

$$\langle \phi_\alpha^- | \phi_\beta^- \rangle = \sum_{\substack{\ell,k\,=\,1 \\ \ell \neq k}}^{n} \langle (-U)^\ell \phi_\alpha | (-U)^k \phi_\beta \rangle = \sum_{\substack{\ell,k\,=\,1 \\ \ell < k}}^{n} \{ \langle (-U)^{k-\ell} \phi_\alpha | \phi_\beta \rangle + \langle \phi_\alpha | (-U)^{k-\ell} \phi_\beta \rangle \}. \tag{27}$$

Those terms in (26,27) with (k − ℓ) odd will cancel in (25) and the remainder is

$$\approx \langle \psi_\alpha | \psi_\beta \rangle = 2 \sum_{\substack{\ell,k\,=\,1 \\ \ell < k,\ (k-\ell)\,\text{even}}}^{n} \{ \langle U^{k-\ell} \phi_\alpha | \phi_\beta \rangle + \langle \phi_\alpha | U^{k-\ell} \phi_\beta \rangle \}, \tag{28}$$

which can be nonzero if $\alpha \neq \beta$. If n = 2, then k − ℓ < 1, and the orthogonality (24) is satisfied. By choosing appropriate real constants a_α we can also normalize ψ_α so that $||\psi_\alpha|| = 1$. For n > 2 the orthogonality (24) is not satisfied. However, because $\langle \psi_\alpha | \psi_\beta \rangle$ is real for any α and β according to (25) we can construct an orthonormal basis from the A-adapted basis by linear combinations of ψ_α with real coefficients. Real linear combinations do not spoil the A-adaptedness, and consequently we obtain a complete, orthonormal A-adapted basis denoted by ψ_α again.

Whenever an ensemble of Hamiltonians possesses an antiunitary symmetry A belonging to one of the two classes, we predict GOE spectral fluctuations. If A belongs to neither of the two classes, I am not able to construct an A-adapted basis, and the representation argument cannot be used.

But there are invariance arguments. The distribution of the Hamiltonians of the given ensemble must be invariant with respect to the canonical group of transformations that preserve the symmetry structure of the ensemble. A similarity transformation SHS^{-1} must preserve the Hermitian property of H,

$$(SHS^{-1})^+ = SHS^{-1}, \tag{29}$$

from which follows

$$SS^+H - HSS^+ = 0. \tag{30}$$

Since H is arbitrary SS^+ = constant and can be chosen unity. Therefore the most general canonical group is the group of unitary transformations.

If every member H of the ensemble has the antiunitary symmetry A = UK, U unitary, then we require that SAS^+ is also of the form U'K, U' unitary i.e.

$$SAS^+ = SUKS^+ = SUS^+K = U'K, \tag{31}$$

which can be satisfied only if

$$KS^+ = S^+K, \tag{32}$$

whence $S^+ = KS^+K = \tilde{S}$ = transpose of S, and thus S is real, i.e.

$$S = S^*. \tag{33}$$

The conclusion is that, if the ensemble has an antiunitary symmetry, S must be real and the canonical group of transformations preserving the symmetry structure is the group of real orthogonal transformations. On these grounds we arrive at <u>the most general classification scheme</u>: the spectral fluctuations of an ensemble with arbitrary antiunitary symmetry and ergodic classical dynamics are described by the GOE statistics, and by the GUE statistics if no antiunitary symmetry exists.

The role of antiunitary symmetries is quite different from that of unitary symmetries. The antiunitary symmetries for which eigenvalues cannot be defined dictate the type of spectral fluctuations. The unitary symmetries have eigenvalues and imply decomposition of energy level sequences into subsequences corresponding to different eigenvalues. They do not affect the type of spectral fluctuations for each subsequence.

Charged Particle in a Magnetic Field: Anticanonical Symmetry

Consider a charged particle with mass m and charge e moving in the $\underline{r}$ = (x,y) plane under the influence of a scalar potential $V(\underline{r})$ and a magnetic field $\underline{n}B(\underline{r})$ directed perpendicularly to the plane. The

Hamiltonian is

$$H(\underline{r},\underline{p}) = \frac{1}{2m} (\underline{p} - \frac{e}{c} \underline{A})^2 + V(\underline{r}),$$ (34)

where the vector potential $\underline{A}$ satisfies

$$\underline{\nabla} \times \underline{A}(\underline{r}) = \underline{n}B(\underline{r}).$$ (35)

It is clear that in discussing the symmetries of (34) we must remove the arbitrariness of gauge. The symmetry of the kinetic energy of (34) can be changed rather arbitrarily by the (physically insignificant) gradient transformations

$$\underline{A} \rightarrow \underline{A} + \underline{\nabla}f,$$ (36)

where $f = f(\underline{r})$ is any function. Therefore we decompose $\underline{A}$ into its curl and gradient parts and require vanishing of the gradient part. So

$$\underline{A} = \underline{\nabla} \times \underline{F},$$ (37)

implying the Coulomb gauge

$$\underline{\nabla} \cdot \underline{A} = 0.$$ (38)

In (38) we can choose

$$\underline{F} = \underline{n}F(\underline{r}),$$ (39)

and obtain the Poisson equation for F,

$$B = - (\frac{\partial^2 F}{\partial x^2} + \frac{\partial^2 F}{\partial y^2}) = -\Delta F.$$ (40)

For the uniform magnetic field we can choose $F = -\frac{1}{2} Bx^2$, so that $A_x = 0$, $A_y = Bx$. But there is further a continuum of other choices, namely e.g. $F = -\frac{1}{2} B(x^2\cos^2\alpha + y^2\sin^2\alpha)$, where α is arbitrary real angle. Thus to make (38) unique we must impose conditions on the solution F of the Poisson equation (40). We require that F is the "purely inhomogeneous" solution (thereby removing the freedom of adding to F any harmonic function F', (ΔF' = 0). Specifically our requirement is that

$$F(\underline{r}) = \frac{1}{2\pi} \int d^2\underline{r}'B(\underline{r}')\ln|\underline{r} - \underline{r}'|.$$ (41)

This is only the natural gauge that we can have. According to (41) F and with it the vector potential (37) vanish if the field B is identically zero. Also F($\underline{r}$) has the same symmetry as the field B($\underline{r}$). In fact the gauge (41) maximizes the symmetry of F relative to B. In the Hamiltonian (34) we shall henceforth assume that $\underline{A}$ is given by (37) and (41).

Normally time reversal T is defined as the reversal of velocities and not of the canonical momenta p. Nevertheless we define T as the reversal of canonical momenta. No difficulty arises because with the gauge (41) the system does or does not have the invariance with respect to the reversal of velocities if H is or is not invariant under T defined as p-reversal. Namely, if B = 0 then $\underline{A}$ = 0 and H has T and also has invariance with respect to the reversal of velocities. If B $\neq$ 0 then $\underline{A} \neq 0$ and H neither has T nor is it invariant with respect to the reversal of velocities.

Now we consider the following symmetry operations in the classical phase space.[22,24]

Inversion P: $(x,y,p_x,p_y) \rightarrow (-x,-y,-p_x,-p_y)$

Time reversal T: $(x,y,p_x,p_y) \rightarrow (x,y,-p_x,-p_y)$

Reflection S_x: $(x,y,p_x,p_y) \rightarrow (-x,y,-p_x,p_y)$.   (42)

These maps satisfy $P^2 = T^2 = S_x^2 = 1$. For any symmetry U we will say that H has symmetry U if HU = H. It follows from (34), (37) and (41) that

   (i)   H has P if both V and B have P;
   (ii) H has T if B = 0;
   (iii) H has S_x if B = 0 and V has S_x;

Remark: if both V and B have S_x this does <u>not</u> imply that H has S_x, but we have instead

   (iv) H has TS_x if B and V have S_x.

There is an important difference between (i, iii) and (ii, iv). The operations (i) and (iii) are canonical, i.e. they leave invariant both the Hamiltonian H and the Hamilton equations, for the Poisson brackets remain unchanged. On the other hand the transformations (ii) and (iv) leave H invariant, but they change the sign of the Hamilton equations, this change of sign being equivalent to the time inversion. We call them _anticanonical_ as they preserve the Poisson brackets apart from the change of sign. The canonical symmetries establish the equivalence between forward orbits, whereas the anticanonical symmetries relate a forward orbit to a backward orbit. Anticanonical symmetries correspond to the antiunitary symmetries in quantum mechanics.

When the periodic orbit sum formula is used to calculate the spectral statistics the degeneracy of classical orbits is very important. Degeneracy is either due to canonical or anticanonical symmetry and implies equivalence of pairs (or of n-tuples) of closed orbits, such that their periods and amplitudes are the same, and in such a case they must be combined coherently, but incoherently in absence of symmetries. At this level of semiclassical approximation the anticanonical symmetries affect the spectral statistics in much the same way as the canonical symmetries. The effect in the spectral rigidity $\Delta(L)$ is much less pronounced and the leading asymptotic term of $\Delta(L)$ is the same as for the GOE ensembles if the Hamiltonian has a discrete symmetry (no matter whether canonical or anticanonical), or the same as for the GUE ensemble if it does not have a symmetry. On the other hand, the distribution of level spacings P(S) is known to depend very strongly on the unitary or antiunitary nature of the symmetry: in the unitary case we get superposition of independent sequencies,[4] while the antiunitary symmetry merely changes the type of fluctuations from GUE to GOE. From this we conclude that the semiclassical theory based on the usage of the periodic orbit sum formula cannot explain the level repulsion laws (P(S) ~ const, S, S^2 as S → 0 for integrable, GOE and GUE, respectively) and even less so the global aspects of level spacings distributions.

There is a large class of systems corresponding to the case (iv), where H has neither T nor S_x, but $TS_x = S_xT$. For all such Hamiltonians we predict GOE spectral fluctuations if the classical dynamics of H is ergodic. Important examples are atoms in uniform magnetic fields. The hydrogen atom was studied by Robnik[17,18] and was shown to become chaotic above a certain critical energy. Other examples have been given by Berry and Robnik,[5] for the case of Aharonov-Bohm billiards, analyzed by a conformal method as developed in Robnik.[19] Due to the limited space of these proceedings I am not able to present these results but the interested reader is referred to Refs. 5 and 23.

Discussion and Conclusions

The conclusion of this lecture is that the discrete energy spectra of Hamiltonians that have ergodic classical dynamics fall into two classes as to their fluctuation properties. The spectral fluctuations of the first class are described by the statistics of the Gaussian Orthogonal Ensemble (GOE) of random matrix theories, and those of the second class are described by the statistics of the Gaussian Unitary Ensemble (GUE). The criterion necessary to classify a given Hamiltonian H is in terms of the antiunitary symmetries A. The representation arguments relying on the existence of an A-adapted basis guarantee that H belongs to the GOE class if there is an antiunitary A such that AH = HA and either $A^n = 1$ for an integer n, or A = UK, and the unitary operator U satisfieds $U^n = 1$ for some even n. The invariance arguments, however, suggest that the existence of an antiunitary symmetry A is sufficient for GOE statistics to apply.

Both assertions – the existence of exactly two universality classes (we ignore spin) – and the classification criterion – are firmly supported by the numerical evidence of Bohigas, Giannoni and Schmit,[7] Seligman and Verbaarschot,[26,27] Berry and Robnik,[5] and Robnik and Berry.[23] In the present lecture I have given theoretical arguments for the classification criterion. In particular, the invariance arguments suggest that an A-adapted basis exists for arbitrary antiunitary symmetry A, but this is yet to be proven and explicitly constructed. I have further pointed out that arguments in the papers by Pechukas[15] and Yukawa[31] lead to the conclusion that there are two universality classes. Let us rediscuss these arguments.

Following Yukawa[31] we study a one parameter family of Hamiltonians

$$H(t) = H_0 + tH_1 \qquad (43)$$

which may be general complex Hermitian. We use the notation: $x_n(t)$ for eigenvalues, $\phi_n(t)$ for eigenfunctions (generally complex), $p_n = \langle \phi_n(t) | H_1 | \phi_n \rangle$, and $f_{mn} = |x_m - x_n| \langle \phi_m(t) | H_1 | \phi_n(t) \rangle$, $m \neq n$, the matrix elements. The equations of motion follow from the (nondegenerate) perturbation theory.

$$\frac{dx_n}{dt} = p_n \qquad (44)$$

$$\frac{dp_n}{dt} = 2 \sum_{m \neq n} \frac{|f_{mn}|^2}{(x_n - x_m)^3} \qquad (45)$$

$$\frac{df_{mn}}{dt} = (x_n - x_m) \sum_{\ell \neq m,n} \frac{f_{m\ell} f_{\ell n}}{|x_n - x_\ell||x_\ell - x_m|} \left\{ \frac{1}{x_n - x_\ell} + \frac{1}{x_m - x_\ell} \right\} \qquad (46)$$

Clearly, x_n and p_n are always real, while f_{mn} are complex in general. It can be easily seen that the flow defined by (44-46) preserves volume in the phase space $\{x_n, p_n, f_{mn}\}$. Further, the "energy"

$$E = \frac{1}{2} \sum_n p_n^2 + \frac{1}{2} \sum_{n \neq m} \frac{|f_{mn}|^2}{(x_n - x_m)} \qquad (47)$$

is a constant of the motion. For a large enough t one can assume that the distribution in phase space is canonical with probability one, so

$$P\{x_n, p_n, f_{mn}\} = \text{const} \exp(-\text{const } E), \qquad (48)$$

and after integrating over p_n and $|f_{mn}|$ we get the probability density for the distribution of the eigenvalues $\{x_n\}$,

$$P\{x_n\} = \text{const} \prod_{m > n} |x_m - x_n|^\beta, \qquad (49)$$

where $\beta = 2$ in the general case of complex f_{mn}, and $\beta = 1$ if f_{mn} are real.

Now, when are f_{mn} real? Suppose there is an antiunitary symmetry A such that $H_0 A = A H_0$ and

$$H_1 A = A H_1 \qquad (50)$$

and suppose that $\phi_n(0)$ is an A-adapted basis, i.e.

$$A\phi_n(0) = \phi_n(0), \qquad (51)$$

for each n. Then f_{mn} are clearly real and their realness is preserved by the flow, for the perturbation theory implies:

$$\frac{d\phi_n}{dt} = \sum_{\ell \neq n} \frac{\langle \phi_\ell | H_1 | \phi_n \rangle}{x_n - x_\ell} \phi_\ell , \qquad (52)$$

which was used to obtain (47). As has been argued by Yukawa, $\beta = 1$ leads to the GOE level spacings distribution, while $\beta = 2$ yields GUE level spacings distribution. The agreement of these results with our previous conclusions will be perfect once we can show that an A-adapted basis exists for any A.

REFERENCES

1. R. Balian and C. Bloch, Ann. Phys. _69_, 76 (1972).

2. R. Balian and C. Bloch, Ann. Phys. _85_, 514 (1974).

3. M. V. Berry, Proc. Roy. Soc. Lond. _A400_, 229 (1985).

4. M. V. Berry and M. Robnik, J. Phys. A _17_, 2413 (1984).

5. M. V. Berry and M. Robnik, J. Phys. A _19_, (1986) in press.

6. O. Bohigas and M. J. Giannoni, Lecture Notes in Physics _209_, 1 (1984).

7. O. Bohigas, M. J. Giannoni and C. Schmit, Phys. Rev. Lett. _52_, 1 (1984).

8. F. J. Dyson, J. Math. Phys. _3_, 1191 (1962).

9. M. C. Gutzwiller, J. Math. Phys. _8_, 1979 (1967).

10. M. C. Gutzwiller, J. Math. Phys. _10_, 1004 (1969).

11. M. C. Gutzwiller, J. Math. Phys. _11_, 1791 (1970).

12. M. C. Gutzwiller, J. Math. Phys. _12_, 343 (1971).

13. M. C. Gutzwiller, in Path Integrals and Their Applications in Quantum Statistical and Solid State Physics ed. G. J. Papadopoulos and J. T. Devreese, (New York, Plenum, 1978) pp. 163-200.

14. H. D. Meyer, E. Haller, H. Köppel and L. S. Cederbaum, J. Phys. A _17_ L831 (1984).

15. P. Pechukas, Phys. Rev. Lett. _51_, 943-946 (1983).

16. C. E. Porter, Statistical Theories of Spectra: Fluctuations (New York, Academic Press, 1965).

17. M. Robnik, J. Phys. A _14_, 3195 (1981).

18. M. Robnik, J. Physique C2 _43_, 45 (1982).

19. M. Robnik, J. Phys. A _17_, 1049 (1984).

20. M. Robnik, J. Phys. A _17_, 109 (1984).

21. M. Robnik, in Proceedings of the NATO Advanced Study Institute Photophysics and Photochemistry in the Vacuum UV, eds. S. P. McGlynn, G. L. Findley and R. H. Huebner (Dordrecht, Reidel, 1985) pp. 579-629.

22. M. Robnik, Regular and Chaotic Billiard Dynamics in Magnetic Fields, in Nonlinear Phenomena and Chaos, ed. S. Sarkar (Bristol, Adam Hilger, 1986).

23. M. Robnik and M. V. Berry, J. Phys. A _19_, (1986) in press.

24. M. Robnik and M. V. Berry, J. Phys. A _18_, 1361 (1985).

25. M. Robnik (1986) to be published.

26. T. H. Seligman and J.J.M. Verbaarschot, Phys. Lett. A _108_, 183 (1985).

27. T. H. Seligman and J.J.M. Verbaarschot, J. Phys. A _18_, 2227 (1985).

28. T. H. Seligman, J.J.M. Verbaarschot and M. R. Zirnbauer, Phys. Rev. Lett. 53, 215 (1984).

29. T. H. Seligman, J.J.M. Verbaarschot and M. R. Zirnbauer, J. Phys. A 18, 2751 (1985).

30. E. Wigner, Group Theory and Its Applications to the Theory of Atomic Spectra (New York, Academic Press, 1959).

31. T. Yukawa, Phys. Rev. Lett. 54, 1883 (1985).

PART II. QUANTUM INTEGRABILITY AND LOCALIZATION OF EIGENFUNCTIONS

Introduction

In this paper I shall discuss the relevance of the quantum integrability[1] for the eigenfunctions of quantum systems, in particular of the classically chaotic systems.[2] By the localization of eigenfunctions in the present context we mean simply concentration of eigenfunctions in regions of enhanced probability density.

Given a Hamilton operator $\hat{H}$, what can we tell about the properties of its eigenfunctions? In general we are helpless and the fact that the Schrödinger equation is in principle solvable is no remedy. On the other hand we would like to have an overview upon the various generic types of eigenfunctions that can arise in order to be able to differentiate between the special and the universal aspects. A similar situation occurred in classical dynamics: the fact that the Hamilton-Jacobi equation is in principle solvable was useless in studying the asymptotic, ergodic properties of classical motion. New techniques and approaches supported (and sometimes guided) by the numerical experiments are necessary to uncover the variety of regular and stochastic motion.

I begin by summarizing the semiclassical picture. There are two cases.

(I) Integrability: the integrals of motion have a finite number of leaves, and consequently the motion is quasiperiodic on the (quantized) invariant torus. When projected onto the configuration space this yields a finite number of possible velocity directions $\hbar \vec{k}_j/m$ (but with the same speed $\hbar |\vec{k}|/m$) at every point $\vec{r}$ inside the torus projection regime. Thus the semiclassical eigenfunction is locally a superposition of a finite number of plane waves

$$\psi(\vec{r}) = \sum_{j=1}^{n} A_j e^{i\vec{k}_j \vec{r}} \tag{1}$$

where $|\vec{k}_j| = |\vec{k}|$ and $\vec{k}$ is determined by $H(\vec{r}, \vec{p} = \hbar\vec{k}) = E$, where $H(\vec{r}, \vec{p})$ is the Hamiltonian and E the energy. The amplitudes are determined by the projection and are nonzero only inside the region of the projection of the torus. This well-known association of invariant tori and of eigenfunctions is originally due to Maslov.[3]

(E) <u>Ergodicity</u>: the nonisolating "integrals" of motion have <u>infinite</u> number of leaves. When projected onto the configuration space this yields an infinite number of possible velocity directions $\hbar\vec{k}_j/m$ (but with the same speed $\hbar|\vec{k}|/m$) at every point inside the energetically allowed region $H(\vec{r}, \vec{p}) = E$. Thus the semiclassical eigenfunction is locally a superposition of an infinite number of plane waves,

$$\psi(\vec{r}) = \sum_{j=1}^{\infty} A_j e^{i\vec{k}_j \vec{r}} \tag{2}$$

Berry[4] has conjectured that ergodicity implies isotropic distribution of $\vec{k}_j$'s on the sphere $|\vec{k}_j|^2 - |\vec{k}|^2$ - constant, and further, that the stochasticity implies loss of memory and therefore random phases implicitly entailed in the amplitudes A_j. Such an assumption would imply that (2) is a Gaussian random function.[5]

The preliminary investigations of the eigenfunctions of classically ergodic systems seemed to confirm this expectation. However, a more careful study by Heller[6] has shown substantial deviations from the Gaussian randomness in the eigenfunctions of the stadium (which is a classically ergodic system). The regions of enhanced probability density (scars), coincide with the classical periodic orbits. According to his arguments the enhancement ratio is

$$\frac{I_s}{I_o} = 1/\lambda\tau, \tag{3}$$

where I_s is the probabilty density on the scar (along an orbit of period τ, having the instability exponent λ), while I_o is the mean probability density according to the hypothesis of Gaussian randomness. As (3) is independent of $\hbar$, the scars never disappear, but can at best become narrower as $\hbar \to 0$. At the same time, however, new scars might emerge as $\hbar \to 0$, so that the Gaussian randomness might possibly be never achieved by at least some of the states when $\hbar \to 0$. In a very recent work Heller, O'Connor and Gehlen[7] show numerical eigenfunctions for states as high as 8390-th eigenstate of the stadium and, still, with increasing energy E there are more and more states exhibiting a scar dominated probability density. It seems indeed as if some states never reach the Gaussian randomness in the semiclassical limit $\hbar \to 0$, or equivalently, $E \to \infty$.

The purpose of this paper is to explain the importance of quantum integrability and of its limiting classical behavior for the understanding of the two types of eigenfunctions. In particular I will give arguments for the breakdown of Berry's random phase hypothesis in (2): at any finite nonzero $\hbar$ the Hamiltonians are quantum integrable and this implies scar structure of the wave functions.

Quantum Integrability

Consider Hamilton systems with purely discrete spectrum. They correspond to the classical Hamiltonians with bounded motion. The question we are addressing concerns the integrability of such quantum systems and its preservation under small perturbations. As such, the problem is to formulate the quantum analogy of the Kolmogorov-Arnold-Moser (KAM) theorem. We will see that the result for the quantum system is stronger[1]: almost all quantum Hamiltonians are (quantum) integrable, but the integrals of motion (the operators representing the observables) generically do not have the classical limit as $\hbar \to 0$.

In analogy with classical mechanics we <u>define</u> a quantum Hamiltonian $H(q, p)$ (of N freedoms) <u>integrable</u> if there exist N operators A_n, $1 \leqslant n \leqslant N$, all of them being functions of the coordinates q_1 and momenta p_k,

$$A_n = A_n(q_1, p_k), \tag{4}$$

such that all commutators vanish pairwise, i.e.

$$[A_1, A_k] = A_1 A_k - A_k A_1 = 0, \quad 1 \leqslant 1, k \leqslant N, \tag{5}$$

where one of A_n's is H, say $A_1 = H$.

One obvious approach is to use the Wigner-Weyl formalism.[8] There is one-to-one correspondence between the operators acting on the Hilbert space and their Weyl symbols. The commutators of operators are mapped onto the Moyal brackets of their Weyl correspondents. Thus if a (q,p) and $b(q,p)$ are the Weyl symbols of the two operators A and B, then their Moyal bracket

$$[a,b]_M = \frac{2}{\hbar} \sin \frac{\hbar \Lambda}{2} a(q,p)b(q,b), \tag{6}$$

is the Weyl correspondent of $-\frac{i}{\hbar} [A, B]$, where

$$\Lambda = \frac{\partial^{(a)}}{\partial q} \frac{\partial^{(b)}}{\partial p} - \frac{\partial^{(a)}}{\partial p} \frac{\partial^{(b)}}{\partial q} \tag{7}$$

and the indices (a) and (b) indicate the action of the partial derivatives. The task of deciding whether $H(q,p)$ is integrable is now in finding N functions $f_n(q,p)$, where $f_1(q,p) = H(q,p)$, such that they mutually commute in the sense of Moyal.

Assertion: every quantum system with purely discrete spectrum is quantum integrable in the sense of the above definition.

The reason is very simple. Let $|\psi_n\rangle$ denote the n-th eigenfunction. The projection operators $P_n = |\psi_n\rangle\langle\psi_n|$ commute pairwise, and with the Hamiltonian. In fact,

$$P_n P_m = P_m P_n = 0, \quad \forall \, n,m \tag{8}$$

and

$$[P_n, H] = 0. \tag{9}$$

Now, the Wigner function

$$\rho_n = \frac{1}{(\pi\hbar)^N} \int \psi_n(q + x)\psi_n^*(q - x)e^{-\frac{2ix\cdot p}{\hbar}} d^N x \tag{10}$$

is the Weyl transform of P_n and therefore the Moyal brackets vanish pairwise

$$[\rho_n, \rho_m]_M = 0, \quad [H, \rho_n]_M = 0. \tag{11}$$

Hence we have <u>infinitely</u> many functions ρ_n, $n = 1, 2, \ldots$, defined in the Wigner phase space, which Moyal-commute pairwise, and with the Hamiltonian H. Q.E.D.

At first glance the quantum integrability is a mere triviality. But this is not so. The reason is that $\rho_n(q,p)$'s are not quite right functions. One way of seeing this is that P_n's are singular (as functions of q and p) reflecting also the fact that the Wigner functions ρ_n become singular as $\hbar \to 0$. To see this recall the orthogonality relation:

$$\text{Tr}(P_n P_m) = \hbar^N \int d^N q\, d^N p\, \rho_n(q,p)\rho_m(q,p) = \delta_{nm}. \tag{12}$$

When $\hbar \to 0$ we have the semiclassical limit

$$\rho_n \to h^{-N} \rho_n^2, \tag{13}$$

showing that ρ_n's become positive definite in this limit. The orthogonality relation (12) then implies that they have (almost) disjoint supports, so they are not defined everywhere in the quantum phase space, but only on a piece of volume h^N. In fact, the condensation takes place near the energy shell $E_n = H(q,p)$. We thus clearly see that the Wigner function ρ_n corresponds to a <u>section of an integral of motion</u> in the classical case rather than to a global integral of motion. To remove this weakness of the definition we impose stronger conditions on the quantum integrals of motion (4): we require that they (or their Weyl correspondents) are globally defined for any $\hbar$, and that they are nonsingular.

Even with these stronger requirements the Assertion remains almost intact, as I have argued using heuristic arguments in a recent paper (Ref. 1). The precise statement is: <u>almost</u> every quantum Hamiltonian (with purely discrete spectrum) is quantum integrable, but the classical limit of the integrals of motion (the operators representing the observables) generically does <u>not</u> exist.

The argument rests upon the perturbation theory: the perturbation series may converge for <u>almost</u> every admissible perturbation. (A perturbation is admissible if it preserves the discreteness of the energy spectrum.) Without discussing the details which can be found in Ref. 1 let me explain the gist of the argument.

Let H_0 be an integrable, and $H = H_0 + \varepsilon H_1$ the perturbed Hamiltonian whose integrals of motion we want to construct by the perturbation expansion starting off H_0. The classical perturbation series diverges, as is well known, due to the existence of the small denominators of the form (for two freedoms, to simplify)

$$F = m\omega_1 + n\omega_2, \quad m,n \text{ integers} \tag{14}$$

As $|m|$, $|n| \to \infty$, F can become arbitrarily small implying the divergence of the series.

In the quantum perturbation series an important change arises: the small denominators obtain a "quantum correction," i.e. a term that is $\hbar$ dependent and is nonzero as a consequence of both the quantum commutation relations and the <u>nonlinearity</u> of H_0. More importantly, F becomes a nonlinear function of n,m,

$$F = m\omega_1 + n\omega_2 + f(\hbar,m,n).$$ (15)

It is this new structure of the "quantum small denominators" which implies convergence of the quantum perturbation series in an appropriate formulation (superconvergence). The rigorous proof is still lacking, but heuristically we have every reason to think that the quantum perturbation series converges for <u>almost</u> every perturbation H_1.

In the classical limit when $\hbar \to 0$ the quantum correction f in (15) vanishes and the series diverges. This implies that the quantum integrals of motion, or their Weyl transforms, <u>generically</u> do not have a classical limit. Classical ergodic systems belong to this class. The nongeneric class consists of classically integrable systems for which the limit of quantum integrals as $\hbar \to 0$ is given by the classical integrals of motion.

The Significance of Quantum Integrability

In the classical mechanics there is the so-called Jeans theorem: a stationary distribution function in the phase space can be any function of the integrals of motion. The reason is that if f Poisson-commutes with $a_1, \ldots, a_N$, then its Poisson commutator with any function $g(a_1, \ldots, a_N)$ of a_j's vanishes as well.

An analogous result does <u>not</u> apply in quantum mechanics: if f Moyal commutes with $a_1, \ldots, a_N$, then in general it does <u>not</u> commute (in the sense of Moyal) with some function of a_j's. I give a simple <u>counter-example</u>[9]: let $f = qp^2$ and $g = f^2 = q^2p^4$. One can calculate in a straightforward manner that the Moyal bracket $[f,g]_M = 2\hbar^2 p^3 \neq 0$, but of course $[f,f]_M = 0$. Q.E.D.

Therefore in quantum mechanics (in Wigner space) there is no reason to expect that the Wigner distribution function $\rho_n(q,p)$ corresponding to the n-th eigenstate $|\psi_n\rangle$ should be a function of the quantum integrals of motion, i.e. of their Weyl symbols.

The famous example[10] of the one-dimensional harmonic oscillator, where Wigner functions are accidentally functions of the total energy $H(q,p)$, is an exception.

One <u>generic example</u>: one dimensional infinite potential well with the potential $V(q) = 0$ inside the interval $0 < q < a$, and $V(q) = \infty$ outside $[0,a]$. The Hamiltonian is simply

$$H(q,p) = \frac{p^2}{2m} + V(q). \tag{16}$$

The eigenfunctions are

$$\psi_n = \sqrt{\frac{2}{a}} \sin \frac{n\pi q}{a} \; , \; n = 1, 2, 3, \ldots \tag{17}$$

and for the Wigner functions we find

$$\rho_n = \frac{n}{qa^2 p} \left\{ \frac{\sin 2M \left(\frac{p}{\hbar} - \frac{n\pi}{a} \right)}{\frac{p}{\hbar} - \frac{n\pi}{a}} - \frac{\sin 2M \left(\frac{p}{\hbar} + \frac{n\pi}{a} \right)}{\frac{p}{\hbar} + \frac{n\pi}{a}} \right\} \tag{18}$$

where

$$M = M(q) = \left\{ \begin{array}{l} q, \text{ if } 0 < q < a/2 \\ a - q, \text{ if } a/2 < q < a \end{array} \right\} \tag{19}$$

The Wigner functions (18) obviously are not functions of the integral of motion (16), but are something more complicated, and depend on q and p. They are concentrated, however, near the contours of constant H, on which they condense when $\hbar \to 0$, namely

$$\rho_n \xrightarrow[\hbar \to 0]{} \frac{n\pi\hbar}{2a^2 p} \left\{ \delta\left(p - \frac{n\pi\hbar}{a} \right) - \delta\left(p + \frac{n\pi\hbar}{a} \right) \right\}. \tag{20}$$

In view of all these generalities and peculiarities, what is the significance of the quantum integrability for the localization of the eigenfunctions? By the localization in this context we mean simply concentration of eigenfunctions in regions of enhanced probability density.

The answer is simple and straightforward. For a given Hamiltonian H we have N integrals of motion with their Weyl symbols $a_1, \ldots, a_N$, which by the definition of quantum integrability Moyal-commute, pairwise. Then to the lowest order in $\hbar$-expansion of the Moyal brackets (5) we have the Poisson brackets

$$[a_1, a_k]_M = \{a_1(\hbar), a_k(\hbar)\}_P + O(\hbar^2) \; O \; (a_1 a_k) \tag{21}$$

where in (21) we have indicated explicitly that a_j's are in general functions of $\hbar$, i.e. $a_j = a_j(\hbar)$. To the lowest approximation, when $\hbar$ is small, the constants of motion a_j do approximately Poisson-commute, and regarding them as classical integrals of motion, this implies the

Lagrangian property and existence of <u>approximate</u> invariant tori. We can now use these approximate invariant tori to construct approximate action-angle variables I, θ. Now, recall that within the semiclassical approximation the quantization commutes with the canonical transformation.[9] Therefore, we can construct the eigenfunctions as plane waves on approximate tori,

$$\psi = Ae^{iI\theta/\hbar} \tag{22}$$

where in evaluating the amplitudes A we must of course take care of caustics. The Wigner functions corresponding to (22) are then associated with these approximate tori. In fact it is easy to show[11]

$$\rho_n \xrightarrow[\hbar \to 0]{} (2\pi)^{-N} \delta\left(I(q, p; \hbar) - I_n\right) \tag{23}$$

where I_n is quantized action, and the approximate torus $I(q, p; \hbar)$ is in general a function of $\hbar$.

We can now distinguish between the two classes explained in the introduction.

(I) The system is classically integrable, and therefore each $a_j(\hbar)$ has a limit as $\hbar \to 0$. The Wigner functions (23) condense on them as $\hbar \to 0$, because $I(q, p; \hbar) \to I_{classical}(q, p)$.

(E) The system is classically nonintegrable, and to be specific we assume that it is classically ergodic. The constants of the motion $a_j(\hbar)$ have <u>no</u> classical limit, but for every finite value of $\hbar$ they do define an approximate torus $I(q, p; \hbar)$ near which the Wigner function (23) is concentrated. As $\hbar \to 0$ the geometry of this torus is continuously changing, because no limit exists. It is expected on natural grounds but yet to be proved, that approximate tori in such a case exist around the classically periodic orbits, since this is the only nontrivial invariant component in phase space, although of measure zero.

It follows from these considerations that quantum integrability sheds new light on the important discovery of scars by Heller,[6] and might supplement his theory of scars. One specific prediction can be made on the basis of my arguments: if we fix the (quantum number of an) eigenstate and change $\hbar$ continuously, then I predict that in a classically ergodic <u>nonscaling</u> system we shall see continuously changing but always present scar structure. It remains to be clarified whether a sufficiently complex and uniform scar structure can simulate a Gaussian random function. A final remark concerns the special case of linear Hamiltonians and their perturbations. For reasons explained in Ref. 1 they might be genuinely quantum nonintegrable. Indeed, Eckhardt[12] finds random ϕ-correlated eigenfunctions in one such case. E. Heller and E. Stechel[13] have another example. Further theoretical and numerical work is necessary for a more detailed understanding of these and similar phenomena.

Note Added in Proof

In the discussion of this talk the question of the relevance of the present work to the problem of a hydrogen atom in a strong magnetic field has been raised, in particular, "as ħ is always finite in real systems." The answer is that the question is slightly wrong, for firstly I have mentioned the problem of a hydrogen atom in a strong magnetic field only as a motivating example (the paradigm of quantum chaos) for more general studies; and secondly the important point of Part II is to see that the quantum integrability is relevant in classically ergodic systems for _any_ value of ħ, and not only in the semiclassical limit ħ → 0. The really important part of this answer to a slightly wrong questions is as follows. If the scars exist in a quantum system, as predicted by the present arguments, and if their structure is correlated over a sufficiently large number of states (i.e. if their geometry does _not_ change too wildly from state to state), then, firstly, their existence implies clustering of energy levels, and, secondly, for an appropriate polarization of the photons the dipole matrix elements are enhanced, so that the modulations show up in photoabsorption cross sections, for which the quasi-Landau resonances are an analogy.

ACKNOWLEDGEMENT

I thank Eric J. Heller for the preprint concerning his recent theoretical and computational work[7] on scars.

REFERENCES

1. M. Robnik, J. Phys. A _19_, L841-847 (1986).
2. E. J. Heller, _Lecture Notes in Physics_ 263, 162-181 (1986).
3. See e.g. review by M. V. Berry, in _Chaotic Behavior of Deterministic Systems_, eds. G. Iooss, R. H. Helleman and R. Stora (North Holland, Amsterdam, 1983).
4. M. V. Berry, J. Phys. A _12_, 2083-2091 (1977).
5. M. S. Longuet-Higgins, Phil. Trans. Roy. Soc. A _249_, 321-387 (1986).

6. E. J. Heller, Phys. Rev. lett. $\underline{53}$, 1515-1518 (1984).

7. E. J. Heller, P. O'Connor and J. Gehlen, Preprint, University of Washington, Seattle (June 1987).

8. One useful review can be found in: S. R. de Groot and L. G. Suttorp, Foundations of Electrodynamics (North Holland, Amsterdam, 1972) pp. 317-364. See also E. J. Heller, J. Chem. Phys. $\underline{65}$, 1289-1298 (1976).

9. M. Robnik, J. Phys. A $\underline{17}$, 109-130 (1984).

10. T. Takabayasi, Prog. Theor. Phys. $\underline{11}$, 341-373 (1954).

11. M. V. Berry, Phil. Trans. Roy. Soc. A $\underline{287}$, 237-271 (1977).

12. B. Eckhardt, J. Phys. A $\underline{19}$, 1823-1831 (1986).

13. E. J. Heller and E. Stechel, private communication (October 1986).

QUANTUM CHAOS: DESTRUCTION AND SCARS OF SYMMETRIES

Dominique Delande and Jean-Claude Gay

Laboratoire de Spectroscopie Hertzienne de l'ENS
Tour 12, Etage 1, 4 Place Jussieu
75252 Paris Cedex 05, France

ABSTRACT

The classical dynamics of the hydrogen atom in a magnetic field evolves from regular at low field to chaotic in the strong-field-mixing regime. In the low-field regime, there exists approximate dynamical symmetries. We study the destruction of these symmetries when the magnetic field is increased, from both the classical and quantum point of views. In the classically chaotic region, remnants of the symmetries scar the quantum dynamics. Furthermore, a part of the quantum states presents a phase-space localization which leads to large deviations from ergodicity.

I. THE HYDROGEN ATOM IN A MAGNETIC FIELD. EQUIVALENCE WITH A SYSTEM OF COUPLED OSCILLATORS

The hydrogen atom in a strong magnetic field is one of the simplest classically chaotic systems. The classical and quantum dynamics (§ III-IV) can be studied very accurately, and a lot of nice experimental results have been obtained in the last twenty years.[1-5] Conjectures on quantum chaos previously established on model systems (billiards,[6] coupled oscillators,[7,8] anisotropic Kepler system[9] ...) or observed on nuclear energy levels[6] can be tested on a "real" completely calculable system.

The Hamiltonian of the hydrogen atom in a magnetic field is (in atomic units):

$$H = \frac{\vec{p}^2}{2} - \frac{1}{r} + \frac{\gamma}{2} L_z + \frac{\gamma^2}{8} (x^2 + y^2) \tag{1}$$

where $\gamma = B/B_c$ is the magnetic field (along z-axis) measured in atomic units ($B_c = 2.35 \times 10^5$ T). The only constants of the motion are L_z (projection of the angular momentum $\vec{L}$ on z-axis) and parity.

In the following, we consider the $L_z = M = 0$ states. Similar conclusions can be obtained for the other series. The paramagnetic term $\gamma L_z/2$ is constant over a given M series and generates a global shift of the energy levels.

The hydrogen atom in a magnetic field is a two-dimensional Hamiltonian system. It belongs to the most simple class of systems which may exhibit a chaotic behavior.[10]

Calculation of the eigenvalues and eigenstates of the Hamiltonian (1) is not straightforward. In contrast with the hydrogenic Stark effect, there is no dynamical symmetry. A careful analysis of the symmetry properties is likely to increase the efficiency of numerical calculations.[11] The diamagnetic interaction $\gamma^2(x^2 + y^2)/8$ actually couples almost any hydrogenic states with any other one. A structure describing the hydrogenic spectrum in its whole would be convenient. It is the dynamical group[12-16] SO(2,2) (subgroup with fixed value of M of the full dynamical group SO(4,2)).

The SO(2,2) dynamical group is easily built using the equivalence of the hydrogen atom with a pair of harmonic oscillators.[11-12] The semi-parabolic coordinates are defined through:

$$\left\{ \begin{array}{l} \mu = \sqrt{r + z} \\ \\ \nu = \sqrt{r - z} \end{array} \right.$$

In this system of coordinates, the Schrödinger equation is (E is the energy):

$$\left\{ H(\mu) + H(\nu) + \frac{\gamma^2}{8} \mu^2 \nu^2 (\mu^2 + \nu^2) \right\} |\psi\rangle = 4|\psi\rangle \qquad (2)$$

with

$$H(\mu) = -\frac{\partial^2}{\partial \mu^2} - \frac{1}{\mu} \frac{\partial}{\partial u} + \frac{M^2}{\mu^2} - 2E\mu^2$$

$H(\mu)$ is the radial part of the Hamiltonian of a 2-dimensional harmonic oscillator with frequency $\sqrt{-2E}$ and angular momentum M. Equation (2) describes two harmonic oscillators coupled by the interaction $\gamma^2 \mu^2 \nu^2 (\mu^2 + \nu^2)/8$.

The dynamical group for the radial motion of the two-dimensional harmonic oscillator is SO(2,1), the hermitian generators of which are[12-13]:

$$S_x^{(\alpha)} = \frac{\alpha}{4} \left(\frac{\partial^2}{\partial \mu^2} + \frac{1}{\mu} \frac{\partial}{\partial \mu} - \frac{M^2}{\mu^2} \right) + \frac{1}{4\alpha} \mu^2$$

$$S_y^{(\alpha)} = \frac{i}{2} \left(1 + \mu \frac{\partial}{\partial \mu} \right) \qquad (3)$$

$$S_z^{(\alpha)} = -\frac{\alpha}{4} \left(\frac{\partial^2}{\partial \mu^2} + \frac{1}{\mu} \frac{\partial}{\partial \mu} - \frac{M^2}{\mu^2} \right) + \frac{1}{4\alpha} \mu^2$$

with α a positive adjustable parameter.

These generators satisfy the commutation relations of a SO(2,1) Lie algebra. With the convenient choice $\alpha = 1/\sqrt{-2E}$, $S_z^{(\alpha)}$ is proportional to the Hamiltonian $H(\mu)$. The eigenvalues of $S_z^{(\alpha)}$ are $n_\mu + \frac{|M| + 1}{2}$, labelled with the non-negative integer n_μ.

The three generators of the SO(2,1) group act like the three components of the angular momentum for the usual SO(3) rotation group. $S_\pm^{(\alpha)} = S_x^{(\alpha)} \pm i \, S_y^{(\alpha)}$ are ladder operators, rising n_μ by ±1. The matrix elements are known from group representation theory and involve simple algebraic expressions.[18]

The three $\vec{S}^{(\alpha)}$ generators play for the two-dimensional radial harmonic oscillator, the role played by the creation and annihilation operators a and a^+ for the one-dimensional oscillator.

The dynamical group of the hydrogen atom (or of the equivalent oscillator problem) is the direct product SO(2,1) ⊗ SO(2,1) = SO(2,2). The six generators are the three components of $\vec{S}^{(\alpha)}$ and $\vec{T}^{(\alpha)}$ (defined as $\vec{S}^{(\alpha)}$ but with the change ($\mu \rightarrow \nu$). Using Eqs. (2) and (3), the Schrödinger equation is written under a pure algebraic form as a function of the generators:

$$\{ S_z^{(\alpha)} + T_z^{(\alpha)} - \alpha +$$

$$\frac{\gamma^2 \alpha^4}{2} (S_z^{(\alpha)} + S_x^{(\alpha)})(T_z^{(\alpha)} + T_x^{(\alpha)})(S_z^{(\alpha)} + S_x^{(\alpha)} + T_z^{(\alpha)} + T_x^{(\alpha)}) \} | \psi \rangle = 0 \quad (4)$$

with $\alpha = (-2E)^{-1/2}$.

Although this equation formally establishes the equivalence of the hydrogen atom in a magnetic field with a pair of coupled harmonic oscillators, it traduces a more general property, namely that the intrinsic group structure of the Coulomb problem (at constant L_z) is SO(2,2).

From Eq. (4), the zero-field eigenstates are those of $(S_z^{(\alpha)} + T_z^{(\alpha)})$ with eigenvalue n, the principal quantum number. The degeneracy of the n-shell is $n - |M|$ and several choices of basis are possible. Three of them are of special interest for the present purpose:

* The parabolic basis. The eigenstates are eigenfunctions of $S_z^{(\alpha)}$ and $T_z^{(\alpha)}$. They separate in semi-parabolic (and parabolic) coordinates. They are labelled with the parabolic quantum numbers n_1 and n_2 which identify with n_μ and n_ν. They fulfill $n = n_1 + n_2 + |M| + 1$. The eigenstates are direct products of functions of μ and ν. The motions of the two oscillators are here uncorrelated. The associated subgroup chain is:

$$SO(2,2) = SO(2,1) \otimes SO(2,1) \quad SO(2) \otimes SO(2)$$

* The spherical basis associated with the standard coupling:

$$\vec{U}^{(\alpha)} = \vec{S}^{(\alpha)} + \vec{T}^{(\alpha)}$$

The eigenstates are eigenfunctions of $U_z^{(\alpha)}$ and $(\vec{U}^{(\alpha)})^2 = (U_z^{(\alpha)})^2 - (U_x^{(\alpha)})^2 - (U_y^{(\alpha)})^2$ which, from Eqs. (3) can be shown[22] to identify with the orbital angular momentum $\vec{L}^2 = (\vec{r} \times \vec{p})^2$. They are labelled with the usual ℓ quantum number and the principal quantum number n (the eigenvalue of $U_z^{(\alpha)}$). The wavefunctions are the spherical ones and separate in spherical coordinates. This scheme expresses a strong correlation between the two oscillators. The associated subgroup chain is $SO(2,2) \quad SO(2,1)_U \quad SO(2)_{U_z}$.

* The λ-type basis associated with the non-standard coupling:

$$V_x^{(\alpha)} = -S_x^{(\alpha)} + T_x^{(\alpha)}$$
$$V_y^{(\alpha)} = -S_y^{(\alpha)} + T_y^{(\alpha)}$$
$$V_z^{(\alpha)} = S_z^{(\alpha)} + T_z^{(\alpha)}$$

The eigenstates are eigenfunctions of $V_z^{(\alpha)}$ and $(\vec{V}^{(\alpha)})^2 = (V_z^{(\alpha)})^2 - (V_x^{(\alpha)})^2 - (V_y^{(\alpha)})^2$. With the choice $\alpha = n = 1/\sqrt{-2E}$, $(\vec{V}^{(\alpha)})^2$ identifies in the n shell with $\vec{\lambda}^2$, where $\vec{\lambda}$ is the non-standard angular momentum[14,22,23] built from the components of $\vec{L}$ and $\vec{A}$ (the Runge-Lenz vector), namely

$\vec{\lambda} = (A_x, A_y, L_z)$. The eigenstates are labelled with n (the eigenvalue of $V_z^{(\alpha)}$) and an integer λ. The eigenvalue of $(\vec{V}^{(\alpha)})^2$ is $\lambda(\lambda+1)$. They are not of any usual type and express, once again, a strong correlation between the two oscillators. The associated subgroup chain is $SO(2,2) \quad SO(2,1)_V \quad SO(2)_{V_z}$.

II. THE LOW-FIELD LIMIT

In the limit of vanishing magnetic field (the so-called inter-ℓ-mixing regime), the eigenstates are obtained by first order perturbation theory. Using either classical mechanics and the secular approximation,[25] or quantum mechanics[26,27] and the SO(4) symmetry group of the hydrogen atom, Solov'ev and, independently, Herrick,[24] showed that an adiabatic invariant, constant of the motion to first order in γ^2, can be expressed in terms of the Runge-Lenz vector $\vec{A}$:

$$\Lambda = 4 \ \vec{A}^2 - 5 \ A_z^2$$

This result, and higher order corrections in classical and quantum[13] mechanics, are easily obtained using the SO(2,2) dynamical group through perturbative development of Eq. (4).

This has a simple interpretation in terms of coupled oscillators dynamics. Rather than using the (μ, ν, p_μ, p_ν) system of coordinates in phase space, we define "phase-amplitude" coordinates:

$$\phi_s = \tan^{-1} \left(\frac{S_y^{(\alpha)}}{S_x^{(\alpha)}}\right) = 2 \ \tan^{-1} \left(\frac{p_\mu}{\mu}\right)$$

$(\phi_S, \ \phi_T, \ S_z^{(\alpha)}, \ T_z^{(\alpha)})$ is a set of <u>canonical</u> coordinates matching the dynamical symmetries of the system. The adiabatic invariant can be expressed as a function of these coordinates just by keeping in the diamagnetic term of Eq. (4), the part which does not change the $n = S_z^{(\alpha)} + T_z^{(\alpha)}$ value. The result is:

$$1 + \Lambda = \frac{4}{n^2} \ S_z^{(n)} \ T_z^{(n)} \ (3 + 2 \ \cos(\phi_S - \phi_T)) \tag{5}$$

with $S_z^{(n)} + T_z^{(n)} = n$ (principal quantum number)

This can be reexpressed using the canonically conjugate amplitude and phase differences:

$$X_z^{(n)} = T_z^{(n)} - S_z^{(n)}$$

$$\Delta\phi = (\phi_T - \phi_S)/2$$

(6)

$X_z^{(n)}$ is diagonal in the parabolic basis (see above) with eigenvalue $(n_2 - n_1)$ and thus identifies with the z component of the Runge–Lenz vector. Figure 1 shows a Poincaré surface of section in the $(X_z^{(n)}, \Delta\phi)$ plane gotten from numerical integration of the equations of motion (the section is taken at $\phi_S + \phi_T = 0$ in order to preserve the $S \leftrightarrow T$ symmetry). The invariant curves are in excellent agreement with the predictions of Eq. (5).

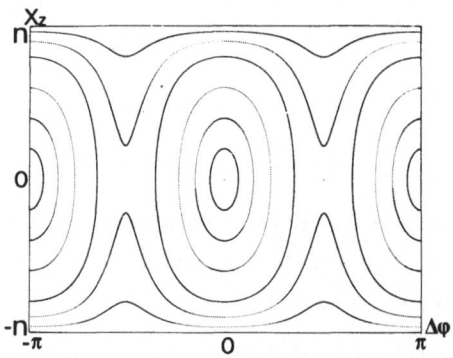

Fig. 1 Poincaré surface of section in the low–field limit ($\beta = 0.01$). The section is in the plane ($X_z^{(n)}$, $\Delta\phi$) at $\phi_S + \phi_T = 0$. The equation of the invariant curves is given by Eq. (5).

In the weak-field limit, the eigenstates are obtained through diagonalization of the Hamiltonian (2) in a given (n, M) manifold. They are labelled with an integer K ranging from 0 (highest state) to $n - |M| - 1$ (lowest state). The wavefunctions are usually very complicated, showing a lot of oscillations and, except for a few of them, no clear spatial localization.

Comparison with the classical dynamics requires to define a distribution function in phase space rather than in real space.[28] The most famous one is the Wigner density. Other definitions are possible, all of them sharing the same semi-classical behavior.

In the present case, the Q-density defined from the coherent states of the SO(2,2) dynamical group[29,30] is used. For each eigenstate, the phase-space distribution function depends on the four dynamical variables

$(\phi_S,\ \phi_T,\ S_z^{(n)},\ T_z^{(n)})$. We verified that, as expected from general arguments,[28], it is localized near the energy surface H = E with a gaussian dispersion. A further section in the plane $\phi_S + \phi_T = 0$ allows us to plot it as a function of $(X_z^{(n)},\ \Delta\phi)$, which is done for the (n = 89, K = 40, M = 0) state on Fig. 2. As expected, the phase-space distribution function localizes near a classical invariant curve Λ = Cst (compare with Fig. 1). This result is very important for understanding the strong field regime. All the eigenstates are localized in phase space, not only the lowest or highest ones as usually inferred.

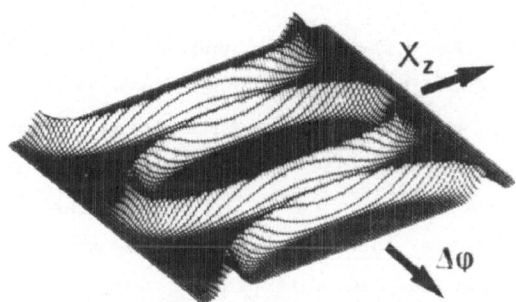

Fig. 2 Phase-space distribution function for the (n = 89, K = 40, M = 0) state represented in the $(X_z^{(n)},\ \Delta\phi)$ plane. It is localized near the invariant curve Λ = Cst of the classical motion.

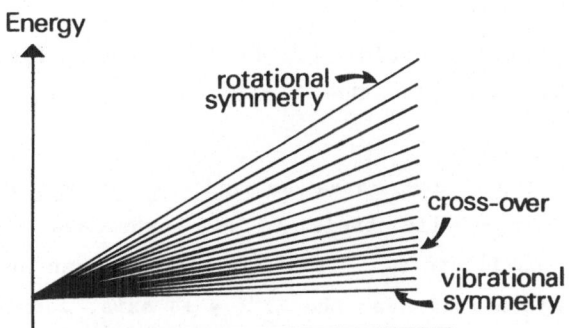

Fig. 3 Structure of the diamagnetic multiplet (n = 36, M = 0, even parity). The upper states have an approximate rotational symmetry. The lower states have an approximate vibrational symmetry.

The breaking of the Coulomb dynamical symmetry by the diamagnetic interaction has been studied by Herrick[24] in the inter-ℓ-mixing regime, and this has been further extended in the inter-n-mixing regime.[13,14] There exists two limiting approximate symmetries (see Fig. 3):

* The high-lying states have an <u>approximate rotational symmetry</u> described by the λ-type basis, expressing a strongly correlated motion in the pair of oscillators. This is not surprising from the classical point of view. The associated phase-space distribution functions are localized near the fixed point $(X_z^{(n)} = 0, \Delta\phi = 0)$ with maximum Λ (and energy). This is also true for the $|n - \lambda| \ll n$ states of the λ-type basis, which is easily seen from the equations of the invariant curves $\chi^2 = Cst$:

$$\chi^2 = \left(n^2 - (X_z^{(n)})^2\right)(1 + \cos 2\Delta\phi)/2$$

Near the fixed point $(X_z^{(n)} = 0, \Delta\phi = 0)$, these curves look very much like the invariant curves (Eq. (5)) found for diamagnetism. The phase-space distribution function of the upper (n = 89, M = 0, K = 0) state represented on Fig. 4a demonstrates this localization. The μ and ν oscillators have nearly the same amplitude and zero dephasing. Hence, $\mu \simeq \nu$ and the electronic motion takes place near the z = 0 plane. Here, the general <u>phase space localization</u> of the motion also leads to a localization in real space near this plane.

How good the approximate rotational symmetry is, can be measured through the overlap between the diamagnetic and λ-type basis. For the upper state, it is[17,31] (for n ≫ 1):

$$|\langle n\ K = 0\ M = 0|n\ \lambda = n - 1\ M = 0\rangle|^2 \simeq 0.998 \qquad (7)$$

This high numerical value proves that the λ-type basis is an excellent approximation to the highest states of the diamagnetic multiplet.

* The low-lying states have an <u>approximate vibrational symmetry</u> described by the parabolic basis. This expresses weak correlation between the two oscillators. This is clearly seen on Fig. 1: the invariant curves localized near the $X_z^{(n)} = \pm n$ axis, corresponding to the lowest values of Λ (and energy) are nearly straight lines characteristic of zero correlation between oscillators. The phase-space distribution function for the (n = 89, K = 88, M = 0) state, plotted in Fig. 4b, is localized near these invariant curves. Classically, only one oscillator has an important amplitude, the other being nearly at rest. This corresponds to an electronic motion localized along z-axis. As a <u>consequence of the phase space localization</u>, the wavefunctions are localized near this axis. The overlap between the parabolic and diamagnetic basis measures the quality of the approximate symmetry. For

$n \gg 1^{17,31}$:

$$|\langle n \ K = n - 1 \ M = 0|(\text{symmetrized } n \ n_1 = 0 \ M = 0 \ \text{state})|^2 \simeq 0.998 \quad (8)$$

* The $\Lambda = 0$ invariant curve is a separatrix[10] between the two types of motion.

The low-field motion presents some analogy with the motion of a pendulum in a homogeneous gravitational field. There exists two limiting types of motion: harmonic oscillation at low energy and rotation at high energy (the separatrix is the situation where the pendulum stops in its upper position). As a matter of fact, the invariant curves are very similar to the ones of Fig. 1.

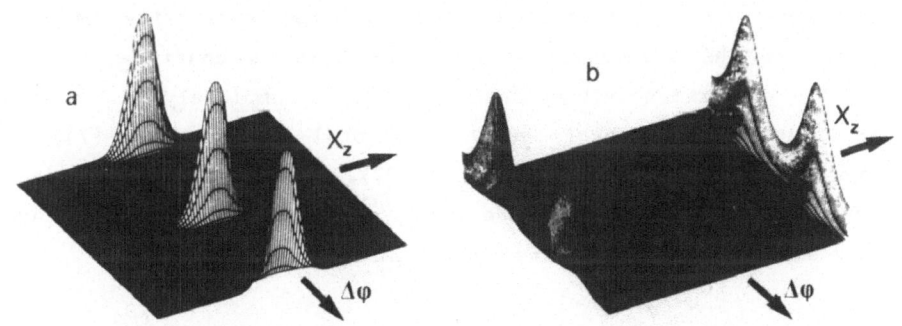

Fig. 4   Phase space-distribution functions:
(a) Rotational (n = 89, K = 0, M = 0) state
(b) Vibrational (n = 89, K = 88, M = 0) state

III. CLASSICAL DYNAMICS

The hydrogen atom and the equivalent system of coupled oscillators share the same classical dynamics. The latter can be studied using either the dynamical variables $(\vec{S}^{(\alpha)}, \vec{T}^{(\alpha)})$ or the canonical variables $(\phi_S, \phi_T, S_z^{(\alpha)}, T_z^{(\alpha)})$. The Hamiltonian of the system is the classical form of Eqs. (2) and (4). Some selected Poincaré surfaces of section are drawn on Fig. 5 (section in the plane $\phi_S + \phi_T = 0$ represented in the $(X_z^{(\alpha)}, \Delta\phi)$ plane).

Because the Coulomb and diamagnetic interaction are homogeneous functions of the electronic position, there is a scaling law for the classical dynamics. It depends on the only parameter:

$$\beta = \frac{\gamma^2}{(-2E)^3}$$

(or on the scaling energy[19] $E \gamma^{-2/3} = -0.5 \beta^{-1/3}$).

The various sets of trajectories associated with different values of γ and E, and same value of β, are simply homothetic.

The $\beta \ll 1$ regime has been studied in the previous section. When $\beta \approx 1$, the Coulomb and diamagnetic interactions with completely different symmetries are of the same order of magnitude and chaos takes place.[32-36] This can be verified on Fig. 5 for $\beta = 1$: a small chaotic region is visible near the separatrix. Such a localization of the chaotic motion is not a surprising feature: in the vicinity of the separatrix, the symmetry of the motion is not well defined which makes it more sensitive to the higher order perturbations. This is the same for the motion of the pendulum close to the separatrix. Starting from its upper position, its evolution is strongly affected by small perturbations. The existence of (approximate) symmetries is thus a limitation to the development of chaos. This is confirmed at higher field: the approximate vibrational symmetry is completely destroyed near $\beta = 3$, while for the nearly exact rotational one (see Eq. (7)) chaos develops around $\beta = 60$ (see Refs. 32 to 37). Above this value and up to the ionization limit ($\beta = \infty$), the phase space is fully chaotic, except for very small regions (relative volume smaller than 10^{-3}).

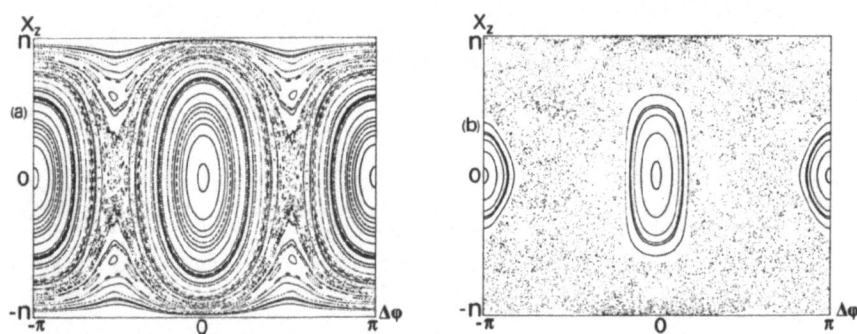

Fig. 5   Poincaré surfaces of section in the $(X_3^{(n)}, \Delta\phi)$ plane.
(a) $\beta = 1$
(b) $\beta = 5$

IV.   QUANTUM SPECTRUM AND EIGENSTATES

Equation (4) can be used to calculate the spectrum and the eigenstates of the hydrogen atom in a strong magnetic field. In the parabolic basis, the matrix elements of the generators $(\vec{S}^{(\alpha)}, \vec{T}^{(\alpha)})$ are known from group representation theory.[13] They only connect a given

$|n_1\ n_2\ M\rangle$ state with its nearest neighbor $|n_1\pm1\ n_2\pm1\ M\rangle$. Consequently, the Hamiltonian matrix (Eq. (4)) written in the parbolic basis has only a few non-zero elements. The resulting matrix has a band structure allowing efficient numerical diagonalization. Depending on the algorithm used,[32,38] we can calculate the spectrum at fixed magnetic field, fixed energy, fixed β or fixed ratio γ/E. The convergency of the computations is usually excellent, except very close to the ionization limit.

The key point for this type of calculation is to use a basis having the appropriate symmetry properties. This is the case for the <u>oscillator</u> parabolic basis defined in Section I. To the contrary, numerical diagonalization of the Hamiltonian (1) in a non-zero field hydrogenic basis only gives very poorly converging in the inter-n-mixing regime.[39]

Depending on the type of states to be studied, one of the three oscillator basis defined in Section I (parabolic, spherical, "lambda") may give faster convergency. For instance, a better convergency is obtained for the rotatational states in the "λ-type" basis and for the vibrational states in the parabolic basis,[13] as can be expected from the low-field limit analysis.

Figure 6 shows some numerical simulations of the spectra from the low-field ($\beta \lesssim 1$, Fig. 6a) regime up to the strong-field regime ($\beta \gtrsim 60$, Fig. 6d).

In the classically regular regime (Fig. 6a), there are only level crossings, or very small avoided crossings not visible at this scale, when inter-n-mixing takes place. This is very clear from the point of view of classical mechanics: the different eigenstates are localized (see Fig. 2) near the invariant curves of the classical motion in phase space. Hence, two different eigenstates are localized in different regions and the energy levels cross.

When classical chaos begins to appear, the energy levels anticross (right part of Fig. 6b and Fig. 6c). This is associated with the destruction of the phase-space invariant curves, leading to delocalization of the eigenstates and strong repulsion between them. Actually, the eigenstates located near the separatrix of the classical motion are the first ones to experience delocalization and strong anticrossings (see Fig. 6b).

As the magnetic field strength is increased, more and more energy levels anticross, corresponding with the increasing size of chaotic regions. On Fig. 6c, the energy levels of the rotational states are the ony ones still exhibiting "crossings," as they correspond to the last invariant curves of the classical motion. Finally, above β = 60, all the energy levels do anticross (Fig. 6d).

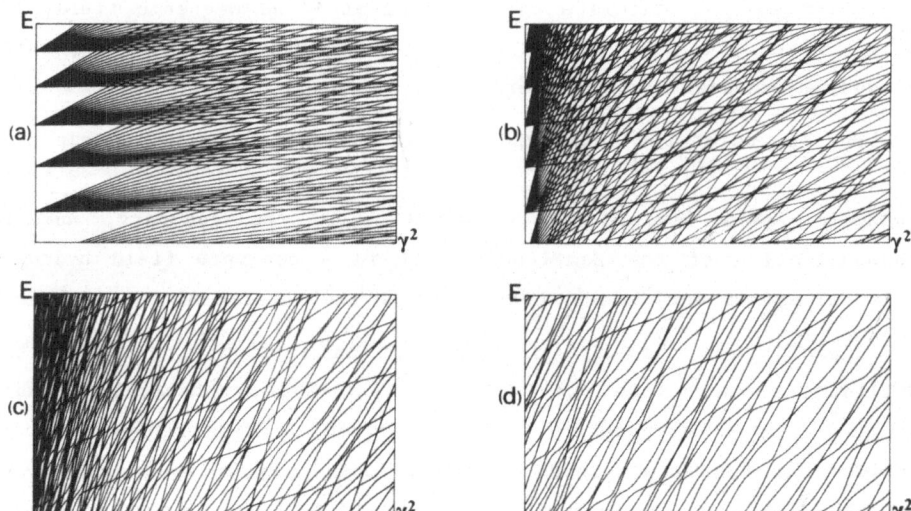

Fig. 6 Energy diagrams of the hydrogen atom as a function of γ^2:

(a) Low field $-4\times10^{-4} < E < -3\times10^{-4}$; $0 < \gamma^2 < 1.4\times10^{-10}$

(b) Moderate field $-4\times10^{-4} < E < -3\times10^{-4}$; $0 < \gamma^2 < 1.4\times10^{-9}$

(c) High field $-4\times10^{-4} < E < -3\times10^{-4}$; $1\times10^{-9} < \gamma^2 < 1.4\times10^{-8}$

(d) High field $-2.5\times10^{-4} < E < -2\times10^{-4}$; $6\times10^{-9} < \gamma^2 < 1.3\times10^{-8}$

The diagrams on Fig. 6 give a qualitative characterization of quantum chaos, that is evolution from level crossing to level anticrossing when chaos develops.

A quantitative measure is provided with the analysis of the statistical properties of the energy levels. It has been numerically shown[19,32,40-42] that, in the regular regime, these statistical properties are well described by a model of non-interacting eigenstates while, in the chaotic regime, they obey a random matrix model (here, the Gaussian Orthogonal Ensemble, see Ref. 6). For instance, the nearest neighbor distribution evolves from a Poisson distribution in the regular case to a GOE distribution in the chaotic case.

However, such a generic random matrix model cannot explain any property of a Hamiltonian system. Indeed, non-generic properties of the system still exist. They appear on the energy levels distribution (see Ref. 19), but more clearly by studying the destruction of the symmetries.

V. DESTRUCTION AND SCARS OF SYMMETRIES

In order to measure the destruction of the low-field approximate symmetries, we numerically calculate the eigenstates of the hydrogen atom in a strong magnetic field when different hydrogenic manifolds are completely mixed. If the rotational symmetry is a good one, the upper rotational states of the n-manifold will be mixed only with the upper rotational states of the $n \pm 1$, $n \pm 2$ ... ones. Hence, we project each eigenstate on the subspace spanned with the pure rotational states $|n \lambda = n - 1 M = 0\rangle$ of each manifold. We then plot this squared projection as a function of the energy (or of the effective principal quantum number $\varepsilon = 1/\sqrt{-2E}$). This builds a stick spectrum containing one line for each eigenstate. In the low field limit, most of these lines are very small as a consequence of the nearly exact character of the rotational symmetry and do not appear in the spectrum. In each diamagnetic manifold, only the upper state is visible with "intensity" 0.998, the other states are nearly vanishing. This procedure allows us to select the upper state among all the other ones.

Figure 7a shows such a spectrum obtained at moderate field strength. This simulation is made at constant ratio γ/E, corresponding to a constant coupling for the equivalent oscillator problem, but the same conclusions are obtained at fixed γ (see Fig. 9). In this low field regime, though different n-manifolds are mixed by the diamagnetic interaction, the rotational symmetry remains valid to more than 98%. This has been previously noticed when studying the convergency of the numerical calculations in truncated basis.[13,19,21] In the right part of Fig. 7a, small deviations from the rotational symmetry are visible: a secondary series of states with intensity around 1% is hardly visible.

At higher field (Fig. 7b), a new phenomenon is displayed: many new lines are growing in the stick spectrum near $\varepsilon = 1/\sqrt{-2E} = 36$ corresponding to the value $\beta = 60$ where the last classical invariant torii with rotational symmetry (located near $z = 0$) are destroyed.[33-36] This indicates that the onset of classical chaos and the destruction of the quantum symmetries just coincide.

Above $\varepsilon = 36$, the rotational symmetry breaks down, but a regular pattern still exists in the stick spectrum. Clusters of levels, approximately equally spaced, carry an important part of the initial symmetry. The positions of the clusters are regularly distributed in continuity with the individual dominant lines of the regular regime and are obeying the semi-classical predictions for $z = 0$ (Quasi-Landau resonances). In each cluster, the individual positions and intensities of the lines are irregular: the rotational symmetry may be concentrated on one dominant line (near $\varepsilon = 62.2$) or redistributed over several lines (near $\varepsilon = 59.1$).

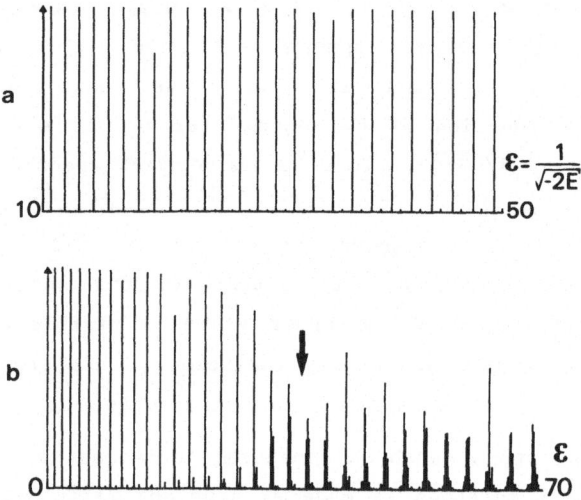

Fig. 7 Squared projections of the eigenstates of the hydrogen atom in magnetic field onto the <u>rotational</u> subspace $\lambda = n - 1$. The heights of the lines measure the fraction of rotational symmetry carried by the eigenstates.
(a) Calculated spectrum for $E/\gamma = -8.678$ (moderate field). The irregularity near $\varepsilon = 20$ is due to accidental degeneracy.
(b) Calculated spectrum for $E/\gamma = -2.311$ (high field). The arrow indicates the classical transition to chaos. The clustering levels above $\varepsilon = 36$ indicates the scar of the rotational symmetry.

The same phenomenon is observed for the vibrational symmetry (Fig. 8). The eigenstates are projected onto the subspace spanned with the parabolic states $|n \; n_1 = 0 \; M = 0\rangle$. A stick spectrum is built from these squared projections. At moderate field strength, this procedure selects the lower state of each manifold: the associated line is dominant and carries 85% of the vibrational symmetry. This symmetry remains valid up to $\varepsilon = 30$ where it breaks down (see Fig. 8a). This precisely corresponds to the destruction of the classical invariant torii with vibrational symmetry (located near the z-axis) near $\beta = 3$. Above $\varepsilon = 30$, clusters of levels, each sharing some part of the initial symmetry, are visible. Their average positions are in continuity with the individual dominant lines of the low field regime; in addition, they approximately obey the semi-classicaly predictions for the motion along z-axis. This phenomenon extends far into the chaotic regime (see Fig. 8b at higher field strength).

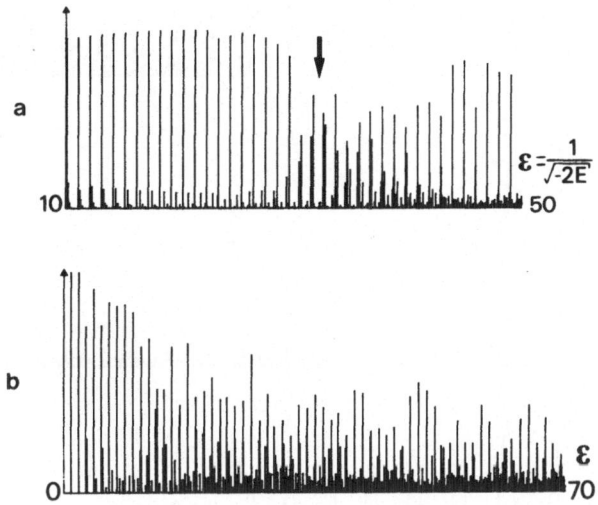

Fig. 8   Squared projections of the eigenstates onto the vibrational subspace $n_1 = 0$.
(a) $E/\gamma = -8.678$. The arrow indicates the classical transition to chaos for the vibrational symmetry. The clustering of levels above $\varepsilon = 30$ indicates the scar of the vibrational symmetry.
(b) $E/\gamma = -2.311$.

Actually, the clusterings of levels manifest the existence of <u>scars of the rotational or vibrational symmetries.</u> By "scars," we mean that some residual symmetry still exists for the system in the chaotic region, in contradiction with Random Matrix Theories expectations. In that sense, they are comparable to the scars found by Heller[43] in the wavefunctions of billiards leading to weak and unexpected localization effects. Above $\varepsilon = 36$ in Fig. 7b and 8b, the classical motion is completely chaotic and most of the energy levels fluctuations are described by a random matrix model (see Section IV or Ref. 19 in this volume). In spite of this, one spectrum of the system contains scars of both rotational and vibrational symmetries, which, in turn, are associated with very different cluster spacings (compare Fig. 7b and 8b).

In the quantum spectrum of a classically chaotic system, the individual positions of the energy levels then seem to carry no information about the dynamics and have eseentially random properties. To the contrast, the properties averaged over many adjacent levels do carry this information.

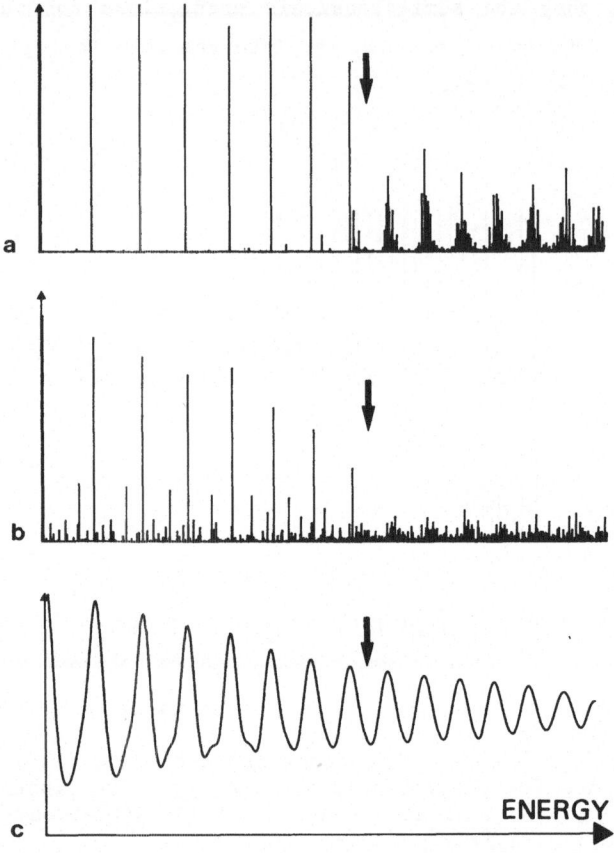

Fig. 9 Simulation of the M = 0, even parity spectrum of the hydrogen atom in a field of 8T ($\gamma = 3.4 \ 10^{-5}$) between −132 and 44 cm^{-1}. The arrow indicates the transition to complete chaos.
(a) Squared projections onto the $\lambda = n - 1$ rotational subspace.
(b) Oscillator strengths from the (2p, M = −1) state.
(c) Same as (b), with a resolution 0.00002 a.u.

These scars of symmetries have very important experimental consequences illustrated on the simulated spectrum of the M = 0 series of hydrogen in a field of 8T. Fig. 9a is a stick spectrum built with the squared projections of the eigenstates onto the rotational subspace. When the classical motion turns chaotic, the rotational symmetry partly breaks down, leading to clusters of levels. As expected, the cluster spacing at E = 0 is 1.5 $\hbar\omega_c$, corresponding to the usual quasi-Landau resonances (as the rotational states are localized near the z = 0 plane, their spacing is the frequency of the z = 0 classical motion).

Figure 9b is a simulation of a real spectrum using optical excitation from the (2p, M = −1) state (which are the experimental conditions of Ref. 5). The clusters of rotational levels are still visible, though very weakened.

The symmetry of optical excitation is essentially spherical (there are selection rules on ℓ and M, but no selection rule on n_1 in the parabolic basis or on λ in the λ-type basis). Hence it does not comply with the internal symmetries of the magnetized atom (parabolic and λ-type): all the eigenstates are then efficiently optically excited. However, depending on the polarization being used, the rotational or vibrational states may be favored.[44] With σ^+ polarization used in Fig. 9b, the rotational states are preferably excited, which explains that the rotational clusters are visible. Figure 9c is a simulation of an experimental spectrum assuming finite resolution (modelled through convolution with a gaussian curve). The short-range fluctuations are smoothed out; the spectrum looks "regular" continuously from the regular to the chaotic regime. The familiar aspect of the Quasi-Landau spectra with 1.5 $\hbar\omega_c$ spacing is rediscovered.[1]

VI. PHASE SPACE LOCALIZATION

In the previous section, we showed that the "chaotic" eigenstates are strongly scarred by the remnants of the low-field symmetries. The associated wavefunctions are only weakly localized, because, especially for highly excited states, they fail to traduce the dynamical correlations (for instance between position and momentum, see Section II in the low-field limit). The phase-space distribution functions directly highlight these correlations (Figs. 2 and 4). We calculate some of them for "chaotic" eigenstates of the system in the strong field conditions of Figs. 7b and 8b. We choose a typical rotational state ($\varepsilon = 41.86$), a typical vibrational state ($\varepsilon = 41.44$) and a typical state with no evident symmetry ($\varepsilon = 43.60$). The phase-space distributions are respectively plotted on Figs. 10a, 10b and 10c. They are normalized with respect to the classical phase-space volume, that is they should be constant for an "ergodic" eigenstate.

Clearly, the eigenstates are very far from being "ergodic." Figure 10a is strongly localized near the classical fixed point ($X_z^{(\alpha)} = 0$, $\Delta\phi = 0$). Such a plot is not very different from the one for a "regular" rotational state (compare with Fig. 4a). This indicates that the phase-space distribution function is strongly scarred by the remnant of rotational symmetry. As a consequence, the wavefunction in real space is scarred (in Heller's sense[43]) close to the origin, leading to an enhanced optical excitation probability. The same feature is obtained for the "chaotic" vibrational state (Fig. 10b to compare with Fig. 4b) which phase-space distribution function is strongly localized near

$X_z^{(\alpha)} = \pm \alpha$ and scarred by the vibrational symmetry.

On the other hand, an important part of the "chaotic" phase-space distribution functions are not clearly localized, though absolutely not "ergodic" (Fig. 10c). Some eigenstates may even have important scars of both limiting symmetries.

The ergodic behavior of the system is rediscovered when averaging the phase-space distribution function over many eigenstates. The short range oscillations are completely smoothed and the distribution is approximately constant (see Fig. 11).

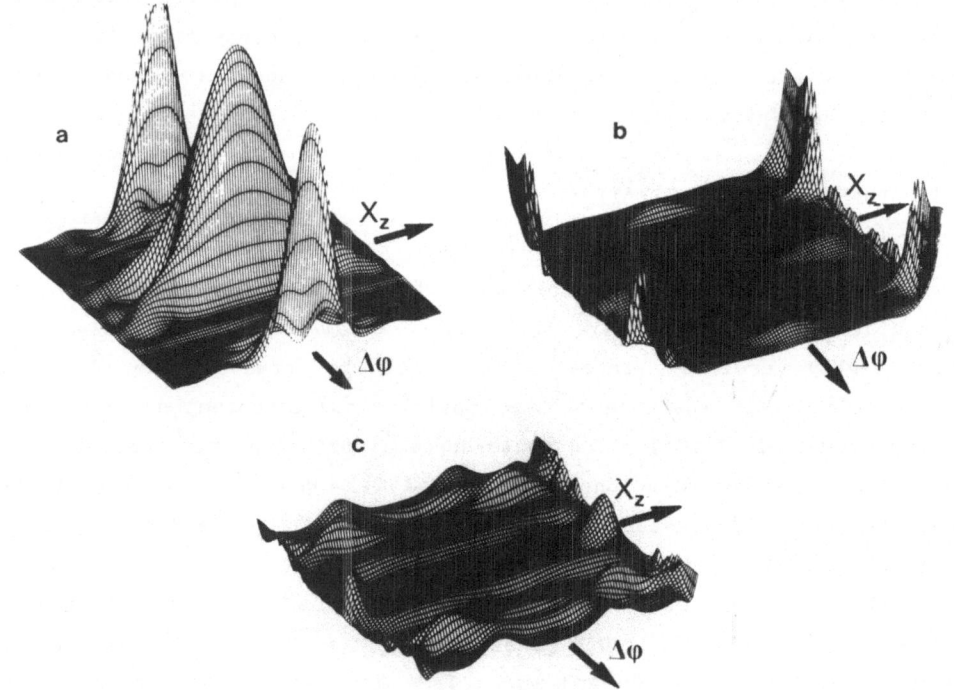

Fig. 10 Phase-space distribution functions for some "chaotic" eigenstates, represented in the $(X_z^{(\alpha)}, \Delta\phi)$ plane.
(a) Rotational state.
(b) Vibrational state.
(c) State with no evident symmetry.

VII. CONCLUSION

In summary, we proved that the classically chaotic magnetized hydrogen atom hs the following quantum characteristics:

* The short range fluctuations of the energy levels and eigenstates are essentially governed by a random matrix model.

* The long range behavior traduces the essential part of the dynamics. At large scale, the spectrum is partly ordered: averaged over

several states, the physical properties are regular and even present some periodicity. This is why low resolution experiments have shown regular features for a long time,[1] and why high resolution experimental results are so difficult to interpret.

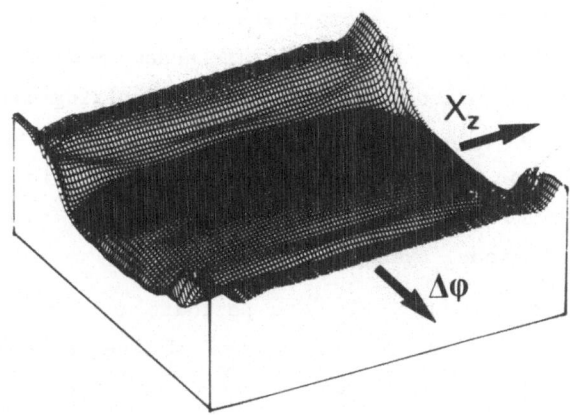

Fig. 11 Averaged phase-space distribution function in the chaotic regime. The average is performed on 77 states of the M = 0 series in the conditions of Figs. 7b and 8b (E/γ = -2.311).

* The onset of chaos is essentially governed, both in classical and quantum mechanics, by the destruction of the internal symmetries of the system. The mechanisms, though not discussed here, are quite general: basically, they are 1:1 resonances.

* The "fully chaotic" eigenstates of the quantum system are strongly scarred by the remnants of the low field symmetries, though the classical system is nearly completely chaotic and probably ergodic. These scars affect the phase-space distribution function, traducing the internal dynamical correlations. Optical spectra reveal the existence of these scars of symmetries. But the effects are weakened and entangled as optical excitation is a process not complying with the internal symmetries of the system.

In conclusion, it is interesting to compare the present approach with the "periodic orbit" approach[19,33,37,45-48] which nicely relates the modulations of the periodic orbits issuing from the origin. The approximate rotational symmetry is of course associated with the z = 0 periodic orbit, while the vibrational one is associated with the trajectory along z-axis. The spectra, scarred by these symmetries, have modulations at the corresponding frequencies of the periodic orbits.

However, the scars of symmetries reflect a global property of the eigenstates, and not only an enhanced excitation probability or localization near the origin. Figure 9 clearly demonstrates that the Quasi-Landau modulation is not only a modulation of the cross section appearing in the Fourier transform of the spectrum, but rather expresses a strong dynamical correlation of the eigenstates in their whole (visible on Fig. 10). This can be exemplified as follows: the scar of the rotational symmetry leads to the Qausi-Landau modulation in the odd z-parity spectrum of hydrogen.[45] This is surprising from the periodic orbit analysis: as the wavefunctions identically vanish on the z = 0 plane, the modulation associated with the z = 0 periodic orbit should not exist! Even far from the z = 0 plane, the wavefunctions are actually scarred by the rotational symmetry.

Obviously, considering only the periodic orbits issuing from the origin cannot explain all the properties of the quantum system. The other periodic orbits probably have a great importance (neglecting these orbits is equivalent to forget one phase-space dimension, that is reducing the distribution functions of Figs. 2, 4, 10, and 11 to their part along $\Delta\phi = 0$). In addition, it is clear that the most interesting aspects of quantum chaos are probably pure quantum effects, not classical ones.

We believe that the existence and shapes of periodic orbits are the consequences of the underlying symmetries of the system. This is clear for the two "fundamental" trajectories along the z-axis (vibrational symmetry) and in the z = 0 plane (rotational symmetry). A great advance would be to classify the periodic orbits from the internal symmetries of the system.

REFERENCES

1. W.R.S. Garton and F. S. Tomkins, Astrophys. J. 158, 839 (1969).

2. J. C. Castro, M. L. Zimmerman, R. G. Hulet, D. Kleppner and R. R. Freeman, Phys. Rev. Lett. 45, 1780 (1980).

3. J. C. Gay in Atoms in Unusual Situations, ed. J. P. Briand (Plenum, 1986).

4. P. Cacciani, E. Luc-Koenig, J. Pinard, C. Thomas and S. Liberman, Phys. Rev. Lett. 56, 1124 (1986).

5. K. H. Welge, this volume; A. Holle, G. Wiesbuch, J. Main, H. Rottke, B. Hager and K. H. Welge, Phys. Rev. Lett. 56, 2594 (1986).

6. O. Bohigas, M. J. Gianonni and C. Schmit, Phys. Rev. Lett. _52_, 1 (1984).

7. T. H. Seligman, J.J.M. Verbaarschot and M. R. Zirnbauer, Phys. Rev. Lett. _53_, 215 (1984).

8. T. Zimmerman, H. D. Meyer, H. Koppel, L. S. Cederbaum, Phys. Rev. A _33_, 4334 (1986).

9. M. Gutzwiller, Phys. Rev. Lett. _45_, 150 (1980).

10. M. Henon in _Chaotic Behavior of Deterministic Systems_, Les Houches Summer School (North-Holland, 1983).

11. C. W. Clark and K. T. Taylor, J. Phys. B, _15_, 1175 (1982).

12. M. J. Englefield, _Group Theory and the Coulomb Problem_ (Wiley, New York, 1972).

13. D. Delande and J. C. Gay, J. Phys. B Lett. _17_, L335 (1984).

14. D. Delande and J. C. Gay, J. Phys. B Lett. _19_, L173 (1984).

15. D. Delande and J. C. Gay, Comm. At. Mol. Phys. _19_, 35 (1986).

16. J. C. Gay in _Spectrum of Atomic Hydrogen_, Part II, (G. W. Series World Publishing, 1987).

17. D. Delande and J. C. Gay, Phys. Rev. Lett. _59_, 1809 (1987).

19. D. Wintgen, this volume.

20. D. Wintgen and H. Friedrich, J. Phys. B _19_, 991 (1986).

21. D. Wintgen and H. Friedrich, J. Phys. B _19_, 1261 (1986).

22. A. O. Barut and R. Raczka, _Theory of Group Representations and Applications_ (PWN, Varsovie, 1980).

23. J. J. Labarthe, J. Phys. B _14_, L467 (1981).

24. D. R. Herrick, Phys. Rev. A _26_, 323 (1982).

25. E. A. Solov'ev, JETP Lett. _34_, 265 (1981).

26. E. A. Solov'ev, JETP _55_, 1017 (1982).

27. T. P. Grozdanov and E. A. Solov'ev, J. Phys. B _17_, 555 (1984).

28. M. V. Berry in _Chaotic Behavior of Deterministic Systems_, Les Houches Summer School (North-Holland, 1983).

29. M. Hillery, R. F. O'Connell, M. O. Scully and E. P. Wigner, Phys. Reports _106_, 121 (1984).

30. D. Delande and J. C. Gay, submitted.

31. D. Delande and J. C. Gay, to be published.

32. D. Delande and J. C. Gay, Phys. Rev. Lett. _57_, 2006 (1986).

33. A. R. Edmonds and R. A. Pullen, Imperial College Preprints, 79-801 nos. 28, 29, 30.

34. M. Robnik, J. Phys. A _14_, 3195 (1981).

35. A. Harada and H. Hasegawa, J. Phys. A _16_, L259 (1983); J. B. Delos, S. K. Knudson and D. W. Noid, Phys. Rev. A _30_, 1208 (1984).

36. W. P. Reinhardt and D. Farelly, J. Physique (Paris) _43_, C2-29 (1982).

37. P. F. O'Mahony, this volume.

38. D. Wintgen and H. Friedrich, Phys. Rev. A 35, 1464 (1987).

39. D. Delande and J. C. Gay, Phys. Lett. 82A, 393 (1981).

40. D. Wintgen and H. Friedrich, Phys. Rev. Lett. 57, 571 (1986).

41. G. Wunner, U. Woelk, I. Zech, G. Zeller, T. Ertl, P. Geyer, W. Schweitzer and H. Ruder, Phys. Rev. Lett. 57, 3261 (1986).

42. G. Wunner, this volume.

43. E. J. Heller, Phys. Rev. Lett. 16, 1515 (1984).

44. G. Wunner, J. Phys. B 19, 1623 (1986).

45. D. Wintgen, Phys. Rev. Lett. 58, 1731 (1987).

46. D. Wintgen and H. Friedrich, Phys. Rev. A 36, 131 (1987).

47. M. L. Du and J. B. Delos, Phys. Rev. Lett. 58, 1731 (1987).

48. J. B. Delos, this volume.

STOCHASTIC FORMULATION OF ENERGY LEVEL STATISTICS:

APPLICATION TO THE HYDROGEN ATOM IN A MAGNETIC FIELD

Hiroshi Hasegawa

Department of Physics
Kyoto University
Kyoto 606, Japan

This article is to give an account of stochastic approach to the energy-level statistics for those quantum systems with varying degree of nonintegrability; those lying between the regular limit adapted for the Poisson statistics and the chaotic limit for the Gaussian ensemble of random matrices. It is a modification of Dyson's Brownian Motion theory for the above purpose by incorporating the original idea into the recent dynamical proposal of Yukawa. The interpolation formula for the nearest-neighbor spacing distribution we propose is discussed in the light of its application to the diamagnetic Kepler problem.

I. GENERAL BACKGROUND

Statistical description of energy levels for complex quantum systems was initiated around 1960 in the field of nuclear physics,[1] and has been extended in these years to other fields such as atomic and molecular physics by the name of "quantum chaos" (for the recent development, see Refs. 2 and 3). According to Bohigas, Giannoni and Schmit,[3] the hydrogen atom in a magnetic field is an extremely interesting system for its fundamental "simplicity": it can hardly be imagined that its fluctuations of levels are of the same feature as those found in compound nuclei. Actually, investigations of energy levels of the hydrogen atom in a magnetic field also have a long history,[4,5] and have now received a good deal of attention as a prototype study of quantum chaos.[6] In this article, we discuss a new formulation[7] of the statistics in more detail, stimulated by the recent development of research into this system.[8] We first outline the classical Brownian-motion theory of Dyson,[9] and then discuss its modification in accordance with a framework of stochastic differential equations later developed.

Consider an ensemble of N × N Hamiltonian matrices which are a) real symmetric, b) complex hermitian and c) those belonging to a direct-product algebra (N' × N' matrix alg.) ⊗ (Pauli spin alg.) (2N' = N) with time-reversal symmetry besides hermiticity. Dyson summarized the main result of the random matrix theory about the Gaussian ensemble[10] E_G in his two theorems:

Theorem 1

If a matrix M is chosen at random in E_G, the probability for finding an eigenvalue of M in each interval $[x_n, x_n + dx_n]$ is $F(x_1, \ldots, x_N) dx_1 \ldots dx_N$, where

$$F(x_1, \ldots, x_N) = Ce^{-\hat{\beta}(\sum_n x_n^2)/2a^2} \prod_{n>m} |x_n - x_m| \tag{1}$$

where C is the normalization constant. It is required that i) every matrix element is statistically independent apart from symmetry specified in each class a) ~ c) and ii) the probability measure F(M)dM is invariant by every transformation relevant to each class i.e. a) orthogonal, b) unitary, c) symplectic transformations, for which

$$\text{a) } \hat{\beta} = 1 \text{(GOE)} \qquad \text{b) } \hat{\beta} = 2 \text{(GUE)} \qquad \text{c) } \hat{\beta} = 4 \text{(GSE)} \tag{2}$$

Theorem 2

Suppose that a matrix M in E_G executes a Brownian motion, each matrix element subject to the Langevin equation

$$f \frac{du}{dt} = -\frac{1}{a^2} u + \sigma r(t), \qquad f > 0 \text{ and } \qquad \langle r(0)r(t) \rangle = \delta(t) \tag{3}$$

hence its distribution given from the Fokker-Planck equation

$$f \frac{\partial P}{\partial t} = \frac{\partial}{\partial u} \left(\frac{u}{a^2} P + \frac{\sigma^2}{2f} \frac{\partial P}{\partial u} \right) \text{ (Smoluchowski equation)}. \tag{4}$$

The random force $\sigma r(t)$ is assumed independent of each other for all the elements, and their strength given by $\sigma^2 = 2f\hat{\beta}^{-1}$ and $f\hat{\beta}^{-1}$ for every diagonal and off-diagonal elementst, respectively. Then all the eigenvalues of M execute a Brownian motion subject to the Langevin equation

$$f \frac{dx_n}{dt} = \frac{\partial W(x_1, \ldots x_N)}{\partial x_n} + \sigma r_n(t) \tag{5}$$

$$\sigma^2 = 2\hat{\beta}^{-1}, \qquad \langle r_n(0) r_m(t) \rangle = \delta_{nm} \delta(t), \tag{6}$$

$$W(x_1, \ldots, x_N) = - \sum_{n>m} \log |x_n - x_m| + \sum_n x_n^2 / 2a^2, \tag{7}$$

whose stationary distribution density is identical to (1).

This second theorem is what Dyson aimed in order to prove the result of random matrix theory (i.e. Theorem 1) by means of the Brownian motions, where the logarithmic part of the potential W in (7) arises as the extra diagonalization for an eigenvalue $x_n(t)$ when "time" t proceeds on to a small amount. Hence, this time-variable is actually to represent a perturbation strength: we shall call this "Dyson's fictitious time."

II. YUKAWA'S DYNAMICAL APPROACH TO THE STATISTICS

Here, we give an account of the recent work of Yukawa along the line of his second paper.[12] For simplicity, the Hamiltonian matrices are confined to real symmetric (class a) in the Gaussian case (i.e. Gaussian orthogonal ensemble (GOE)).

Let H(t) denote such a Hamiltonian, N × N real symmetric matrix, of the form

$$H(t) = H(0) + tV, \tag{8}$$

namely, the unperturbed Hamiltonian H(0) is modified by the perturbation V whose strength is expressed by a parameter t. H(0) is supposed to represent the Hamiltonian of an integrable system whose spectrum is regular, and the perturbation V to modify the spectrum of H(t) irregularly depending on the strength t. This parameter t can then be identified with Dyson's fictitious time discussed in Sec. I. Accordingly, our task will be to examine the "motion" of quantities associated with H(t), specifically the n-th eigenvalues of H(t), in the lapse of this time.

We assume that the matrix is represented in a fixed basis labelled by i, j, ... from 0 to N. There must exist an orthogonal matrix depending on t, R(t), which diagonalizes H(t) at every instant (the diagonal representation of H(t) will be labelled by n, m, ...):

$$R^T(t)H(t)R(t) = E(t) \equiv \begin{bmatrix} x_1(t) & & & 0 \\ & x_2(t) & & \\ & & \ddots & \\ 0 & & & x_N(t) \end{bmatrix} \tag{9}$$

$$\text{with } R^T(t)R(t) = R(t)R^T(t) = 1.$$

By time differentiation of (9) and (10),

$$\frac{d}{dt} E(t) = R^T(t)VR(t) + [M(t),E(t)]M(t) = \left(\frac{d}{dt} R^T(t)\right)R(t) = -R^T(t) \frac{d}{dt} R(t). \tag{10}$$

Let A_t denote a matrix of the form $R^T(t)AR(t)$ where A is time-dependent. Then the equation of motion for A_t can be rewritten as $(d/dt)A_t$ can be rewritten as $(d/dt)A_t = [M(t), A_t]$, which is called the Lax form. If we define $F(t) \equiv [E(t), V_t]$, then it can be rewritten as $[R^T(t)H(t)R(t), \ R^T(t)VR(t)] = R^T(t) \ [H(t), V]R(t) = R^T(t)[H(0), V]R(t) = F_t$ with $F = [H(0), V]$, for which the equation of motion is of a Lax form. Thus we have

$$\frac{d}{dt} E(t) = V_t + [M(t), E(t)] \tag{11}$$

and two Lax forms

$$\frac{d}{dt} V_t = [M(t), V_t]; \quad \frac{d}{dt} F_t = [M(t), F_t] \tag{12}$$

where

$$F_t = [E(t), V_t]. \tag{12a}$$

It can be shown that the three matrix-equations in (11) and (12) fully determine the motion of all the eigenvalues $\{x_n(t), n = 1, \ldots N\}$: the diagonal and off-diagonal parts of Eq. (11) yield

$$\frac{dx_n(t)}{dt} = V_{nn}(t) \text{ and } M_{nm}(t) = \frac{V_{nm}(t)}{x_n(t) - x_m(t)} \quad (n \neq m),$$

respectively, which can be substituted into the diagonal and off-diagonal parts of the first equation in (12), closing the equations for $x_n(t)$ and $V_{nm}(t)$:

$$\frac{dx_n}{dt} = P_n \ (\equiv V_{nn}) \tag{13}$$

$$\frac{dp_n}{dt} = \sum_{n' \neq m} \frac{2V_{nn'}^2}{x_n - x_{n'}} \tag{14}$$

$$\frac{dV_{nm}}{dt} = \sum_{n' \neq n} V_{nn'} V_{n'm} \left(\frac{1}{x_n - x_{n'}} + \frac{1}{x_m - x_{n'}}\right) - (p_n - p_m)\frac{V_{nm}}{x_n - x_m} . \tag{15}$$

The initial conditions for them must be

$$x_n(t = 0) = E_i, \quad V_{nm}(t = 0) = V_{ij} \tag{16}$$

i.e. the unperturbed energy and the matrix elements with respect to the unperturbed eigenvectors.

Yukawa argued that the dynamical system subject to the equations of motion $(13 \sim 15)$ is a completely integrable one with a number of constants of motion, which are prescribed by

$$\mathrm{Tr} \{V_t^{n_1} F_t^{n_2} V_t^{n_3} \ldots\} \tag{17}$$

i.e. the trace of any polynomial generated by V_t and F_t because of the fact that such a polynomial is subject to its equation of motion in the same Lax form and hence its trace must be time-independent. Yukawa's basic idea is to apply the standard method of statistical mechanics to this well-defined dynamical system: a canonical equilibrium distribution can be introduced dependening on each constant of motion, normally in the form $e^{-\beta K}$, if K (the Hamiltonian, not to be confused with H(t)) is the unique constant of motion.

Yukawa noticed, among many (precisely, $\frac{1}{2}N(N + 1)$) independent constants of motion associated with the above dynamics, two relevant ones, which are explicitly given by

$$K = \frac{1}{2} \mathrm{Tr} V^2 = \frac{1}{2} \sum_n p_n^2 + \sum_{n>m} V_{nm}^2 \tag{18}$$

$$= \frac{1}{2} \sum_n p_n^2 + \sum_{n>m} \frac{F_{nm}^2}{(x_n - x_m)^2} , \tag{18a}$$

$$Q = \frac{1}{2} \mathrm{Tr} F^2 = \sum_{n>m} F_{nm}^2 = \sum_{n>m} (x_n - x_m)^2 V_{nm}^2 . \tag{18b}$$

Note that (12a) implies the relation between F_{nm} and V_{nm} given by

$$F_{nm} = (x_n - x_m) V_{nm}. \tag{19}$$

Yukawa showed that the joint distribution for GOE, (1) with $\hat{\beta} = 1$, can be deduced as a consequence of certain procedures of the canonical distribution associated with the two constants K and Q,* namely a partial integration and taking a limit of the large-space distribution $\rho_{\beta,\gamma} = e^{-\beta K - \gamma Q}$. To explain this, we first establish the volume element to measure the phase space of the dynamics (13 ~ 15), which is shown to be given by

$$d\Gamma = \prod_n dx_n dp_n \prod_{n>m} dF_{nm} \psi(\alpha) \prod_{i=1}^{N(N-1)/2} d\alpha_i. \tag{20}$$

We note that in the fixed representation of $H(t)$ the dynamics is a $\frac{1}{2} N(N+1)$ dimensional free motion $\dot{x}_{ij} = p_{ij} (= V_{ij})$, $\dot{p}_{ij} = 0 (i > j)$ for which $d\Gamma = \prod dx_{ij} p_{ij}$, and that the transformation from this to the diagonal representation by the rotation $R(t)$ gives rise to the Jacobian $\prod |x_n - x_m| \psi(\alpha)$, where α represents $\frac{1}{2} N(N-1)$ angle variables to characterize the rotation $R(t)$. Since these are cyclic variables in the dynamics absent in K and Q, it is unnecessary to specify them. Consequently, the integration of $\rho_{\beta,\gamma}$ over all the irrelevant variables i.e. over those other than the first $\prod dx_n$ in (20) yields, by virtue of (18a,b) and the cyclicity of α's,

$$\int e^{-\beta K - \gamma Q} d\Gamma(\text{irrelevant}) = C \prod_{n>m} [1 + \frac{\gamma}{\beta}(x_n - x_m)^2]^{-1/2} |x_n - x_m|. \tag{21}$$

This expression, if a limit of $\gamma \to 0$ is considered asymptotically for $N \gg 1$ such that γN is kept constant, becomes identical to the GOE joint distribution where the variance a^2 is equal to $\beta/N\gamma$. This seems to have a significance from a viewpoint of the central limit theorem: GOE distribution represents a chaotic limit of the energy levels valid only for matrices with large dimension.

Thus, the above dynamical formulation opens a possibility that by changing the value of γ/β in the expression (21) an extent of the nonchaoticity of the random Hamiltonian $H(t)$ could be incorporated into this distribution. The possibility is indeed pointed out in Yukawa's paper[11] (although his argument about it is rather obscure). We plan to realize this possibility by modelling the above dynamics to a stochastic process. This, in turn, will clarify how Dyson's Brownian motion theory can be related to Yukawa's approach.

---

*Of course, a question may be asked why the other constants are irrelevant. The question is a matter of the fundamental ergodicity in dynamical systems: for the present system, one can impose some boundary conditions on the dynamics (such as to confine the trajectories in a container) to examine how three basic expressions survive as constant i.e. the above K and Q and also $P = \sum p_n$ (total momentum, which can be assumed to be zero) (private communication from T. Yukawa).

III. FORMULATION BY STOCHASTIC DIFFERENTIAL EQUATIONS

Our plan is to transform the deterministic equations of motion discussed in Sec. II into stochastic ones. Such problems often occur in statistical physics, but there exists no established prescription for such: we must depend largely on wisdom. A reinspection of our abject-equations (13 ~ 15) suggests that it is reasonable to assume all V_{nm}'s (n > m) to be random variables. This means that we assume a legitimacy to discard the third equation (15). It then enforces the other two equations to be indeterministic. We thus have

$$\frac{dx_n}{dt} = p_n, \qquad \frac{dp_n}{dt} = \sum_{n'(\neq n)} \frac{2v_{nn'}^2}{x_n - x_{n'}} \qquad n = 1, \ldots N. \tag{22}$$

This dynamical system consists of $\frac{1}{2} N(N + 1)$ variables, of which 2N are just like the usual canonical system $\{x_n, p_n\}_{n=1}^N$ and the rest $\frac{1}{2} N(N - 1)$ play the role of random variables acting on the momenta of this subsystem. The structure reminds us of the Ornstein–Uhlenbeck Brownian motion.[13]

A typical Ornstein–Uhlenbeck process (in one dimension) is expressed in a form of Langevin equation

$$\frac{dx}{dt} = p, \qquad \frac{dp}{dt} = -\Gamma p - \frac{\partial \phi}{\partial x} + \sigma r(t) \qquad \Gamma > 0. \tag{23}$$

Namely, a one-dimensional particle moves, being subject to a Newtonian mechanics with Hamiltonian $\frac{1}{2} p^2 + \phi(x)$, which is influenced also by a random force $\sigma r(t)$ from its environment. Its randomness is expressed only by the stationary time-correlation property, $\langle r(t)r(t + \tau)\rangle$ = some function of τ, and the constant σ represents the strength of the force. If the probability density with which the above correlation average is calculated is a Gaussian, then the random variable $r(t)$ is said to be a Gaussian stationary process, and if δ-correlated i.e.

$$\langle r(t)r(t + \tau)\rangle = \delta(\tau) \tag{24}$$

it is called a Gaussian white noise. The presence of such random force causes the friction of the momentum, $-\Gamma p$ in Eq. (23), the friction constant Γ being related to the strength σ of the force, in the ideal white-noise situation (24), through

$$\sigma^2 = 2\beta^{-1}\Gamma, \tag{25}$$

where β^{-1} is the variance of the momentum in the equilibrium distribution (i.e. the temperature of the kinetic energy) to be attained for $t \to \infty$.

An important results elucidated by the work of Uhlenbeck and Ornstein[13] was that the friction constant Γ for the momentum of the Brownian particle appears _inversely_ in the process of its coordinate, as was known (prior to O-U) by Smoluchowski in his Fokker-Planck Eq. (4), the friction constant being denoted by f so that Smoluchowski's f is precisely equal to the above Γ. Presently, this fact is well understood as a typical example of "elimination of fast variables,"[14] sometimes called "adiabatic elimination." Namely, in (23) if the decay of motion of p is much faster than the motion of x, then the p-motion can be eliminated from the whole dynamics, and the remaining x-motion in a gross time scale $\gg 1/\Gamma$ can be described correctly by getting an equation for x with the substitution of p from $(dp/dt) = 0$:*

$$\frac{dx}{dt} = \frac{1}{\Gamma} \left(- \frac{\partial \phi}{\partial x} + \sigma r(t) \right). \tag{23'}$$

The above story of the reduction of the Ornstein-Uhlenbeck process to the Smoluchowski process can be exploited in the present study of the dynamics (22): upon introduction of such a single Γ and an extra random force $R_n(t)$ of the Gaussian white characteristic, which are subject to (25), we replace (22) by

$$\frac{dx_n}{dt} = p_n, \qquad \frac{dp_n}{dt} = - \Gamma p_n + \sum_{n(\neq n)} \frac{2v_{nn'}^2}{x_n - x_{n'}} + R_n(t) \tag{22'}$$

and eliminate p_n to get

$$\frac{dx_n}{dt} = \frac{1}{\Gamma} \left(\sum_{n(\neq n)} \frac{2v_{nn'}^2}{x_n - x_{n'}} + R_n(t) \right) \qquad n = 1, 2, \ldots, N. \tag{26}$$

This is our starting stochastic equation for an eigenvalue of the random Hamiltonian matrix. If this equation were treated following Dyson's prescription stated in Theorem 2, then a Langiven equation of the form (5) would be obtained with the logarithmic potential W, in which the harmonic part to yield the Gaussian exponent would be lacking. We shall

---

*In Ref. 14 a comprehensive discussion of the subject of elimination of fast variables is given, and this simple procedure of p-elimination conforms to the author's first case. However, it is only in the framework of deterministic equations. In the stochastic framework generally much complexities arise, but still the simple procedure holds in the present example (see Ref. 15).

not go on this way, but deal with the equation by taking the fluctuations more efficiently in accordance with the theory of stochastic differential equations (SDE) to conform to Yukawa's approach. The rest of this section is devoted to a summary of SDE for this purpose.

A typical SDE for one variable x_t is expressed as

$$dx_t = a(x_t)dt + \sigma(x_t)dB(t), \tag{27}$$

where $a(x)$ and $\sigma(x)$ are smooth functions of x, and $B(t)$ denotes a standard Brownian motion defined by the Gaussian process with

$$\text{mean } (B(t) - B(s)) = 0, \quad \text{variance } (B(t) - B(S)) = |t - s| \tag{28}$$

besides $B(t = 0) = 0$. If σ is a constant, then Eq. (27) is equivalent to the Langevin equation, $(d/dt)x_t = a(x_t) + \sigma r(t)$, with the Gaussian white noise $r(t)$ defined in (24). A mathematical significance of writing SDE(27) instead of the conventional Langevin equation consists in the case of nonconstant $\sigma(x_t)$, because then the multiplication of $dB(t)$ by such functions requires a precise rule; a framework of calculus for stochastic differentials elucidated by Itô.[16]

The stochastic differential to which the calculus applies is an object, beyond the ordinary sense of differentials, which obeys the following multiplication table:

	dt	dB(t)
dt	0	0
dB(t)	0	dt

Every such differential is expressed uniquely as a sum of dt part and dB(t) part as on the right-hand side of Eq. (27). SDE(27) in this sense is precisely the Itô SDE, for which the distribution of x_t can be determined from

$$\frac{\partial}{\partial t} P_t(x) = \frac{\partial}{\partial x} (-a(x)P_t(x)) + \frac{1}{2} \frac{\partial^2}{\partial x^2} (\sigma^2(x)P_t(x)) \tag{29}$$

that is the Fokker-Planck equation.

Itô SDE is an ideal object which cannot be reached by the ordinary limiting process of nonwhite noise: roughly speaking, if a Gaussian white noise is simulated by a limit of nonwhite series of noises, then the resulting Langevin equation corresponds to an SDE not in the Itô sense: instead, one obtains the Stratonovich SDE of the form (for preciseness, see Ref. 17)

$$dx_t = \tilde{a}(x_t)dt + \sigma(x_t) \circ dB(t). \tag{30}$$

However, one can transform this equation into the equivalent Itô SDE by a rule $Y \circ dX = YdX + \frac{1}{2}dYdX$: thus $\sigma(x_t) \circ dB(t) = \sigma(x_t)dB(t) + \frac{1}{2}\sigma'(x_t dx_t dB(t)(\sigma'(x) \equiv d\sigma(x)/dx)$, and the substitution of (30) into the last part, $dx_t dB(t)$, yields (by a use of the above table)

$$dx_t = [\tilde{a}(x_t) + \frac{1}{2}\sigma'(x_t)\sigma(x_t)]dt + \sigma(x_t)dB(t)$$

which is identical to (27) by setting $\hat{a} + \frac{1}{2}\sigma'\sigma = a$ in the above.

The great facility of the $(\circ dx)$-differential, we remark, is that it obeys all the ordinary rules of the calculus (see Ref. 18). Hence, we get a theorem which can be confirmed easily:

<u>Consider a Stratonovich SDE of the form</u>

$$dx_t = \sigma(x_t) \circ dB(t), \quad \text{or in general} \quad \sum_i \sigma^i(x_t) \circ dB_i(t) \tag{31}$$

<u>where $\{B_i(t); i = 1, \dots r\}$ is a set of mutually independent standard Brownian motions. Then, the associated Fokker-Planck equation is given by</u>

$$\frac{\partial}{\partial t} P_t(x) = \frac{\partial}{\partial x} D(x)\left(\frac{1D'(x)}{2D(x)} P_t(x) + \frac{\partial}{\partial x} P_t(x)\right) \equiv LP_t \tag{32}$$

$$D(x) = \frac{1}{2}\sigma(x)^2, \quad \text{or in general} \quad \frac{1}{2}\sum_i\{\sigma^i(x)\}^2. \tag{32a}$$

We add that the stationary distribution of the process x_t whose density satisfies $LP = 0$ in the above is given by

$$P(x) = \text{const} \times \{D(x)\}^{-\frac{1}{2}} \tag{33}$$

for some generic class of the processes subject to Eq. (31) (the fact that (33) satisfies the stationarity $LP = 0$ is easy to see, but it is difficult to specify the condition under which (33) is the unique stationary density). In what follows we assume the validity of (33).

IV. INTERPOLATION FORMULA FOR LEVEL SPACING DISTRIBUTION

Consider any two of the eigenvalues (x_n, x_m), and set up SDE for

$$x_{nm}(t) \equiv x_n(t) - x_m(t) \qquad \text{(spacing of the n and m levels).} \qquad (34)$$

We first combine two equations for them from Eq. (26) in Sec. III to write dx_{nm}, which is still a multi-dimensional SDE. Therefore, we try to simplify it by regarding all the variables other than x_{nm} as external parameters so that the resulting SDE can be of the form (31). To avoid ambiguity, we list the prescriptions as follows:

(i) Every eigenvalue is labelled such that $x_m < x_n$ for $m < n$, and after rearrangement the indices in x_{nm} can be assumed as $m < n$.

(ii) All the $\frac{1}{2} N(N - 1)$ random variables $V_{nn'}^2$, $n' \neq n$, are statistically independent of each other, each regarded as a Brownian motion defined, apart from normalization, in (28).

(iii) The multiplication of $dB(t)$ by any stochastic variable is defined as in the sense of Stratonovich i.e. as $\circ dB(t)$.

(iv) In setting up SDE for dx_{nm}, all the $x_{n(m),n'}$ other than x_{nm} in the denominator of the right-hand summand are regarded as being absorbed in the $V_{nn'}^2$'s entering the numerator.

Consequently, we get SDE for the particular spacing x_{nm} as follows:

$$
\begin{aligned}
dx_{nm} &= \sum_{n' \neq n,m} \sigma_{nn'} \circ dB_{nn'}(t) + \frac{\lambda}{x_{nm}} \circ dB_{nm}(t) \\
&= \sigma d\hat{B}_{nm}(t) + \frac{\lambda}{x_{nm}} \circ dB_{nm}(t).
\end{aligned}
\qquad (35)
$$

Note that a linear combination of dB's with constant coefficients yields another dB. It shows that the two kinds of independent fluctuations, $\hat{B}(t)$ and $B(t)$, affect the spacing with the respective strength σ^2 and λ^2: the latter has its origin from itself exhibiting the effect of level repulsion, hence the former represents just a normal fluctuation accumulated from all other spacings. We show that the ratio of these two strengths is precisely identical to Yukawa's nonchaoticity parameter so that

$$\sigma^2/\lambda^2 = \gamma/\beta \qquad (36)$$

This is because the theorem about the Stratonovich SDE (31) together with the statistical independence of all the Brownian motions in (35) for different pairs assures the joint distribution (21), provided the relation (36) holds. Our aim is to exploit this result to obtain an analytic expression for the <u>nearest-neighbor spacing</u> (NNS) distribution.

It has been known that a simple relation exists between the distribution of an eigenvalue (the level density) and that of an eigenvalue spacing:[1a] if $\mu(x)dx$ denotes the probability of a level found in the interval $(x, x + dx)$, then $e - \int_0^S \mu(x)dx$ represents the conditional probability that given a level at 0 there exists no level in the open interval $(0,S)$, and hence $e - \int_0^S \mu(x)dx_{\mu S}$ yields the density of the NNS distribution. This result can be supplemented from the present stochastic version:

<u>Let the variable S of NN-spacing be regarded as a stochastic process with increase of Dyson's fictitious time subject to the following SDE:</u>

$$d \{e^{-\int_0^{S_t} \mu(x)dx}\} = \sigma B(t), \qquad \sigma = \text{const.} \tag{37}$$

<u>Then, the stationary distribution of the process S_t is identical to</u>

$$P(S) = e^{-\int_0^S \mu(x)dx}\mu(S) \qquad \text{(density of NNS distribution).} \tag{38}$$

The proof can be seen from rewriting (36) in the standard form

$$dS_t = \{\frac{1}{\mu(S_t)} e^{\int_0^{S_t} \mu(x)dx}\} o\sigma dB(t) \tag{37'}$$

i.e. the form of (31) with a single $B(t)$ for which $D(S) = \frac{1}{2} \{\sigma(S)\}^2$. We note the facility of the (odX)-differential for deriving this result. Two such examples are, of couse, $\mu(x) = $ const. to yield the Poissonian and $\mu(x) = $ const. $\times x$ to yield the Wigner surmise, and we see that the level-density functions for both enter SDE (35) just inverse-additively. Hence we propose to write SDE for the NN-spacing S_t adequately as

$$dS_t = e^{\rho S_t} od\hat{B}(t) + \frac{\lambda}{\rho S_t} e^{\frac{\alpha^2}{2} \rho^2 S_t^2} odB(t), \tag{39}$$

where three parameters have been introduced to cover the most general circumstance i.e. ρ, α and λ: ρ and $\alpha\rho$ are the scaling parameters of the average spacing associated with the Poisson part and the Wigner part, respectively, and λ yields the relative strength of the two kinds of fluctuations. The resulting NNS distribution (density) is given by

$$P(S) = N_\rho S[\rho^2 S^2 e^{2\rho S} + \lambda^2 e^{\alpha^2 \rho^2 S^2}]^{-1/2} \qquad 0 \leq S < \infty \tag{40}$$

where N is the normalization constant. For $\lambda = 0$, it reduces to the Poissonian:

$$P(S) = Ne^{-\rho S} \text{ with } N = \rho \qquad \text{by the normalization} \tag{41}$$

and $\rho = 1$ by the scaling normalization. For $\lambda \to \infty$, it reduces to the Wigner surmise:

$$P(S) = \frac{N}{\lambda} \pi S e^{-1/2 \alpha^2 \rho^2 S^2} \text{ with } N = \lambda \alpha^2 \rho \text{ and } \alpha^2 \rho^2 = \frac{\pi}{2} \tag{42}$$

by the normalization and the scaling normalization, respectively, where the scaling normalization means $\int_0^\infty P(S) S dS = 1$. In the general intermediate case, these two normalizations can be analytically performed with ρ and N as functions of λ and α given by

$$\rho = \frac{F_2(\lambda, \alpha)}{F_1(\lambda, \alpha)} , \qquad N = \frac{F_2(\lambda, \alpha)}{\{F_1(\lambda, \alpha)\}^2} \tag{43}$$

where

$$F_S(\lambda, \alpha) = \int_0^\infty [x^2 e^{2x} + \lambda^2 e^{\alpha^2 x^2}]^{-1/2} x^S dx. \tag{43a}$$

In Fig. 1 graphs of the one-parameter family of the densities $\{P_\lambda(S), 0 < \lambda < \infty\}$ with a fixed value of α i.e. $\alpha = 1$ are exhibited. A characteristic feature of the graphs is the nonanalytic behavior near $\lambda = 0$: $P_\lambda = 0$ is identical to the Poissonian, but for $\lambda > 0$ $P_\lambda(S)$ is proportional to S for $0 < S < \lambda$, which implies that the level repulsion is effective for any small λ. This has been anticipated by Robnik[19] as the unique universality of the NNS distribution for nearly regular quantum systems (GOE case).

V. APPLICATION TO THE HYDROGEN ATOM IN A MAGNETIC FIELD

Systematic analyses of surface-of-sections for the diamagnetic Kepler motion have been made by several authors[20] to elucidate the KAM tori, their destruction and the transition to a complete chaos. The connection of such analyses to the random-matrix theory has become and important discipline in the subject of quantum chaos[2] in order to locate the precise correspondence between the irregularity of the classicl trajectories and that of the quantum spectra. The work of Harada and

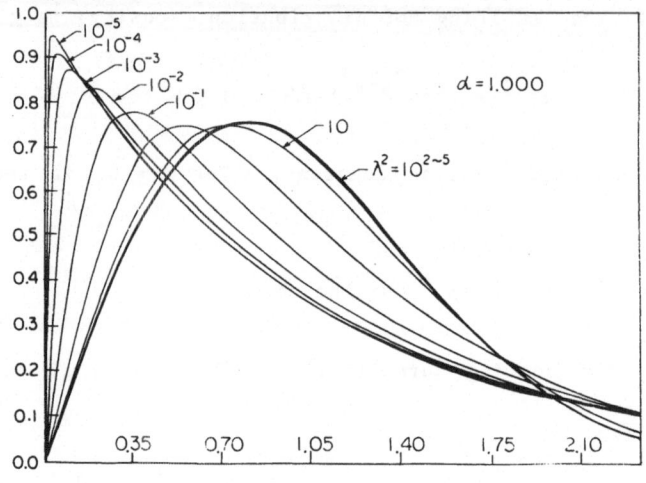

Figure 1

Hasegawa[20] demonstrated the most direct such correspondence: they computed the surface of section (p_ρ, ρ; $z = 0$) in the cylindrical coordinate system of the diamagnetic Kepler motion and compared the weight of its irregular portion with the irregular (optical) spectrum in the so-called inter-n mixing regime of the Lyman series calculated by Clark and Taylor,[21] thus locating the range of transition on the energy scale (Fig. 2). This is just the range of interest also for the study of the NNS distribution, and has been rechecked witha higher precision by Wunner and his collaborators.[6] There have been three groups of investigators[6] who computed the NNS frequency histogram from their secular determinants for the quantum diamagnetic Kepler Hamiltonian and examined it to confirm the Poisson-Wigner transition.

Two formulas have been used for the interpolation purpose: the Brody[22] and Berry-Robnik[23] formulas, both of the one-parameter family of NNS distribution density functions which interpolate the two opposite limits (41) and (42). The Brody formula is just intuitively to interpolate the level-density function in the power form $\mu(x) = \text{const.} \; x^q$; $q = 0$ for the Poisson and $q = 1$ for the Wigner. One sees in the recent

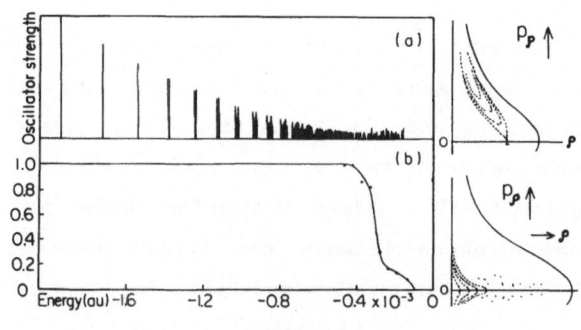

Figure 2

literatures that this simple idea of interpolation has been accepted to simulate the actual histograms for various systems rather well. However, it has been devised on no physical basis. The Berry-Robnik formula, on the other hand, has been proposed on a very clear physical basis: the phase space is assumed to be divided into two parts, a regular part and a chaotic part which are statistically uncorrelated to each other. This assumption of statistical independence of the two parts seems to be the reason that the derived NNS density formula has generally nonvanishing value for S = 0 except in the chaotic limit (which is contradictory to the Brody formula and also to the present one). The work of Wintgen and Friedrich[8] has indeed exhibited inadequacy of this formula in comparison with their histogram near S = 0. In this respect the present formula (40) can be looked upon as a rescue of the Berry-Robnik formula, in that it agrees with the idea of dividing the phase space into a regular part and a chaotic part represented by the two kinds of Brownian motions B(t) and B̂(t), respectively. These are statistically independent objects themselves, but their fluctuations affecting the NN spacing certainly give rise to a mixture of both parts that is the background idea behind our formula (40).

Figure 3 shows an example of numerical result best fit of the NNS histogram to the formula (4) computed in Tübingen. It should be noted that the formular depends on the extra parameter α, showing a nonuniversality of the distribution for such intermediate regime.[19] Thus the fit has been made by the process of least-square to determine the optimum α for each λ (private communication from Wunner). It is conjectured that the degree of nonintegrability for a given system is represented solely in the fluctuation ratio λ and not affecting the scaling ratio, α. This should be one of exploring points in computational studies of the NNS histogram for individual systems. More detailed results are expected in the articles by Wunner and by Wintgen.

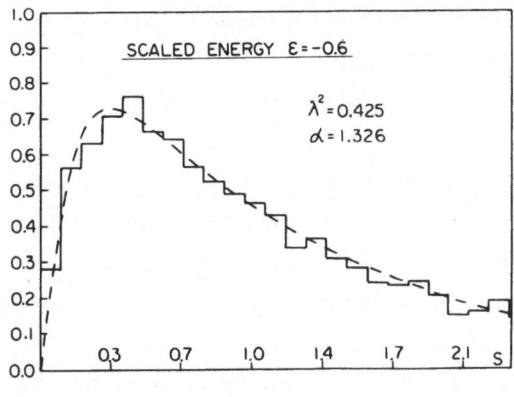

Figure 3

VI. CONCLUDING REMARKS

Two questions must be cleared up in order for the present stochastic formulation to be effective. The first question is how to extend the stochastic differential equation (35) to GUE and GSU cases, and the second one, how the formulation to predict the correlations between different spacings for the study of the spectral rigidity.[2,3] Both questions are not simple to answer. It is remarked that the diamagnetic Kepler Hamiltonian is <u>not</u> time-reversal invariant: however, one need not worry about the GOE applicability to the system because the Hamiltonaian is still anti-unitary invariant (here, time-reversal times a reflection; private communication from Robnik). Nevertheless, the formulation should be satisfactorily extended to GUE case in parallel with Yukawa's theory, but at present I cannot answer this question. As to the second question, it is possible to extend the formulation by removing the prescription (iv) stated in IV so that an SDE can be set up for two spacing variables. The study in this direction is a promising one.

ACKNOWLEDGEMENT

I wish to express hearty thanks to G. Wunner and U. Woelk for their interest and earnest collaboration to my proposal.

REFERENCES

1. For a review, see <u>Statistical Theories of Spectra: Fluctuations</u>, ed. by C. E. Porter (Academic Press, New York, 1965); C. E. Porter, <u>Fluctuations of Quantal Spectra</u> in the above references, VI.

2. O. Bohigas and M.-J. Giannoni, "Chaotic Motion and Random Matrix Theories," <u>Lecture Notes in Physics 209</u> (Springer-Verlag, Berlin, Heidelberg, New York, Tokyo, 1984) 1.

3. O. Bohigas, M.-J. Giannoni and C. Schmit, "Spectral Fluctuations of Classically Chaotic Quantum Systems," <u>Lecture Notes in Physics 263</u> (Springer-Verlag, 1986) 18.

4. J. C. Gay, <u>New Trends in Atomic Diamagnetism</u> (D. Reidel Publishing Co., 1985) 631.

5. H. Hawegawa, <u>Quantization of Nonintegrable Systems; the Hydrogen Atom in a Magnetic Field</u> (Plenum Press, New York and London, 1985) 193.

6. D. Wintgen and H. Friedrich, Phys. Rev. Lett <u>57</u>, 571 (1986); D. Delande and J. C. Gay, Phys. Rev. Lett. <u>57</u>, 2006 (1986); G. Wunner, U. Woelk, I. Zeck, G. Zeller, T. Ertl, F. Geyer, W. Schweitzer and H. Ruder, Phys. Rev. Lett. <u>57</u>, 3261 (1986).

7. H. Hasegawa, H. J. Mikeska and H. Frahm,

8. J. Main, G. Wiebusch, A. Holle and K. H. Welge, Phys. Rev. Lett. 57, 2789 (1986); M. A. Al-Laithy, P. F. O'Mahony and K. T. Taylor, J. Phys. B: At. Mol. Phys. 19, L773 (1986); D. Wintgen and H. Friedrich, Phys. Rev. A 35, 1465 (1987).

9. F. J. Dyson, J. Math. Phys. 3 1191 (1962); J. Math. Phys. 13, 90 (1972).

10. M. L. Mehta, Random Matrices and the Statistical Theory of Energy Levels (Academic Press, New York, 1967).

11. T. Yukawa, Phys. Rev. Lett. 54, 1883 (1985).

12. T. Yukawa, Phys. Lett. 116, 227 (1986).

13. G. E. Uhlenbeck and L. S. Ornstein, Phys. Rev. 36, 823 (1930).

14. N. G. van Kampen, Phys. Rep. 124, 69 (1985).

15. H. Hasegawa, M. Mabuchi and T. Baba, Phys. Lett. 79A, 273 (1980). H. Hasegawa, M. Mizuno and M. Mabuchi, Prog. Theor. Phys. 67, 98 (1982).

16. K. Itô and S. Watanabe, Introduction to Stochastic Differential Equations, Proc. Intern. Symp. SDE Kyoto 1976, ed. by K. Itô Kinokuniya, (Tokyo, 1978) ixxx.

17. E. Wong and M. Zakai, Ann. Math. Stat. 36, 1560 (1965).

18. H. Hasegawa and H. Ezawa, Prog. Theor. Phys. Suppl. 69, 41 (1980).

19. M. Robnik, J. Phys. A: Math. Gen. 20, L495 (1987).

20. M. Robnik, J. Phys. A: Math. Gen. 14, 3195 (1981); W. P. Reinhardt and D. Farrelly, J. Physique 43, C2, 29 (1982); A. Harada and H. Hasegawa, J. Phys. A: Math. Gen. 16, L259 (1983); J. B. Delos, S. K. Knudson and D. W. Noid, Phys. Rev. A 30, 1208 (1984).

21. C. W. Clark and K. T. Taylor, J. Phys. B: At. Mol. 13, L737 (1980); J. Phys. B: At. Mol. 15, 1175 (1982).

22. T. A. Brody, J. Flores, J. B. French, P. A. Mello, A. Pandey and S. M. Wong, Rev. Mod. Phys. 53, 385 (1981).

23. M. V. Berry and M. Robnik, J. Phys. A 17, 2413 (1984).

HIGHLY EXCITED HYDROGEN ATOMS IN STRONG MICROWAVES

J. E. Bayfield and D. W. Sokol

Department of Physics and Astronomy
University of Pittsburgh
Pittsburgh, Pennsylvania 15260 USA

INTRODUCTION

Consider an atom that is highly excited to a state with principal quantum number n near 60. Such an atom's electron binding energy is a few meV. External electric fields of modest field strength, in the range 10-100 V/cm, can have such a strong influence on such atoms that they can be rapidly ionized, even within a few classical electron orbit periods. For microwave fields at near-ionizing strengths, many quantum states are coupled and the problem of energy absorption by the atom from the external field can be addressed as a nonlinear dynamics problem. As both quantum numbers and numbers of photon absorption/emission events can be large, a semiclassical picture based upon an underlying classical electron dynamics is expected to be useful. The problem when viewed classically is in the class of externally-driven nonlinear oscillators. This class of problems has much in common with the class of two coupled nonlinear oscillators, as one of the two is just replaced by an external oscillator of fixed amplitude an frequency. An excited hydrogen atom in a sufficiently strong static magnetic field exhibits electron motion involving a strong competition between Larmor precession in the plane perpendicular to the field and Coulombic motion along the field. This is two coupled nonlinear motions, as in two coupled nonlinear oscillators. Thus the highly excited atom in strong magnetic filds and in strong microwave fields are two systems of study in the same subfield of quantum nonlinear dynamics.[1]

The underlying classical dynamics of the microwave ionization process is that of a diffusion in action space that is the result of electron trajectories that exhibit deterministic chaos. Interest in the quantum analog of diffusive classical time evolution has generated a great deal of recent work on our problem.[2-6]

Classically our system exhibits bifurcations in the loci for periodic electron trajectories in the parameter space of field strength versus frequency.[7] Thus a period-doubling path to chaos is one apparent feature of the nonlinear classical dynamics system. Yet it is believed that microwave ionization predominantly occurs via another path to classical chaos, via quasiperiodic orbits that become replaced by chaotic ones at field strengths determined by the Chirikov classical resonance overlap criterion.[8] This belief is based upon the larger volumes of phase space occupied by quasiperiodic and chaotic orbits rather than periodic ones, and assumes that quantum interference does not greatly enhance the role of the periodic orbits. However, in several model problems, some quantum enhancement of this type does occur.[9,10]

Suppose that our highly excited atom is hydrogenic, and is exposed to a microwave pulse that produces a final probability for ionization in the range 5 to 90%. Then a primary signature of the quantized diffusion that leads to the ionization is an accompanying smooth population of final quantum bound-state energy values that begins near the initial value and extends up close to the ionization limit. The distribution decreases exponentially with increasing change in energy value,[3] and decreases approximately so with increasing change in quantum number value for small changes in quantum number.[2] By identifying quantum number with classical action, the final bound-state distributions can be calculated assuming classical physics, either numerically or by using the Fokker-Planck equation.[3] Quantum modifications of the classical results show up as added resonance peaks in the distributions, produced by two-state multiphoton transitions, and also as modifications resulting from quantum tunneling.

The ionization probabilities themselves also can be calculated assuming classical physics.[4] Here quantum effects can again show up via resonance phenomena.[6] Microwave field strength thresholds for observable (1% or higher) ionization probabilities approach static field quantum tunneling values only as the microwave frequency is decreased to 1/10 the initial electron orbit frequency and lower.[5] At higher frequencies, the presence of the ladder of bound-state energy levels that underlies the classically-chaotic ionization mechanism plays a major role in reducing ionization field strength thresholds by as much as an order of magnitude.[5]

When the microwave frequency becomes larger than the initial electron orbit frequency, theory predicts that quantum and classical physics give increasingly different ionization probabilities. The origin of the underlying quantum suppression of the diffusion is of considerable interest. It might be a special wave interference effect generated by (near) randomness, in an analogy with Anderson localization in condensed matter physics. Experiments to resolve this question are not yet carried out.

Our hydrogen-in-microwaves system is fascinating partly because it lies at the interface of two different areas of research, one being multiphoton absorption physics and the other being quantized nonlinear dynamics. Our problem can be approached using the methods of either area, with different views of the system's characteristics being the result.

In the present paper we primarily view the photon absorption in terms of a ladder of quantum energy levels that become strongly coupled at ionizing field strengths. New experimental results on the character of the transitions between rungs of the ladder will be presented. The evidence for semiclassical behavior just below and above the classical threshold field for chaos will be discussed.

PHOTON ABSORPTION PROCESSES

Observability thresholds for four different photon absorption processes in the excited hydrogen atom are sequentially traversed as the strength of the microwave field is increased from zero on up through the threshold for rapid ionization. In order, these processes are the time-oscillating Stark effect, two-state quantum transitions, perturbative electron delocalization and diffusive electron delocalization. Let us briefly consider each of these.

In problems involving linearly polarized strong electric fields, it is customary to introduce the parabolic stationary states of the hydrogen atom, as these are the zeroth order solutions when the atom is in a constant static electric field. They contain the physics of the presence of a very small electric field that does provide a field direction and does break the degeneracy in the atom to produce states of mixed parity that have large electric dipole moments. These states also pertain in nonhydrogenic atoms when the applied electric field is strong enough for the Stark interaction to exceed the Rydberg electron interactions with the atom's core. Let us for concreteness consider the extremal parabolic state with parabolic quantum numbers n, n_1, n_2, m = n, 0, n-1, 0. Then the electric dipole operator has diagonal matrix elements z_{nn} of magnitude $1.5n^2$, a factor of five or more larger than its off-diagonal matrix elements $z_{nn'}$. For this reason, the time-oscillating Stark effect becomes observable at field strengths where the other photon absorption processes are still unimportant. This effect involves the phase modulation of the atom's wavefunction via a phase factor whose time derivative is proportional to the instantaneous diagonal electric dipole interaction of the atom with the microwave field:[11]

$$\exp\ [iz_{nn}F\ \sin(\omega t)/\omega] = \sum_{K=-\infty}^{+\infty} J_K\ (z_{nn}F/\omega) \tag{1}$$

As in the phase modulation of radio waves, sidebands in the absorption spectra of Rydberg atoms are induced by strong microwave fields.[12] The Bessel function sideband amplitudes of equation 1 reach their first maxima when the argument is about equal to the order K. Thus the K = 1 sideband observability threshold in the absorption spectroscopy experiments is:

$$F_1 = 0.1/n^5 \text{ when } n^3\omega = 0.3. \tag{2}$$

At large field strengths and adequately separated Rydberg atom energy levels, well-defined envelopes of sidebands occur about each stationary-state energy level.[13] These envelopes have an energy width equal to twice the peak instantaneous first order Stark interaction, which can be comparable to the energy of many microwave photons. The presence of such wide envelopes containing many important sidebands is associated with many photon absorption and emission events occuring every microwave oscillation cycle, without however any change of atom quantum state or net change in atom energy. However, if the widths of adjacent sideband envelopes are large enough for a significant overlap, then real transitions can occur if there is adequate coupling between the states of adjacent values of n. The condition for envelope overlap is that envelope width equal the energy level spacing leading to:

$$F_2 = 0.3/n^5. \tag{3}$$

Note that this involved the diagonal matrix element. Equation 3 gives a lower bound for the field thresholds for many photon absorption processes in all atoms, and gives the ionization threshold for the case of nonhydrogenic atoms in states of low angular momentum.[14] In hydrogen,

318

however, the thresholds for absorption are higher than this lower bound.

For electrically polarized hydrogen, n, n_1, m = n, 0, 0, the off-diagonal matrix elements are largest for coupling between adjacent states and have the value $0.3n^2$.[15] Both the diagonal and off-diagonal elements enter crucially in the perturbative evaluation of the transition probability for multiphoton n-changing by one unit.[15,16] Said another way, the strong phton absorption and emission associated with the oscillating Stark effect and the more perturbative photon absorption associated with the off-diagonal elements are both important, in a way involving the coherent inerference of all possible time-ordered, energy-conserving elementary absorption processes. The perturbative evaluation of the fields necessary for observable n-changing transition probabilities are in reasonable agreement with experimental data [15] and give the threshold condition:

$$F_3 = 0.5/n^5 \text{ when } n^3\omega = 0.3. \tag{4}$$

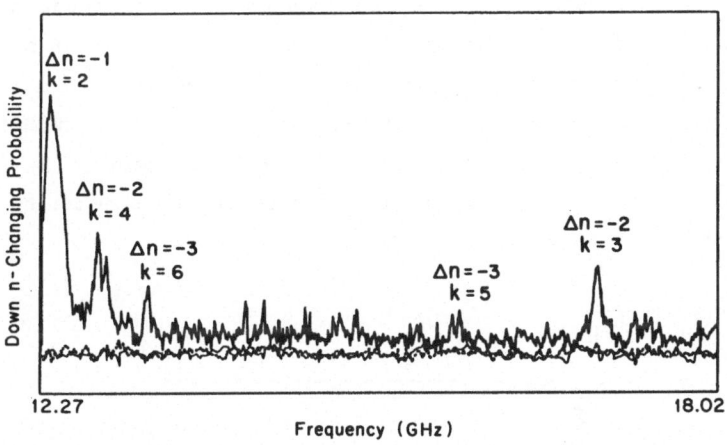

Fig. 1. The dependence on microwave frequency of n-changing microwave multiphoton transitions in electrically polarized hydrogen atoms with principal quantum number n = 63. Displayed is the relative probability for transitions to states with n less than 63, summed over such states. A static electric field of 8.0 V/cm was present. The microwave field was at an intensity of 22 mW/cm², which corresponded to a peak field of about 7.9 V/cm. Broadband microwave noise with an intensity of 0.03 mW/cm² was also present. Five quantal multiphoton resonances are identified, with their final states being verified by field ionization studies. Also seen are n-changing signals at most frequencies away from the resonances; these contain contributions primarily from atoms with n = 62 and 61. One of the zero reference level curves was obtained by turning the microwaves off, and the other by ionizing the n = 63 atoms just before their entrance into the microwave region, instead.

As the microwave field strength is further increased, the lowest order resonant n-changing transitions saturate, and higher order resonant processes become observable. Fig. 1 shows a spectrum in this regime for n-changing down from the initial state n = 63. Unlike earlier work carried out with a microwave exposure of 3000 field oscillations at frequencies in the 6 to 8 GHz region, this new data is for 100 oscillations in the region 12 to 18 GHz. The scaled frequency $n^3\omega$ now reaches 0.7, thrice that of the earlier work. In this region, we have observed large resonance shifts with increasing microwave field strength, see Fig. 2. Such shifting of quantum resonances may account for previously observed steps in the ionization probability as a function of microwave field strength.[2,17] Once these shifts are accounted for, the observed resonance frequencies are in reasonable accord with expected values contained in Fig. 3.

Identified in Fig. 3 are bundles of resonant quantum transitions of constant ratio of quantum number change to photon absorption order, as well as bundles of such transitions of constant ratio of increments in these quantities. As the field strength is further increased, the number and widths of saturated resonances increase, increasing the probability for the bundles to produce bands of strongly coupled quantum states. The population of such bands is conjectured to somehow reflect semiclassical quantum behavior when the threshold field strength for perturbative electron delocalization is reached. This threshold is defined by the coupling between states of adjacent values of n being about equal to one-third their energy level spacing. At $n^3\omega = 1$:

$$F_4 = (\Delta E_n/3)/(z_{n,n+1}F) = 1.0/n^5. \tag{5}$$

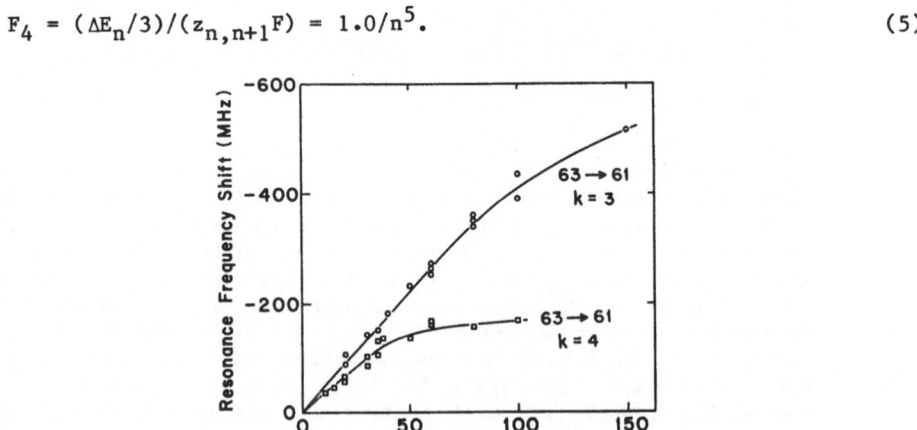

Fig. 2. Shifts in two of the quantal n-changing resonances shown in Fig. 1, as a function of microwave waveguide power. A power of 100 mW corresponded to a peak microwave field of about 11.8 V/cm.

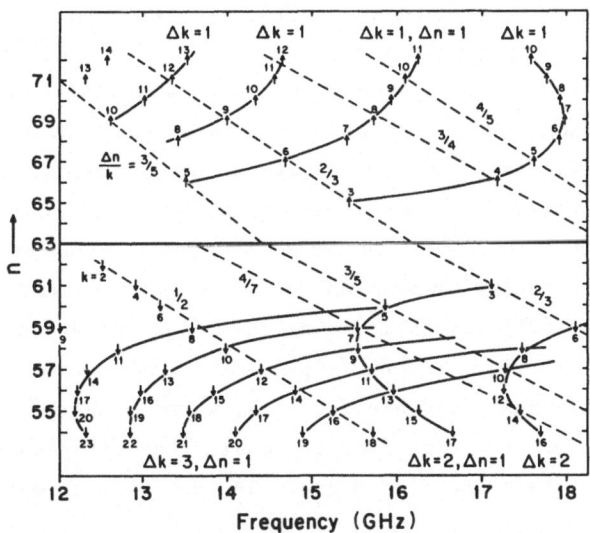

Fig. 3. Locations of some resonance frequencies for n-changing transitions from the n = 63 initial state and ending in various final states, at a static field strength of 8.0 V/cm. Straight lines give resonances of constant ratio of change in n to number k of photons absorbed. Curved lines connect resonances with the same ratios of increments in these quantities.

Note that this condition involves the off-diagonal matrix element, rather than (just) the diagonal one. When perturbative delocalization occurs, the original energy level structure of the atom is blurred out by large energy level widths produced by rapid state-changing microwave photon absorption and emission processes. This Rabi-broadened atom-in-microwaves is really one inseparable system that is free to demonstrate its own structure and dynamics. In view of the large values of actions involved, the composite system is expected to be described by semiclassical quantum mechanics and to have underlying classical behavior.

The question now is, what kind of semiclassical dynamics occurs just above the threshold fields given by equation 5? Those thresholds are some 30% lower than thresholds for 10% ionization probability.[2] Experimentally what is observed is significant amounts of n-changing to nearby states, with changes of n of one or two, up or down.[2] This occurs at most microwave frequencies, not just at those resonant for such n-changing; some of the evidence for this is shown in Figs. 1, 4, and 5. Below the classical threshold for chaotic time evolution, classical electron trajectories are quasiperiodic and evolve within a region of phase space having a range of action values that corresponds to roughly

five values of quantum number.[2] Thus it is tempting to ascribe the
observed local n-changing without ionization to a strong coupling of a
finite number of quantum states that produces a finite number of important frequencies in the quantum time evolution. Over long times the
energy of the electron remains within a small band of values, and the
time-averaged energy is close to a constant. Thus the observability of
quasiperiodic semiclassical evolution must be connected with energy-
nonconserving switching on and off of the microwave field; the switching
must not be quantally adiabatic. One then does not expect a large dependence of the n-changing signals upon microwave exposure time. Fig. 4
shows the experimental evidence for this, where a change in exposure time
of a factor 30 does not significantly modify the off-resonance n-changing
probabilities. The n-changing signals in the quasiperiodic semiclassical
regime thus depend strongly only on microwave field strength, but not on
either exposure time or microwave frequency. We should point out that
the experimental evidence does not in itself rule out unrelated transition probabilities for population transfer to the different final
states. The production of a coherent linear superposition of quantum
states during the field switching-on process is expected on theoretical
grounds.

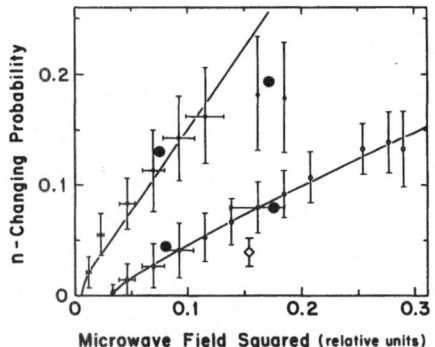

Fig. 4. Measured n-changing probabilities for n = 63 to 64 and to 62, as
a function of the square of microwave field strength, averaged
over frequencies away from observable resonance. The lines are
drawn through data taken in the 6-8 GHz region with a microwave
exposure time of 3000 field oscillations, see reference 2. The
four special points are for data in the 12-18 GHz region with an
exposure time of 100 oscillations, with broadband microwave
noise levels of 0.12-0.15 mW/cm^2. The single point is for
12-18 GHz at the reduced noise level of .03 mW/cm^2, as in Fig.
1.

Additive broadband microwave noise produced by travelling wave tube
microwave amplifiers also induces local n-changing over wide frequency
ranges, since the noise can provide a photon at the needed frequency to

make an otherwise nonresonant multiphoton n-changing transition quantally resonant. We have observed and studied such noise effects. These are particularly easy to detect in our recent experiments as the cutoff frequency of our waveguide microwave system produces a corresponding sharp frequency cutoff near 15 GHz in the noise-enhanced portion of the signals for n-changing down shown in Fig. 1. At the noise power level for the data in this figure, noise enhancement is reduced to about 20% or less of the off-resonance n-changing produced by the monochromatic microwave field alone. Fig. 4 shows a point for this low noise data. It appears that the other n-changing data may have suffered noise enhancement of a factor of two.

We finally consider the fourth process, at the highest range of microwave field strengths, where ionization probabilities become large enough to be observable. At such field strengths, our past studies have uncovered regions of microwave frequency where extended n-changing occurs to final states with quantum numbers above the narrow band ascribed to quasiperiodic dynamics.[2,18,19] The final state distribution for this additional n-changing is often exponentially degreasing with increasing change in quantum number, that is, with increasing change in classical action. The exponential behavior is expected for a classical diffusion process and has been seen in numerical quantum calculations.[20] Diffusion is expected to weaken as the normalized microwave frequency $n^3\omega$ is increased, with quantum resonance effects becoming stronger. Fig. 5 shows some recent data taken at a frequency about three times that of earlier data. Outside of the central quasiperiodic peak, quantum resonance peaks may dominate this final state distribution.

In view of the evidence given above that some broadband microwave noice enhancement may have been present in the quasiperiodic peaks of past final state distributions, one may ask whether such enhancement played a role in the rest of those distributions and in the ionization probabilities that are the end product of the quantized diffusion process. So far we have not found any evidence of this. Fig. 6 shows curves of ionization probability as a function of microwave power, for low, typical and considerably larger levels of microwave noise power. At low noise, the value of n^4F for 10% ionization probability is 0.026 a.u., in excellent agreement with other data.[5] The typical level of added noise power does not change the ionization curve at all. We tentatively ascribe this to the fact that the bandwidth of the noise in all experiments so far is insufficient to enhance ionization except at high noise levels.

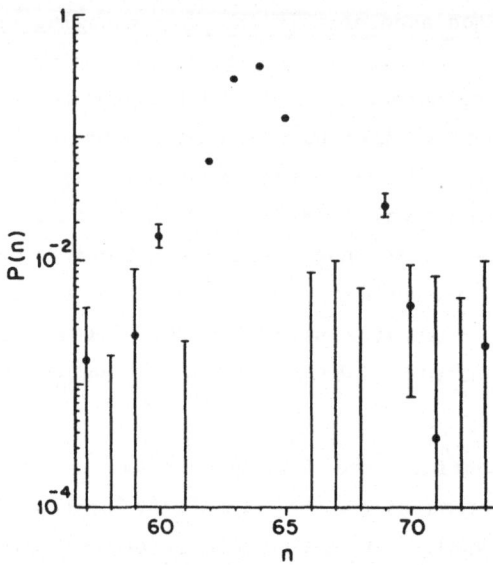

Fig. 5. Preliminary n-changing final state distribution for n = 63 atoms in a microwave field of strength 11.8 V/cm and frequency 17.337 GHz. The static field strength was 8.0 V/cm and the noise level 0.12 mW/cm^2. The ionization probability was measured to be 15%.

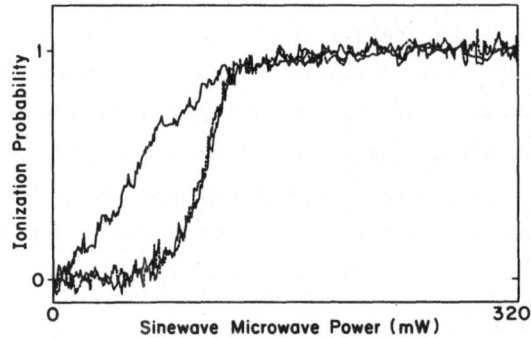

Fig. 6. Ionization probability as a function of microwave power, for n = 63 and a frequency of 17.501 GHz. Curves for three levels of microwave noise are shown, with those at intensities of 0.25 and 1.0 mW/cm^2 being identical within experimental errors. A noise level of 9 mW/cm^2 enhances the ionization. The full scale value of microwave power coresponded to a peak microwave field of 21 V/cm.

Broadband microwave noise of sufficient bandwidth is expected to enhance ionization probabilities at low noise power levels, as the enhancement observed in the local n-changing would then occur for photon absorption throughout the entire ladder of energy levels leading up to the continuum. Indeed, noise-induced diffusion and ionization should occur without application of a monochromatic field. This different type of diffusion process in a quantum system is being pursued in our laboratory, and may provide new information on stochastic interactions in quantum systems.[21]

The authors thank the U.S. National Science Foundation for support in this research.

REFERENCES

1. J. E.Bayfield, Comments At. Mol. Phys. 20, 245 (1987).

2. J. E. Bayfield, in Quantum Measurement and Chaos, E. R. Pike, ed. (Plenum Press, New York, 1987).

3. G. Casati, I. Guarneri, and D. L. Shepelyansky, Phys. Rev. A36, No. 5 (1987), and references therein.

4. R. V. Jensen, in Atomic Physics 10, H. Narumi and I. Shimamura, eds., (North-Holland, Amsterdam, 1987), pages 319-322.

5. P. M. Koch, K. A. H. Van Leeuwen, O. Rath, D. Richards and R. V. Jensen, in The Physics of Phase Space, Y. S. Kim and W. W. Zachary, eds., Lecture Notes in Physics, No. 278, (Springer-Verlag, Berlin, 1987), pages 106-113.

6. R. Blumel and U. Smilansky, in Proceedings of the Conference on Chaos and Related Nonlinear Phenomena, I. Procaccia, ed. (Plenum Press, New York, 1987).

7. J. G. Leopold and D. Richards, J. Phys. B18, 3369 (1985).

8. B. V. Chirikov, Phys. Reports 52, 263 (1979).

9. E. J. Heller, Phys. Rev. Lett. 53, 1515 (1984).

10. E. J. Heller, Phys. Reve. A35, 1360 (1987).

11. P. W. Langhoff, S. T. Epstein, and M. Karplus, Rev. Mod. Phys. 44, 602 (1972).

12. J. E. Bayfield, L. D. Gardner, Y. Z. Gulkok and S. D. Sharma, Phys. Rev. A24, 138 (1981).

13. L. A. Bloomfield, R. C. Stoneman and T. F. Gallagher, Phys. Rev. Lett. 57, 2512 (1986).

14. T. F. Gallagher, in Photons and Continuum States of Atoms and Molecules, N. K. Rahman, C. Guidotti and M. Allegrini, eds. (Springer-Verlag, Berlin, 1987), pages 2-7.

15. J. N. Bardsley and B. Sundaram, Phys. Rev. A32, 689 (1985).

16. W. J. Meath and E. A. Power, J. Phys. B17, 763 (1984).

17. P. M. Koch, in Rydberg States of Atoms and Molecules, R. F. Stebbings and F. B. Dunning, eds. (Cambridge University Press, New York, 1983).

18. J. E. Bayfield and L. A. Pinnaduwage, Phys. Rev. Lett. 54, 313 (1985).

19. J. N. Bardsley, B. Sundaram, L. A. Pinnaduwage and J. E. Bayfield, Phys. Rev. Lett. $\underline{56}$, 1007 (1986).

20. G. Casati, B. V. Chirikov, D. L. Shepelyansky and I. Guarneri, Phys. Reports, (1987).

21. T. Dittrich and R. Graham, Europhys. Lett. $\underline{4}$, 263 (1987), and reference therein.

CHAOS IN ONE-DIMENSIONAL HYDROGEN

D. C. Humm and Munir H. Nayfeh

Department of Physics
University of Illinois at Urbana-Champaign
1110 West Green Street
Urbana, IL 61801

We analyze the effect of a DC electric field on classical chaos in one-dimensional hydrogen in a microwave field. We show that, under ordinary experimental conditions, the system is expected to undergo a transition from a regime of classical behavior to a regime of uniquely quantum behavior as the DC field is increased, for an unclamping DC field. We also show that ionization by classical chaos competes favorably with ionization by tunneling in the transition region, and that tunneling allows very sensitive spectroscopy of this region.

The flourishing of the study of non-integrable classical systems has stimulated interest in the equivalent quantum systems, especially because quantum mechanics was developed and tested from analogies with and observations of systems which classically are integrable.[1] A great deal of theoretical effort has been expended in an attempt to understand the quantum dynamics of systems which are classically chaotic.[1-6] A wide variety of hypotheses have been advanced. It has been widely stated[3] that the mathematical structure of quantum mechanics is incapable of producing chaos in certain physical situations in which classical mechanics leads to chaos. Since chaos is now accepted by some as the correct deterministic foundation of the laws of classical statistical mechanics, the viability of the standard laws of quantum mechanics as a foundation of quantum statistical mechanics has been called into question.[1] Computer simulations have shown that, in some situations, quantum correlation functions decay more slowly than their classical counterparts.[4] Other simulations have exhibited the diffusive energy growth expected for classical chaos, but it stopped after a break time.[3] These results have led to the idea of a quantum "quenching" of chaos. However, studies of other situations have shown that, under certain conditions, there exists a classical limit in which the quantum motion mimics the classically chaotic behavior for a finite time. This classical limit can depend not only on the principal quantum number n but also on whether the frequency of the external driving force is greater than or less than the frequency inherent in the system[5], and on the number of quantum states "trapped" in a given nonlinear resonance[4]. Other analyses have predicted a profound alteration in the spectrum of the quantum system at the point at which the equivalent classical system becomes chaotic.[4,6]

Given the wide variety of theoretical predictions for different quantum systems which classically exhibit chaos, it would seem useful to do experiments on a quantum system which not only exhibits classical chaos, but also has an easily varied parameter that takes one between regimes of different types of behavior, so they can be compared. It would be especially useful to have a parameter that, when changed, would bring the system from a regime in which it is expected to behave classically and exhibit classical chaos to a regime in which effects peculiar to quantum mechanics would be predicted. It would also be best if the system were as simple as possible, so as to be amenable to theoretical analysis and numerical simulation. Finally, it would be even more valuable if, in addition to measuring the time evolution or at least the end result, we could measure the spectrum of the system. The "one-dimensional" hydrogen atom in both DC and AC electric fields, or the surface state electron in an unclamping DC electric field and an AC electric field, is such a system, with the DC electric field the described parameter. We present the following analysis to show that this system does satisfy the above requirements, and to facilitate the design of such experiments and the comparison of their results with the classical theory.

Both the surface state electron and the "one-dimensional" hydrogen atom in AC electric fields have been extensively studied, both theoretically and experimentally.[5,7,8] Even a DC field has been used, but only to prepare the one-dimensional hydrogen atoms, not as a variable parameter to alter the system.[8] The one-dimensional hydrogen atoms are prepared by selecting atoms in a quantum state in which the motion of the electron is chiefly in one direction. However, there always remains a small residual motion in the other directions, and the validity of the one-dimensional approximation is always an important question in the experiments on hydrogen.[5,9] The surface-state electron does not suffer from this problem. Its purely one-dimensional motion perpendicular to the liquid helium surface can be compeletely decoupled from its motion in the other two directions. However, the "one-dimensional" hydrogen atoms are more accessible to experiment, and so we will gear our results to describe them, and give scaling laws to allow the application of our results to surface state electrons. The current experiments all seem to be in the classical limit in which the theories of classical chaos and quantum mechanics agree, and they agree with experiment.[10,11,12]

The unclamping DC field, as opposed to a clamping field (previously analyzed for a high DC field[13]), is important for a number of reasons, including ease of detection of the onset of chaos. With a clamping field, surface state electrons never ionize, and the electron in a one-dimensional hydrogen atom can only ionize by a process violating the one-dimensional approximation. It is precisely ionization which has been used as an indication of the onset of chaos as the AC field is increased. The measurement of this threshold has been a major basis for comparison with theory, both classical and quantum mechanical.[7,10]

With an unclamping field, one could not only measure the onset of chaos by measuring the onset of ionization, but also get spectra of the system near the threshold of chaos by resonant ionization spectroscopy (RIS). One can excite to a given energy with a tunable laser, and observe the ions produced when an electron tunnels away. The unclamping field gives an avenue for the electron to tunnel away. Since current detectors are capable of single-ion detection, this technique is more sensitive than any other spectroscopic technique. It is also well-established as a tool, and has been used to study the spectrum of hydrogen in a strong DC field[14,15] (one would need only to add the microwaves). The technique does have a few limitations. The atoms are excited within the microwave field, which may disturb the production of the one-dimensional states and thus invalidate the one-dimensional approximation. In current experiments[8], the states are prepared first and then the microwaves are applied. However, although the technique does not, apparently, allow selective excitation, it may allow selective detection. One would expect the extreme one-dimensional unclamped states to be the first to ionize by tunneling, as well as the first to ionize by chaos.[10] Therefore, one would expect there to be a range of DC and AC fields for which only the one-dimensional results would be seen. One would be able to study the spectrum using the one-dimensional approximation within this range. It is also important to note that the spectrum even outside this range, though perhaps not one-dimensional, would only contain m=0 states if all external fields are in the z direction. This is a considerable simplification over the general three-dimensional problem, and may be amenable to theoretical analysis or computer simulation.[9] Since one would expect to see structure in the spectrum with peak separations on the order of the frequency of the applied microwaves[9], a laser bandwidth of less than the microwave frequency is necessary. This is consistent with current work[14,15], and improvements are possible. Finally, since one is looking for chaos effects, one has to work in a region in which the atoms are not ionized by tunneling before chaos has a chance to take place. The atoms must tunnel, but they shouldn't tunnel too much. This effect is analyzed later in this paper; the region that one can study is precisely the region that is most interesting.

The classical nonlinear oscillator model is

$$H(r,p,t) = p^2/2 - Z/r + F_0 x + F_1 x \cos\Omega t \qquad (1)$$

where F_0 is the strength of the DC field, F_1 is the amplitude of the microwave field, and Z is the nuclear charge. We can use cylindrical symmetry about the direction of the fields to reduce the system from six to four dimensions in the action-angle phase space.

In the case of strong F_0 one can approximate the Hamiltonian of the nonlinear oscillator to

$$H(r,p,t) = p_x^2/2 - Z/x + F_0 x + F_1 x \cos\Omega t \qquad (2)$$

which reduces the system to two dimensions in the action-angle phase space. Z=1 for the hydrogen atom; it is 7.1×10^{-3} for the surface state electron.[7] The validity of the one-dimensional approximation for selectively excited hydrogen atoms has been studied to some extent. It appears to be valid in at least some situations.[5,8,9] In fact, the 1D approximation can even be used to estimate the AC field chaos threshold for hydrogen atoms not selectively excited, because the nearly one-dimensional states are expected to be the the first to ionize, at the lowest AC fields.[10] Here we will assume that the one-dimensional approximation can be applied. We expect it to apply well for high n with AC fields in the region of the chaos threshold.[9]

This system, in the absence of the external DC field (F_0=0) was extensively analyzed, by numerical simulation and the nonlinear resonance overlap method.[7] It describes the system of an electron located over a liquid-helium surface and interacting with an oscillatory field polarized perpendicular to the surface. Recently some aspects of the effect of a positive clamping DC field on the electron-surface system have been studied using the resonance overlap criterion. Specifically, the studies were oriented towards the determination of the number of quantum levels that the nonlinear resonances trap, and their dependence on the clamping field. Asymptotic results were found giving a quadratic dependence in the limit where the clamping field dominates the coulomb field.[13] Here we analyze the system for positive and negative clamping fields.

We use a canonical transformation to transform the Hamiltonian given in eqn. 2 to

$$K=\alpha(I)+F_1 x(\theta,I)\cos\Omega t \tag{3}$$

where I and θ are the action and angle variables. The constant α is

$$\alpha=p_x^2/2-1/x+F_0 x \tag{4}$$

The variables θ and I are chosen such that $x(\theta,I)$ is periodic in θ with a period of 2π. Thus we expand $x(\theta,I)$ in the Fourier series[18]

$$F_1 x=\sum_{m=-\infty}^{\infty} V_m(I)\ e^{im\theta} \text{ with } V_m=V_{-m} \tag{5}$$

which when substituted in eqn. 3 and rearranging the terms while using the fact that $V_m=V_{-m}$ because x is symmetrical about $\theta=0$ gives

$$K=\alpha\ (K)+\sum_{m=0}^{\infty} V_m(I)\cos\ (m\theta-\Omega t) \tag{6}$$

The problem is first solved for weak time dependent fields. The effect of higher intensities will be determined as a distortion to the weak field solution. If V_m is weak, the microwave field can only affect the system significantly when the phase $m\theta-\Omega t=0$. This condition is called the mth resonance. Thus, when the system is excited to near the mth resonance, it can only interact strongly with this resonance in the weak field limit, and consequently the expansion in eqn. 6 can be separated as follows:

$$K=\alpha(I_m)+V_m(I_m)\cos(m\theta-\Omega t)+f(t) \tag{7}$$

where

$$f(t)= \sum_{n=m} V_n(I_n)\cos(n\theta-\Omega t) \qquad (8)$$

contributes very little to the interaction in the weak field limit, and hence can be neglected. The system of eqn. 7 with $f(t)$ neglected can now be solved by making another transformation to get rid of the explicit time dependence. The new variables Δ and ψ are defined as follows:

$$I-I_m=m\Delta \quad \text{and} \quad \psi=m\theta-\Omega t \qquad (9)$$

The new corresponding generating function is

$$F(\Delta,\theta,t)=(I_m+m\Delta)(\psi-\Omega t/m) \qquad (10)$$

with the transformation equations:

$$I=\partial F/\partial\theta \ , \quad \psi=\partial F/\partial\theta \qquad (11)$$

The transformed Hamiltonian becomes

$$K'=\alpha(I)-\Omega I/m+V_m(I)\cos\psi \qquad (12)$$

It is to be noted that the new variable $m\Delta$ is just the small change in the action of the mth resonance caused by the weak microwave field. The new variable ψ measures the phase difference between the phase of the system $m\theta$ and that of the external field Ωt. In order to see the nature of K' further, we can examine it in the limit of very small V_m, that is in the limit of small Δ and ψ. Taking

$$V_m(I) \approx V_m(I_m) \qquad (13)$$

$$\alpha(I) \approx \alpha(I_m) + \left.\frac{d\alpha}{dI}\right|_{I_m} m\Delta + \frac{1}{2} \left.\frac{d^2\alpha}{dI^2}\right|_{I_m} m^2 \Delta^2 \qquad (14)$$

$$\frac{d\alpha}{dI} = \frac{d\theta}{dt} = \omega \qquad (15)$$

we get:

$$K' = C_1 (\Delta^2 + V_m \cos \psi) + C_2 \tag{16}$$

where $C_1 = \frac{1}{2} m^2 \frac{d\omega}{dI}$ and $C2 = \alpha(I_m) - \frac{\Omega}{m} I_m$ are constants.

Hamiltonian (16) is a pendulum Hamiltonian, where Δ and ψ can be associated with the angular momentum and the pendulum angle respectively.

From K' we can determine the phase and the angular momentum of the system near a resonance. Fig. 1 shows a typical plot of Δ versus ψ (pendulum response). Actually I is plotted instead of Δ. In fact, the plotting of I versus ψ is somewhat misleading, since the ψ variable is different for each resonance, but it is qualitatively correct. The seperatrix divides two kinds of motions where on side the pendulum always swings in the same direction, while on the other side it swings back and forth. Moreover, each seperatrix is symmetric and its width is given in terms of the maximum Δ. Since n maximizes for $\psi=0$ and $\psi=\pi$, one can show that near the mth resonance:

$$(I-I_m)_{max} = 2 \sqrt{V_m/(d\omega/dI)} \tag{17}$$

We can now discuss the case in which the amplitude of the microwave field F_1 is not weak and the term f(t) in eqn. 7 becomes significant. First, the width of the resonances given by eqn. 17 will increase since V_m is proportional to F_1. Second, the f(t) term acts as an additional time dependent perturbation embodying the interaction between the otherwise non-interacting resonances.

What happens to this system when we include the time-dependent perturbation f(t)? Trajectories far away from the separatrix are affected very little. Trajectories near the separatrix are strongly affected, because the time-dependent external force can push them to one side of the separatrix or the other, changing their motion completely. For trajectories very close to the separatrix, the side of the separatrix they end up on in the next cycle depends not at all on which side of the separatrix they start on, but rather which way the force happens to be pushing them along the way. Since different trajectories take different amounts of time to complete a cycle, especially near the separatrix, this pushing can be completely different for two very close trajectories. Very close to the separatrix, the trajectories get all twisted up. Two nearly the same trajectories may end up going in opposite directions, and two different ones in the same direction.

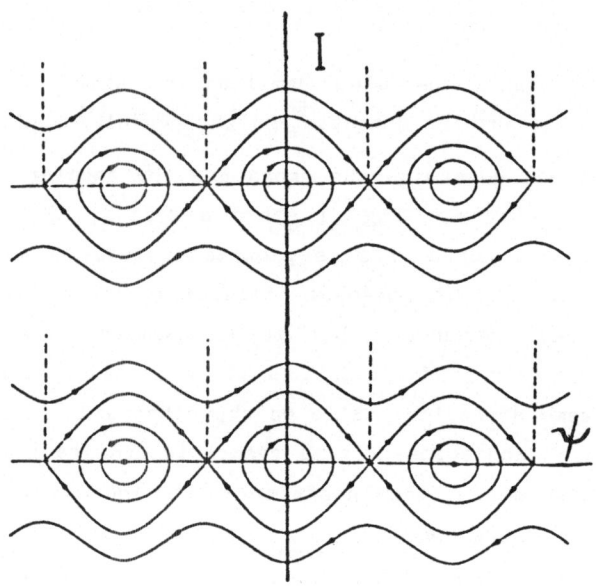

Figure 1.   Schematic of nonlinear resonances.

Chaos Thresholds at Given DC Fields

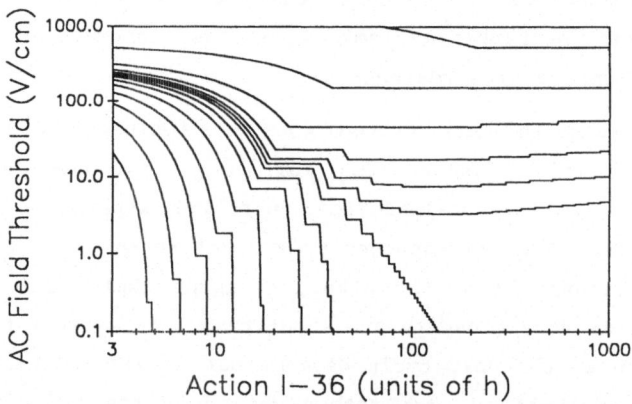

Figure 2.   Chaos thresholds. From the lower left corner to the upper right: E_{DC} = -240, -200, -160, -120, -80, -40, -20, 0, 10, 20, 40, 100, 300, 1000 V/cm. This figure scales in the following way as Ω changes: $I' = (\Omega'/\Omega)^{-1/3}I$, $F'_{AC} = (\Omega'/\Omega)^{4/3} F_{AC}$, $F'_{DC} = (\Omega'/\Omega)^{4/3}F_{DC}$. It scales like the following when Z is changed: $\Omega'=\Omega$, $I' = (Z'/Z)^{-2/3}I$, $F' = (Z'/Z)^{-1/2}F$.

If one simulates this motion on a computer, it looks irregular and disordered. Researchers have searched through their computer results for some kind of correlation and order in this motion but found none. All correlations that have been examined die out within a fairly short time. It appears that a random motion has arisen from a deterministic system. This is called "chaos".

As the external field is increased, the system undergoes a change from a situation in which most orbits are regular to one in which most are chaotic. A reliable estimate to this threshold of chaos is the external field at which two adjacent resonances overlap.[18] This estimate has proven correct in varying situations.[7,17] The transition to global chaos happens within a factor of about 2 in AC field of the resonance overlap estimate in the problem with no DC field.[7]

The pendulum approximation is not as good when the transition to global chaos has occurred for all resonances but one is trying to see if the m=1 separatrix extends down low enough to encompass states of low I (otherwise, those states will still be stable, even if global chaos occurs). In this case, we compute the actual separatrix of the Hamiltonian K', following Jensen.[7]

To make this estimate of the threshold of chaos, we need to calculate $V_m(I_m)$ and $d\omega/dI$ at I_m for all m. In the absence of the DC field, it is possible to get simple analytical expressions for these quantities. However, in the presence of the DC field it is not possible to do so except in some asymptotic regions where the DC field dominates the Coulomb field or vice-versa. Since we are interested in field strengths that are comparable to the Coulomb field strength, then we calculated these numerically.

One can calculate α and I_m by taking (from Eq. 4)

$$H_0(x,p_x)=H_0(x, \frac{\delta W}{\delta x}) = \alpha \qquad \text{or} \qquad (18)$$

$$\frac{1}{2} \left(\frac{\delta W}{\delta x}\right)^2 - \frac{1}{x} + Fx = \alpha \qquad (19)$$

where $Px= \frac{\delta W}{\delta x}$ is evaluated for constant α and

$$I = \frac{1}{2\pi} \oint P_x dx = \frac{1}{2\pi} \oint \frac{\delta W}{\delta x} dx \qquad (20)$$

Moreover we have the fact that $p_x=0$ when x reaches its maximum value x_{max}. From this, we have $\partial W/\partial x=0$ at x_{max}, or $\alpha-Fx+1/x=0$ which can be solved for x_{max}. In addition, at a resonance $m\theta - \Omega t = 0$ or $m\partial\theta/\partial t = \theta$ which gives $m\omega = \Omega$. But $\omega = d\alpha/dI$, thus $d\alpha/dI = \Omega/m$. Thus by solving the first half of the cycle in which x goes from 0 to x_{max} (corresponding to $0 < \theta < \pi$) we can get the result for the full cycle from symmetry. We first calculate $I(\alpha)$ and then invert it to get $\alpha(I)$.

We now present our results on the effect of the DC field on the dynamics of the system. In Fig. 2 we plot the classical chaos AC field threshold for an electron prepared in a state of given initial action for a number of different DC fields at an AC frequency of 30.5 GHz. All plots in the paper can be scaled to any frequency; the scaling laws are given in the captions. We make the following observations making use of the curve corresponding to $F_0 = 0$. It is apparent that the threshold drops essentially in steps as the binding energy of the electron decreases (or equivalently as I increases). As I increases from low values, the threshold drops dramatically till I approaches the center of the first resonance I_1, where it levels off and a plateau develops. As I increases to a value between I_1 and I_2, the center of the second resonance, the threshold drops again, and levels out until approaching I_3, and so on.

Also, note that for negative or zero DC fields, the plateaus get narrower as I increases, whereas for positive fields they get wider as I increases. Thus, if the resonances get closer to each other as I increases, once overlap takes place between the lowest two resonances, the higher resonances will also overlap and the electron will ionize. But for a positive clamping field, the plateaus get wider and eventually the thresholds stop dropping and start increasing. Although local stochasticity over a few resonances can be achieved, the electron will never ionize (without breaking the one-dimensional approximation) because it is bound in an infinitely high well.

There are good theoretical reasons to believe that the number of quantum states trapped in the first nonlinear resonance determines whether the system is in the classical or purely quantum limit, at least in the low-frequency region (below the first resonance).[4] The system is expected to behave classically when the number of states trapped is $\gg 1$, and non-classically when the number of states trapped is $< \approx 1$. With this in mind, we plot in Fig. 3 the number of states trapped in the first resonance for a clamping DC field, and in Fig. 4 the same for an unclamping field. These quantities can be read off the Fig. 2 graph; they are simply the widths of the first nonlinear resonances for the different DC fields. The action measured in units of h is equal to the principal quantum number n in the Wilson-Sommerfeld semiclassical theory

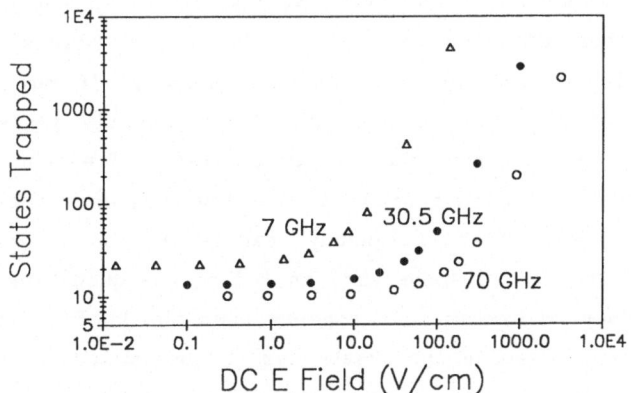

Figure 3. Number of states trapped in the m=1 nonlinear resonance as a function of positive DC E-field. Scaling: $I'=(\Omega'/\Omega)^{-1/3}I$, $F' = (\Omega'/\Omega)^{4/3}F$. As a function of Z: $\Omega'=\Omega$, $I' = (Z'/Z)^{-2/3}I$, $F' = (Z'/Z)^{-1/2}F$.

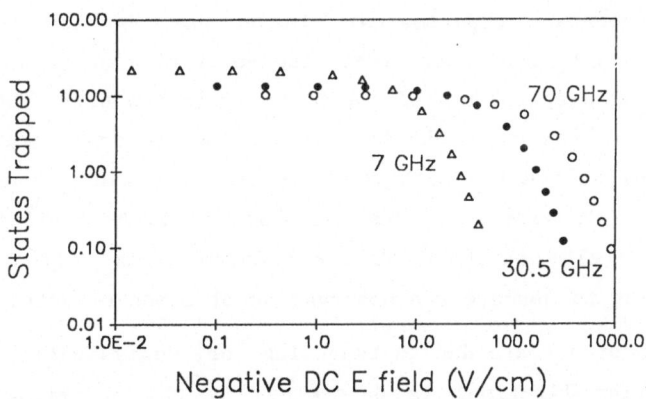

Figure 4. Number of states trapped in the m=1 nonlinear resonance as a function of negative DC E-field. The scaling is the same as in previous graphs.

of quantum mechanics; it is approximately equal for large n in the WKB approximation.[19] Thus, the width of a resonance is a good approximation for the number of states trapped. It is clear that the unclamping field qualifies as a parameter that allows us to go from the classical limit to the uniquely quantum limit. One can see from the graphs that this limit can be reached with easily available microwave frequencies and DC fields. Another advantage of the unclamping DC field is that, by reducing the inherent frequency of the 1D system, it may allow one to more easily reach the regime in which the externally imposed microwave frequency is greater than the inherent system frequency. Uniquely quantum effects which do not occur in the low-frequency regime have been predicted for this high frequency regime by a quantum-mechanical numerical simulation of the system[5], including the quenching of classical chaos by the quantum system. It appears that the DC field allows one to quickly and easily examine the system for various values of the number of trapped states, down to the quantum regime, and helps one to reach a new regime in the frequency ratio as well. Note that if the hypothesis of the quantum quenching of chaos is true, than the application of the unclamping field, a field trying to rip the electron away from the atom or surface, might actually be a stabilizing factor because it brings the system into the quantum regime, a surprising result.

The unclamping field has many advantages as an experimental method to study the quantum manifestations of classical chaos. There is, however, an important experimental limitation. We are detecting the presence of chaos by measuring either the onset of ionization or enhanced line widths in the spectrum. If the DC field is too high, then the atoms will ionize by quantum-mechanical tunneling before they ionize by classical chaos, obscuring both the enhanced line widths and the chaos-produced ions. We have done some calculations on this effect to see if there are DC fields which reach the quantum regime without producing enough tunneling to obscure the observation of chaos effects.

The width of a state due to tunneling (or, equivalently, the inverse of the tunneling lifetime) is determined by the tunneling integral τ given by

$$\tau = \int_a^b \sqrt{2(V-E)} \ dx \tag{21}$$

Parameters a and b are the inner and outer boundaries of the classically forbidden region. The tunneling integral τ is simply the imaginary part of the wavefunction's accumulated phase in the classically forbidden region. In the WKB approximation[19] (which will be sufficiently accurate for our purposes provided n>>1 and τ>1), the width of a state due to tunneling Γ depends only on τ and the energy separation between adjacent states (inverse of the density of states). It is closely approximated by (full width at half maximum)

$$\Gamma \frac{dn}{dE} = \frac{1}{2} e^{-2\tau} \tag{22}$$

for τ>1.[16,19] This expresses Γ as a fraction of the state spacing, or, equivalently, the characteristic period of the system divided by the tunneling lifetime. When τ=1.5, the system on average goes through 40 cycles before ionizing by tunneling, and that number increases very rapidly as τ increases. Since one would generally expect, in usual experimental situations involving this type of system, a system ionizing by chaos to do it within about 40 cycles[7], then when τ=1.5 one would expect the chaos lifetime to be similar to or shorter than the tunneling lifetime, and so the observed effects of chaos (including ionization and widening of line widths) would be at least as strong as those of tunneling, and therefore observable in the presence of tunneling. The τ>1.5 limitation on observing chaos is a conservative one and of course assumes that one's equipment is sufficiently good to measure tunneling. It also makes the most conservative estimate of the effect of classical chaos on the quantum spectrum; it broadens the lines, which it must do if it creates ionization within a certain period of time. If τ>1.5, this effect can be detected against a tunneling background. Of course, chaos is likely to also have other effects on the spectrum[6,9,20], in which case chaos would be even easier to detect.

We calculated τ as a function of unclamping DC field for the three frequencies 7 GHz, 30.5 GHz, and 70 GHz and with initial electron energy at the bottom of the m=1 and m=2 resonances at the threshold of global chaos. It is plotted in Fig. 5. We use the bottom of the m=1 resonance to see the probability of tunneling from the electron's initial state. Of course, chaos cannot be observed if the electron tunnels away from its initial position before the chaos has a chance to take place. However, there is another question to be asked. The nonlinear resonances form before the onset of global chaos. Perhaps an electron will be carried higher in energy by the first nonlinear resonance, and then tunnel away from that new, higher energy. It is only when the electron reaches the second resonance, implying that the first and second resonances have overlapped, that one can be sure the threshold of global chaos has been reached. Therefore, it may be more sensible to look at the tunneling

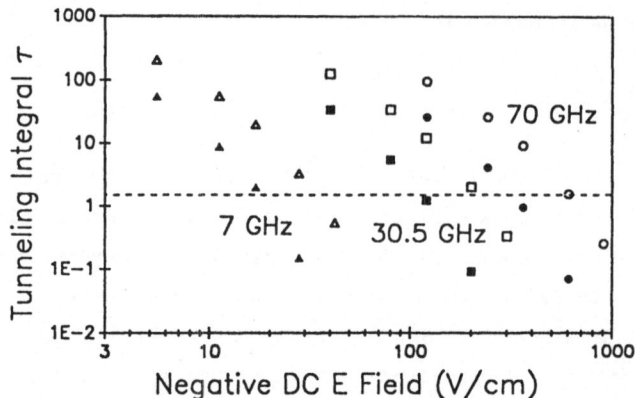

Figure 5.   Tunneling integral τ as a function of negative DC E field, for states at the bottom of the m=1 nonlinear resonance (open symbols) and the bottom of the m=2 nonlinear resonance (filled symbols) for three frequencies of AC field at the threshold of chaos. The tunneling integral scales as follows: $\tau' = (\Omega'/\Omega)^{-1/3}\tau$ at constant Z, and $\tau' = (Z'/Z)^{-2/3}\tau$ at constant Ω. The dotted line at τ=1.5 is a boundary below which chaos would be obscured by tunneling.

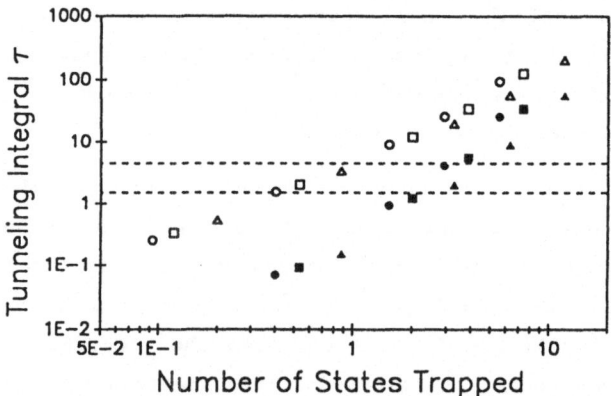

Figure 6.   Tunneling integral τ as a function of the number of states trapped in the m=1 nonlinear resonance, from the bottom of the m=1 (open symbols) and m=2 (filled symbols) resonances at the threshold of chaos. The triangles are 7 GHz, the squares 30.5 GHz, and the circles 70 GhZ. The dotted lines delineate the area within which resonant ionization spectroscopy is possible. Notice that it stretches from fewer than one state trapped to more than one state trapped.

from the bottom of the m=2 resonance, because one can say the electron experienced chaos if it reaches m=2 and tunnels from there. The tunneling integral τ, like Ω, F_0, and F_1, scales as a function of frequency, as it is entirely classical except for a <u>factor</u> of h; the scaling law is given in the caption to Fig. 5.

Of course, the question we are interested in is not how high a DC field can be reached without tunneling obscuring the results, but rather how few states trapped can be reached without tunneling becoming a problem. With that in mind, we combine Figs. 4 and 5 and plot in Fig. 6 the tunneling integral τ as a function of the number of states trapped in the m=1 resonance at the threshold of chaos. Comparing with tunneling from the initial state, one clearly reaches the quantum regime at τ=1.5, with about 0.45 states trapped at 30.5 GHz. The m=2 criterion puts one in the very beginning of the quantum regime at τ=1.5, with about 2 states trapped at 30.5 GHz. Note that the number of states trapped at τ=1.5 is fairly insensitive to frequency.

If one is trying to obtain a spectrum using RIS, instead of just looking at ionization by chaos, then there is also a maximum τ criterion. For a typical experimental setup (ref. 14 plus microwaves), the atom must ionize by tunneling within a microsecond in order to be detected. This criterion gives a maximum τ of about 4.5 for the RIS experiment at 7 GHz (the τ criterion, calculated from eqn. 22, is slightly larger for the other frequencies plotted). This maximum τ criterion, as well as the minimum, is plotted in Fig. 6. Although the band of observable spectra is fairly narrow in τ, it spans a wide range in the number of states trapped. If one looks at the spectrum low in the first resonance, one can go down to about 0.5 states trapped, and if one looks high in the first resonance, one can see the spectrum for as many as 5 states trapped. Since agreement with classical chaos predictions has been observed for as few as 20 states trapped[10], one would expect 5 states to show at least some classical chaos character. Thus, the spectrum measurement would span the region from the clearly quantum-mechanical to the classical.

In conclusion, we find the unclamping field to be a promising technique for bringing the system, be it selectively excited hydrogen atom or surface state electron, from the currently studied classical regime to the uniquely quantum regime. Ionization by classical chaos can compete favorably with ionization by tunneling in usual experimental situations, and tunneling allows very sensitive measurements of the quantum spectrum.

References

1. J. Ford, in The New Physics, eds. S. Capelin and P. C. W. Davies Cambridge Univ. Press 1986).

2. For example, see Stochastic Behavior in Classical and Quantum Hamiltonian Systems, Lecture Notes in Physics V. 93, eds. G. Casati and J. Ford, (Springer-Verlag 1979); Chaotic Behavior in Quantum Systems, ed. G. Casati, (Plenum 1985).

3. G. Casati, J. Ford, I. Guarneri, and F. Vivaldi, Phys. Rev. A 34:1413 (1986).

4. G. P. Berman, G. M. Zaslavsky, and A. R. Kolovsky, Phys. Lett. 87A:152 (1982).

5. G. Casati, B. Chirikov, D. Shepelyansky, and I. Guarneri, Phys. Rep. 154:79 (1987).

6. I. C. Percival, J. Phys. B 6:L229 (1973), and N. Pomphrey, J. Phys. B 7:1909 (1974).

7. R. V. Jensen, Phys. Rev. A 30:386 (1984).

8. J. E. Bayfield and L. A. Pinnaduwage, Phys. Rev. Lett. 54:313 (1985).

9. M. H. Nayfeh, D. C. Humm, and M. S. Peercy, presented to the APS March meeting, K3 3 (1988).

10. R. V. Jensen and S. M. Susskind, Photons and Continuum States of Atoms and Molecules, eds. N. K. Rahman, C. Guidotti, and M. Allegrini, 13 (Springer-Verlag 1987).

11. J. E. Bayfield, Photons and Continuum States of Atoms and Molecules, eds. Rahman, Guidotti, and Allegrini, 8 (Springer-Verlag 1987).

12. P. M. Koch, Fundamental Aspects of Quantum Theory, eds. V. Gorini and A. Frigerio (Plenum 1986).

13. G. P. Berman, G. M. Zaslavsky, and A. R. Kolovsky, Zh. Eksp. Teor. Fiz. 88:1551 (1985) [Sov. Phys. JETP 61:925 (1985)]. Also see M. H. Nayfeh and D. Humm, Photons and Continuum States of Atoms and Molecules, eds. N. K. Rahman, C. Guidotti, and M. Allegrini, 28 (Springer-Verlag 1987).

14. W. L. Glab, K. Ng, D. Yao, and M. H. Nayfeh, Phys. Rev. A 31:3677 (1985).

15. H. Rottke and K. H. Welge, Phys. Rev. A 33:301 (1986).

16. D. A. Harmin, Phys. Rev. A 26:2656 (1982).

17. B. V. Chirikov, Phys. Rep. 52:263 (1979).

18. G. M. Zaslavsky and B. V. Chirikov, Usp. Fiz. Nauk. 105:3 (1971) [Sov. Phys. Usp. 14:549 (1971)].

19. L. I. Schiff, Quantum Mechanics, 3rd ed., 268 (McGraw-Hill 1968).

20. O. Bohigas, M. J. Gianonni, and C. Schmit, Phys. Rev. Lett. 52:1 (1984).

TWO-FREQUENCY MICROWAVE QUENCHING OF HIGHLY EXCITED HYDROGEN ATOMS

L. Moorman, E. J. Galvez, K.A.H. van Leeuwen,[a]
B. E. Sauer, A. Mortazawi-M, G. von Oppen,[b] and P. M. Koch

Physics Department
SUNY at Stony Brook
New York, New York 11794-3800

[a]Technical University, Eindhoven, NL
[b]Technical University, Berlin, FRG

1. INTRODUCTION

This contribution is a preliminary report on the study of ionization of hydrogen atoms in Rydberg states by the simultaneous interaction of these atoms with a microwave field with two frequency components. Since the original work of Bayfield and Koch,[1] the ionization of a highly excited atom by a single microwave field has been studied experimentally in increasing detail. Recently[2] there has been considerable progress in understanding theoretically not only some overall characteristics of the ionization process but also more detailed structures in the ionization curves. The relevant theories have exploited the periodicity of the Hamiltonian to calculate those characteristics. This has motivated us to set up experiments to break this periodicity by applying a second microwave field with a different frequency.

The structure of this contribution is as follows: Sec. 2 present the relevant experimental details. Sec. 3 gives briefly the main conclusions thus far gained from comparisons of single-frequency experiments with theory. Sec. 4 shows preliminary results of the two-frequency experiments discussed in the light of what has been learned from the single-frequency experiments. Sec. 5 compares our two-frequency microwave experiments with two-color laser experiments. Sec. 6 gives some general conclusions.

2. EXPERIMENT

The atoms were prepared in a fast beam, and the interaction with the microwaves took place while the atoms were traversing a cavity. The cavity was specially designed to produce a superposition of parallel, linearly polarized electric field components for microwave fields at two different frequencies. A very simplified picture of the apparatus is given in Fig. 1. The Rydberg states were populated via double resonance excitation using two CO_2 lasers where the atoms were tuned into resonance using the DC-Stark effect in the field regions F_1 and F_3. Charge exchange of the initial proton beam of 14 keV with Xe produced atoms in excited states including those with principal quantum number n = 7. In field region F_1, atoms with parabolic quantum numbers $(n, n_1, n_2, |m|) = (7,0,6,0)$ were excited to $(10,0,9,0)$. Typical field values for F_1 were of order 10^4 V/cm. One example of a spectroscopic scan obtained by tuning the field in the F_3 region is shown in Fig. 2. The labels at the bottom give the principal quantum numbers associated with each group of peaks; for n = 67 identifications are made for various $(67, n_1, n_2, 0)$ substates. Because the linear polarization of the laser light was parallel to the field direction in F_3, only $\Delta m = 0$ transitions were driven. The unique substate that was produced by tuning the fields in F_1 and F_3 was altered by stray fields before the atoms entered the cavity into a statistical distribution of all substates (see Ref. 3).

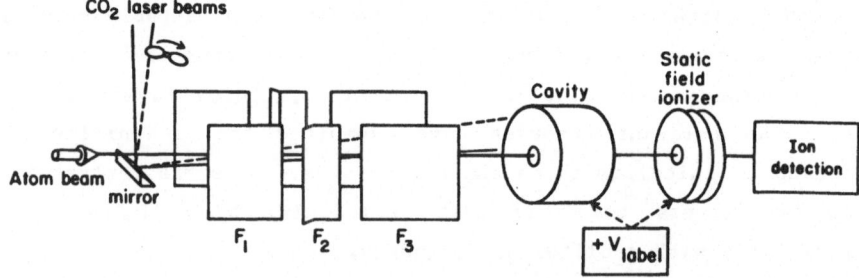

Fig. 1 Apparatus schematic. Atoms in highly excited states are produced via double resonance excitation using two CO_2 lasers and Stark tuning of the atoms in field regions F_1 and F_3. The highly excited atoms that survive the microwaves in the cavity are ionized by a static field and the ions subsequently detected.

In the following we give a detailed analysis of the field distribution inside and just outside the cavity. A cross section of one half of the right circular cylindrical copper cavity is given in Fig. 3. In Fig. 3a the simulation of the cavity and the result of a calculation of the TM_{020} mode disturbed by the collimator and the tapered entrance and exit holes for the beam are shown. The beam particles traveled from left to right, parallel to the horizontal axis. In front of the cavity there was a beam collimator that limited the radial extent of the beam to a region smaller than the radii of the tapered holes. The curves in Fig. 3a, representing contour lines of the product of radial distance and azimuthal magnetic field component or, equivalently, for this mode, the direction of the electric field, were calculated using the computer program "Superfish."[4] The radial dependence of the electric field of a pure TM_{020} mode is a J_0 Bessel function whose second node is on the inside boundary of the cavity. The first radial node is represented by the dashed line in Fig. 3a. As can be seen, the electric field has also a radial component, E_ρ near the entrance and exit holes, where the electric field changes its direction towards the metal boundary. Inside the cavity the maximum deviation between the electric field vector and the beam axis was calculated to be less than 4.5 ± 1.0 degrees for field amplitudes exceeding 90% of the maximal amplitude value. For field values between 50-90%, the maximum angle was calculated to be 23 ± 1 degrees, which occurs where the beam gets closest to the sharp edge of each tapered hole. Near the sharp edges of the tapered holes, there are large amplitudes. In our experiment, the collimator before the cavity prevented particles in the beam from passing too close to the sharp edges of the tapered holes, so they did not experience these large amplitudes. The calculated electric field amplitudes along the beam direction (E_z) actually seen by the atoms traversing the cavity are given in Fig. 3b. The various curves represent the amplitude components along

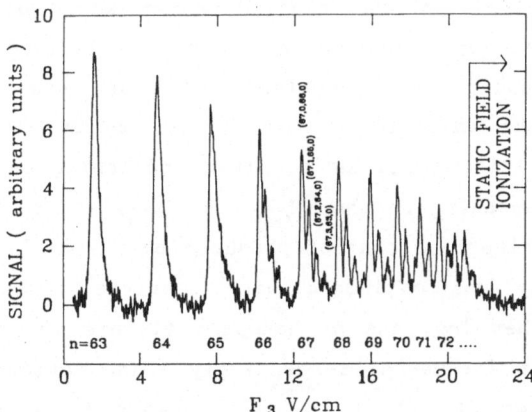

Fig. 2 Laser double resonance production of H Rydberg states. The spectroscopic signal is shown due to Stark tuning for the atoms in field region F_3. The initial state is (10,0,9,0) in parabolic quantum numbers.

straight line trajectories through the cavity at various distances from the symmetry axis. Therefore, each line may be considered the E_z amplitude that was seen by different atoms in the beam ensemble. The various paths through the cavity had a spread of 4% in the maximum electric fields. The typical time the atoms traveled from 5% to 95% in the turn-on (or vice versa in the turn-off) of the maximum field amplitude was 50 cycles for 7.5816 GHz, and the electric field amplitude remained nearly constant during about 200 cycles. Similarly, for the TM_{030} mode at 11.8893 GHz, the field amplitude had a spread of 10% over the beam cross section and turned on (and off) in about 80 cycles, staying near its maximal value for about 300 cycles. For runs on different days the frequencies of the two modes deviated not more than 0.6 MHz from the values mentioned above. The atoms that remained in a highly excited state after passing through the cavity were ionized by a longitudinal static field. The ions created in this part of the apparatus were labeled with a higher kinetic energy and, after energy selection, they were detected in a particle mulitplier. Since the whole method is based on quenching a signal that was present in the absence of microwaves in the cavity, this detection method is called the "quenching mode." Because of fields present in the apparatus used for this detection method, the quenching signal also included atoms that were excited in the microwaves into extremely highly excited final states with n > 87. It was shown previously in our 9.9 GHz experiments[2] for n < 70 that these quench curves give the same information as "direct" microwave ionization curves. The dimensions of the cavity were chosen so that no other modes were degenerate with the two desired ones.

The frequencies for the desired modes were produced with two separate microwave oscillators having no phase relationship. The output powers of each source were combined and sent into a single X-band traveling wave tube amplifier. The cavity, which was coupled to the waveguide via an iris, acted as a filter, selecting only those desired frequencies corresponding to the TM_{020} and TM_{030} modes, rejecting source harmonics and combinations thereof produced by the amplifier. The power at each frequency was monitored with separate directional couplers equipped with either low pass or bandpass filters tuned to pass either the lower or upper frequence, respectively. The reflected power signal from the cavity was fed back to each sweeper in order to lock each frequency to its respective resonance to the order of $1:10^6$. From measured properties of each resonance curve we determined the quality factor, Q, and the coupling factor, K, for each mode and, thereby, were able to calculate each field strength as a function of microwave

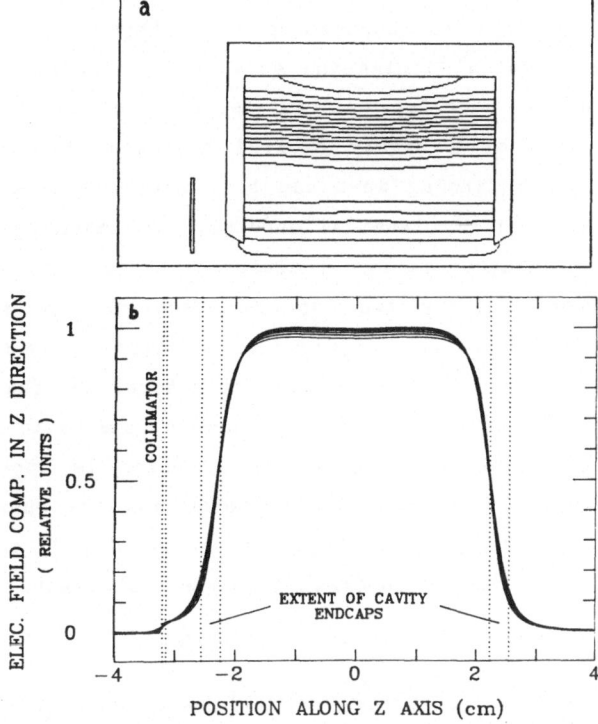

Fig. 3 Half cross section of a collimator and the cylindrical cavity with tapered entrance and exit holes. a) Direction of electric field lines for the TM_{020} made on straight lines parallel to that axis. The various curves correspond to radial distances: 0.0, 0.41, 0.82, 1.24, 1.65 and 2.06 mm from the symmetry axis; and the radius of the collimator hole is 2.5 mm. Vertical dotted lines show the extent of the collimator and endcaps, respectively. Notice that the collimator influence the field only where it is very weak.

power.[5] The accuracy of the power to field calibration in the measurements presented in this article is estimated to be 5%.

3. UNDERSTANDING OF SINGLE-FREQUENCY MICROWAVE IONIZATION

Single-frequency experiments and theoretical calculations that used the periodicity of the associated Hamiltonian have taught us to make connections between different aspects of the experimental ionization curves and characteristics found in different theoretical approaches. In principle the experiments should be described by a fully quantum mechanical theory. However, keeping the correspondence principle in mind, we may ask how well a classical model would describe the inter-action of a highly excited atom with some number of periods of an oscillating classical field. The temporal evolution of the system could be followed in a classical phase space that can be built up in different ways. For instance, it could describe one-, two-, three-dimensional atoms[6] according to the type of experiment. Furthermore, for the field strengths where classical ionization sets in the driven electron motion can exhibit chaotic behavior. This results in the conclusion that studying the system of a highly excited hydrogen atom in a microwave field in detail, and finding what can be understood as genuine quantum mechanical, as opposed to classical aspects, may teach us how to understand a quantum mechanical analog of classical chaos. The (semi-) classical treament suggests two important scaling relationships, most easily stated for circular, unperturbed orbits: (A) for the field amplitude, the ratio of the external electric force and the Coulomb force on the electron, n^4F, and (B) for the frequency, the ratio of the applied frequency and the orbital frequency, $n^3\omega$, which is approximately the ratio of the energy of the microwave photon and the energy difference between the states with principal quantum number $n(\gg 1)$ and $n \pm 1$.   The following conclusions have been drawn from the comparisons of experiments and theories:

1. For the range of parameters studied so far <u>classical</u> calculations have been able to reproduce surprisingly well the <u>overall threshold</u> values of the electric field amplitude at which ionization occurs in the experiments.[1-3,6-10]

2. <u>Quantal</u> calculations using the Floquet approach have explained some <u>structures</u> in some experimental ionization curves[11] via <u>clustering of avoided crossings</u> of Floquet eigenvalues.[12,13]   They have reproduced[14] experimental thresholds as in 1. Also a simpler quantal approach[15] has reproduced structures at the scaled frequencies where they were observed experimentally.   The structures have not been reproduced by classical calculations.

3. <u>Local stability</u> of the hydrogen atom is found near certain <u>rational ratios of the classically scaled frequency</u>.[3,8,11]   This behavior is reproduced very well by <u>classical calculations</u>,[9,10] interpreting this phenomenon as being associated with the stability of some classical orbits that are trapped in islands in the phase space. These inhibit chaotic diffusion towards ionization.   Quantally, note that this relative stability occurs when the energy of a number of photons is near the energy difference between the initial Rydberg state and the nearest neighboring states.   Clearly this involves many strongly coupled levels.

Because relevant two-frequency calculations are not yet available, we shall attempt in the next Section to discuss our two-frequency measurements in the light of these single-frequency results.

4. TWO-FREQUENCY MICROWAVE IONIZATION

We have begun two-frequency experiments with the cavity described in Sec. 2. The approximate Hamiltonian in atomic units for this system is:

$$H = \frac{p^2}{2} - \frac{1}{r} + zA_1(t)F_1\sin(\omega_1 t + \phi_1) + zA_2(t)F_2\sin(\omega_2 t + \phi_2)$$

where, compared to the single frequency Hamiltonian, the last term is new. The linearly polarized electric fields at frequencies ω_i for $i = 1,2$ were both aligned along the z axis. Also, $A_i(t)$ represent the amplitude envelope functions that turned on and off each field as the atoms moved through the cavity (see Fig. 3b) with constant velocity. In the experiment the frequencies were fixed, while the two amplitudes, F_1, and the value of the principal quantum number were external parameters to be varied. Averaging of future theoretical calculations over the experimental field amplitudes and envelope functions will be possible because of our detailed knowledge of the field distribution inside the cavity and the fringe fields near the holes. The phases of the two fields are defined with respect to some arbitrary reference point; for instance if the atom crosses at t = 0 the z = 0 plane, the electric field values are fixed by the phases ϕ_1, ϕ_2 of the two fields. Our experiment does not discriminate between the various phases, and the net result is that theoretical calculations should also be averaged over all possible phases.

For each n-value studied, the experimental curves were obtained in the quenching mode of operation (see Sec. 2). The amplitude of one microwave field (hereafter called the "first field," F_1) was swept, while the amplitude of the second microwave field (hereafter called the "second field," F_2) was kept at a fixed value. We interpret resulting curves as representing the loss of signal from Rydberg states which is due to ionization by the two microwave fields. For each of the limited number of n-values studied we generally observed the microwave quench threshold field to decrease as the amplitude of the second field was increased from run to run. The detailed behavior of the quench curves was different for each case. We present the results for principal quantum number values 43 to 49.

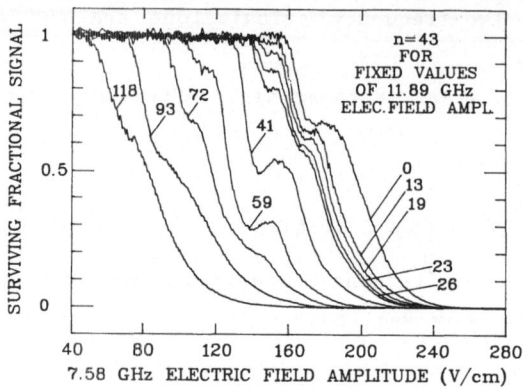

Fig. 4 Probability for n = 43 H atoms to survive the two-frequency
microwave field versus the (scanned) 7.58 GHz electric field
amplitude. The labels of the curves represent the (fixed)
11.89 GHz electric field amplitude (V/cm). These curves are
called two-frequency microwave quench curves.

For n = 43, the "single-frequency" threshold curve (furthest to the
right in Fig. 4; F_2 = 0 V/cm) for $\frac{\omega_1}{2\pi}$ = 7.58 GHz present a non-monotonic
structure about one-third down the curve. As mentioned in Sec. 3,
structures previously seen in our single-frequency experiments have been
shown to be of quantal origin. As F_2 is fixed at increasingly higher
values we observed a surprising evolution of the structure. It is
gradually suppressed until it finally disappears. Furthermore, new
structures appear at nearly fixed, lower values of the swept field,
finally dying out for even larger values of the second field. At this
moment we have no quantitative explanation for the new structures. They
could be related to the type described above, i.e. simply unobserved
"single-frequency" structures made visible by the threshold being shifted
downward in amplitude by the presence of the second frequency, or they
could be caused by a more complicated, combined effect of the two
frequency fields.

The local stabilities at values of n for which the classically
scaled frequency was near rational fractions (e.g. for $n^2\omega$ =
$\frac{1}{6}$, $\frac{1}{5}$, $\frac{1}{4}$, $\frac{1}{3}$, $\frac{2}{5}$, $\frac{1}{2}$, 1) had been observed at 9.92 GHz.[2,3,6] At both new
frequencies alone we have observed consistently the same kind of
effect. These stabilities could in principle disappear with the
inclusion of the second field even at very small amplitudes[16] of the
second field amplitude. At 11.89 GHz the scaled frequency values for
n = 45 and n = 48 are 0.1648 and 0.1999, respectively. These are close
to $\frac{1}{6}$ and $\frac{1}{5}$ frequency ratios for the 11.89 GHz field. The figure shows
classically scaled fields n^4F, at which 50% of quenching takes place, as

a function of the scaled frequency $n^3\omega$ for a range of n values (43-49), for a swept field frequency of 11.89 GHz. The top curve corresponds to zero value of the second field amplitude, i.e. a single-frequency result, and the peaks in the 50% quench threshold curves near scaled frequencies $\frac{1}{6}$ and $\frac{1}{5}$ clearly show the local stabilities (i.e. higher 50% threshold fields). For increasing values of the second field amplitude, both peaks are seen to be suppressed. At about 15 V/cm of the second field amplitude, only 10% of the overall threshold of the first field, the relative stability associated with the $\frac{1}{6}$ and $\frac{1}{5}$ frequency ratios has been lost. Fig. 6 shows the sequence of quench curves for n = 45. As the second field amplitude increases from 0 to 15 V/cm two things happen, but we do not know the relationship between them: the structure which drops in the quench curves and the loss of local stability near scaled frequency $\frac{1}{6}$. The sudden change in the n = 45 overall thresholds is more obvious when this sequence is compared to the corresponding sequence for n = 44($n^3\omega$ = 0.1540), as shown in Fig. 7. Here the quench curves happen to be much smoother, and unlike n = 45, no sudden shift of the threshold is observed. Obviously, the details of the dynamics measured in the two-frequency quench curves may depend sensitively on some or all of the many parameters in Eq. 1. For the n-values studied thus far it is apparent that the two-frequency quench curves can change significantly from one value of the principal quantum number to the other. One example is

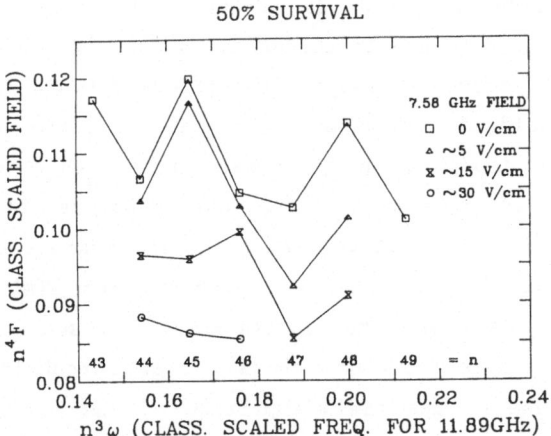

Fig. 5 Classically scaled field for 50% survival vs. classically scaled frequencies near $\frac{1}{6}$ and $\frac{1}{5}$, for 11.89 GHz. The principal quantum numbers are given in the lower part of the diagram. The destruction of the local stability near $\frac{1}{6}$ can be followed as the 7.58 GHz field amplitude is increased.

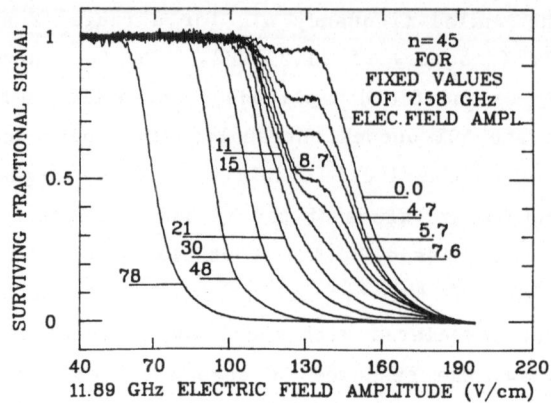

Fig. 6 Two-frequency microwave quench (see Fig. 4) for n = 45 H atoms. The classically scaled frequency is near $\frac{1}{6}$.

n = 43, where we see much structure; at the other extreme is n = 44, where there seems to be only little structure. It is clear that much theoretical work, both classical and quantal, will need to be done.

It is interesting to ask how much an increase in the (fixed) second field decreases the swept field threshold for quenching. This might lead to increased understanding of the interaction of the electron in a highly excited state with the oscillatory microwave fields. We found thus far no single, simple, consistent behavior (e.g. linear, quadratic); some n-values exhibit a linear behavior, but others such as n = 44 deviate strongly from linearity. This is displayed in Fig. 8, where the strength of the two fields required to produce different fixed quenching percentages along the threshold curves are shown. The upper half and the lower half sets of curves correspond to different types of runs, that is, 7.58 GHz swept field with 11.89 GHz fixed field for the upper half curves and vice versa for the lower half curves. The degree of match of the two sets of curves is consistent with our absolute field calibration uncertainty of 5%. The relative uncertainties within each set of curves is lower than this because they were taken under similar experimental conditions. Furthermore, the nonlinearity at small values of the amplitude of the upper frequency shows a slow threshold variation with respect to increase of the field amplitude of the lower frequency (cf.

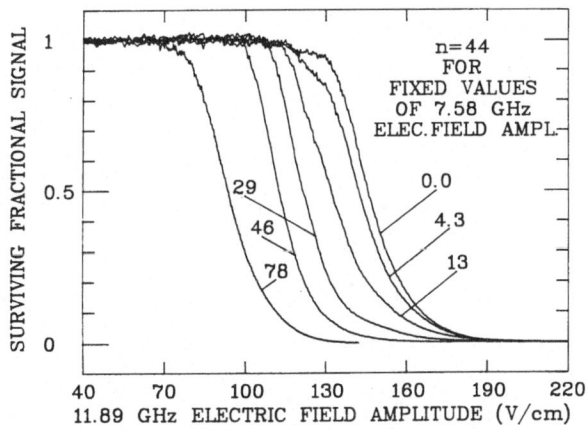

Fig. 7 Two-frequency microwave quench curves for n = 44. Compare to
n = 45 in Fig. 6.

upper half curves), whereas the converse occurs at small values of the
field amplitude corresponding to the lower frequency (cf. lower half
curves). So far this is unexplained.

Finally we want to show how instructive are these contour plots of
the survival probability of the atoms, by showing one in the neighborhood
of the local stability that occurs near n = 45 (see Figs. 5 and 6). In
Fig. 9 we see that there are two regions of different slopes in each
contour line. The interpretation is as follows. If we follow a contour
line, first we see the generic behavior where with a decrease of the
fixed field (on the vertical axis), the swept field increases in order to
maintain the same quenching probability. Then where the slope of the
contour line rather suddenly changes, a given decrease of the fixed field
requires a larger increase of the swept field. This means that the atom
becomes rapidly more stable. Interpreting the slope change in the other
direction, it seems that a relatively stable atom exists for points on
the horizontal axis ($\frac{\omega}{2\pi}$ = 11.89 GHz, $n^2\omega$ close to $\frac{1}{6}$) whose stability is
rapidly destroyed by the presence of the other frequency.

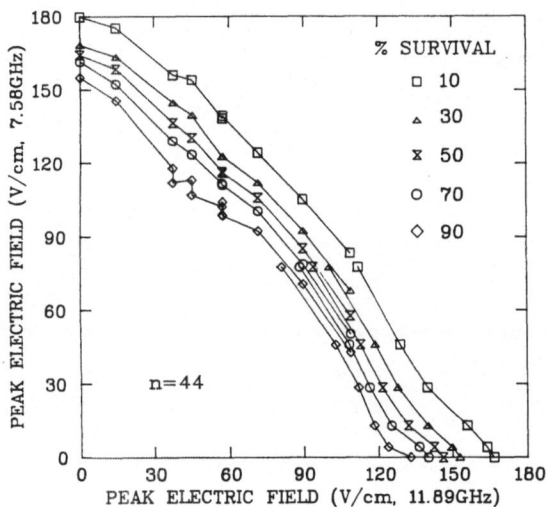

Fig. 8 Contour plots of equal survival probabilities for n = 44 H atoms. On each axis is the field amplitude for the relevant frequency.

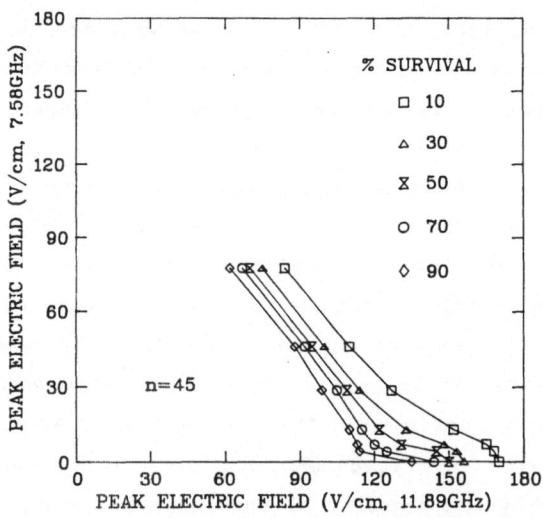

Fig. 9 Contour plots of equal survival probability for n = 45 H atoms. Here one sees the same loss of stability near $\frac{1}{6}$, as is diplayed differently in Figs. 5 and 6.

5. ANALOGY TO A TWO-COLOR LASER EXPERIMENT

If we use the "language" of laser experiments, we have been studying a multiphoton ionization process where the "pulses" of the electromagnetic field seen by the atoms are about 200-400 oscillations of the microwave fields (with slow turn-on and turn-off) and more than one hundred photons are absorbed.[17] In that respect, we could characterize our experiment as the analog of a multiphoton ionization process by a two-colored, pulsed laser system. Such multiphoton ionization experiments with two lasers have been done with Na[18] and with Xe.[19] Another type of distantly related experiment that has been done is the study of dissociation of molecules by two infrared lasers.[20]

In the following, we consider the atomic experiments, which focussed respectively, on resonances in multiphoton ionization spectra[18] and on improving understanding of the above threshold ionization (ATI) process.[19] Neither experiment aimed toward the detailed dependence of the studied processes on the power of both laser fields. Both experiments probed a completely different part of the (external) parameter space with frequencies five orders of magnitude higher, number of field oscillations about four orders of magnitude larger, peak laser electric field amplitudes four to five orders of magnitude larger, and initial atomic binding energies about three orders magnitude larger than in our microwave experiment. Another difference is how accurately those electric field amplitudes were known: in Ref. 19, for example, estimated to about a factor two, whereas in our experiment, to about 5%. Those laser experiments and our microwave experiment can be considered as two limiting cases of the interaction of an atom with an electromagnetic field consisting of two different frequencies and incoherent phases. The two limits are that we study the multiphoton ionization of a weakly bound electron by photon fields that are comparatively weak in laboratory units, whereas the laser experiments explore ionization by intense laser fields. However, because in each type of experiment the interaction of the (active) electron with the external fields is comparable to its Coulomb binding to the nucleus, both type of experiments probe "strong-field" physics.

6. CONCLUSIONS

New experiments with two new frequencies (7.58 and 11.89 GHz) alone have added to and confirmed the results of previous single-frequency (9.92 GHz) microwave ionization and quenching experiments. We continue to find relative stability near certain rational scaled frequency ratios (below 1 in the present case). The first studies with simultaneous application of two microwave frequencies have produced experimental curves with rich new structures and phenomena. Certainly new theoretical work is called for to explain these results, but we are already able to interpret some of our "two-frequency" results. Only a small amplitude of a second driving frequency is needed to destroy the local stability found near rational scaled frequency ratios of the first driving field. More experimental work will pursue this problem in more detail, extending the limited number of cases covered so far.

ACKNOWLEDGEMENTS

This work was supported by the National Science Foundation. We thank Dr. H. G. Kirk of Brookhaven National Laboratory for giving us the opportunity to calculated the fields inside the cavity. We also thank R. Jensen and D. Richards for useful discussions.

REFERENCES

1. J. E. Bayfield and P. M. Koch, Phys. Rev. Lett. 33, 258 (1974).

2. P. M. Koch, to be published in the Invited Papers of the XV Intl. Conf. on the Physics of Electronic and Atomic Collisions (North-Holland, Brighton, 1987).

3. K.A.H. van Leeuwen, G. von Oppen, S. Renwick, J. B. Bowlin, P. M. Koch, R. V. Jensen, O. Rath, D. Richards and J. G. Leopold, Phys. Rev. Lett. 55, 2231 (1985).

4. K. Halbach and R. F. Holsinger, Part. Accel. 7, 213 (1976).

5. B. E. Sauer, K.A.H. van Leeuwen, A. Mortazawi-M and P. M. Koch, in preparation.

6. Note: a difference with the contribution of J. E. Bayfield in this book is that his experiments are done with one parabolic substate, corresponding to quasi-one-dimensional atoms, whereas our experiments are done with a statistical distribution of all substates for a given principal quantum number, corresponding to three-dimensional atoms. P. M. Koch in Fundamental Aspects of Quantum Theory, ed. V. Gorini and A. Frigerio (Plenum, 1986); P. M. Koch, K.A.H. van Leeuwen, O. Rath, D. Richards and R. V. Jensen in Lecture Notes in Physics 278, The Physics of Phase Space, ed. Y. S. Kim and W. W. Zachary (Springer-Verlag, Berlin, 1987).

7. J. G. Leopold and I. C. Percival, J. Phys. B: Atom. Mol. Phys. 12, 709 (1979).

8. D. A. Jones, J. G. Leopold and I. C. Percival, J. Phys. B: Atom. Mol. Phys. <u>13</u>, 31 (1980).

9. R. V. Jensen, Phys. Rev. A <u>30</u>, 386 (1984).

10. J. G. Leopold and D. Richards, J. Phys. B: Mol. Phys. <u>18</u>, 3369 (1985).

11. See P. M. Koch and P. M. Koch, et al. in Ref. 6.

12. R. Blümel and U. Smilansky, Z. Phys. D: Atoms, Molecules and Clusters <u>6</u>, 83 (1987).

13. R. Blümel and U. Smilansky, Phys. Scr. <u>35</u>, 15 (1987).

14. R. Blümel and U. Smilansky, Phys. Rev. Lett. <u>58</u>, 2531 (1987).

15. D. Richards, J. Phys. B: Atom. Mol. Phys. <u>20</u>, 2171 (1987).

16. Note: we think here e.g. of Arnold diffusion in many-dimensional oscillator systems, B. V. Chirikov, Phys. Rep. <u>52</u>, 263 (1979). See also D. Shepelyansky, in <u>Chaotic Behavior in Quantum Systems</u>, ed. G. Casati (Plenum, New York, 1985), who reports preliminary two frequency calculations for $n^3\omega > 1$, which is above the range of scaled frequencies in the present work.

17. For n = 45 the number of photons is $\dfrac{E_t}{\hbar\omega} = \dfrac{1}{(2n^2\omega)}$ (a.u.) = 214 for 7.58 GHz and 137 for 11.89 GHz.

18. D. Feldman, G. Otto, D. Petring and K. H. Welge, J. Phys. B: Atom. Mol. Phys. <u>19</u>, 269 (1986).

19. H. G. Muller, H. B. van Linden van den Heuvell and M. J. van der Wiel, J. Phys. B: Atom. Mol. Phys. <u>19</u>, L733 (1986).

20. S. S. Alimpiev, B. O. Zikrin, B. G. Sartakov and E. M. Khoklov, Zh. Eksp. Teor. Fiz. <u>83</u>, 943 (1982) [Sov. Phys. JETP <u>56</u>, 943 (1982)].

ATOMS IN INTENSE OPTICAL FIELDS: PONDEROMOTIVE FORCES

AND ABOVE-THRESHOLD IONIZATION

P. H. Bucksbaum

AT&T Bell Laboratories
Murray Hill, NJ 07974, USA

ABSTRACT

The electric field strength in a tightly focused high-powered laser beam can approach an atomic unit. Such intense light dramatically alters electron dynamics, causing elastic and inelastic scattering of free electrons by light, and above-threshold ionization of atoms. This paper reviews some recent atomic photoionization experiments in this new regime.[1]

1.   INTRODUCTION

Photoionization of tightly bound atoms by optical wavelength radiation requires the simultaneous absorption of many photons. The rate is negligible at ordinary light levels, but increases rapidly with intensity, so that at 10^{15}W/cm^2, even the most tightly bound atoms are unstable to ionization in less than a few nanoseconds.[2] Two new phenomena, above-threshold ionization (ATI) and ponderomotive scattering, dominate photoelectron spectra at high intensity.[3] These signal the breakdown of minimum order perturbation theory, and can be directly related to the physics of wiggling electrons in the periodic laser field.

1.1 Intensity Scaling

In perturbation theory, the rate for nonresonant multiphoton ionization (MPI) scales as $\sigma_n \rho^n$, where ρ is the photon flux, n the number of photons absorbed, and σ_n is the nth order ionization cross section. Even though ρ^n scaling has been reported for intensities as high as 10^{15}W/cm^2 (22 photon ionization of helium by Nd:YAG radiation[2]), there are a number of reasons to reexamine perturbation theory above 10^{13}W/cm^2.

Perturbation theory is expected to fail when the perturbation expansion parameter approaches unity. For nonresonant MPI, the nth order cross section may be estimated as the product of n, one photon absorptions all occuring within a dephasing time given by the reciprocal of the average detuning $\hbar/\delta E$:

$$\sigma_n \rho^n = \sigma \left(\frac{\hbar\sigma}{\delta E}\right)^{n-1} \left(\frac{I}{\hbar\omega}\right)^n \qquad (1)$$

where I is the laser intensity. Written in this way, it's clear that the dimensionless scaling parameter is simply $\sigma I/\delta E\omega$. For reasonable values $\sigma = 10^{-17} cm^2$ and $\hbar/\delta E = 10^{-16} sec$, this parameter becomes unity as ρ approaches 10^{33} photons/($cm^2 sec$), or for 1 μm photons, $I \approx 10^{14} W/cm^2$.

Laser fields of this intensity may be strong enough to tear atoms apart, even in the static limit. A simple semi-classical estimate for the electric field needed to ionize an atom is $E_0 \approx (2n^*)^{-4}$ atomic units, where n^* is the effective quantum number of the bound state.[4] The atomic unit of electric field, e/a_0^2, is equivalent to the laser peak field at $5.5 \times 10^{16} W/cm^2$. This formula is known to work well for Rydberg states. If it is extrapolated to the ground state of a tightly bound atom such as xenon (an admittedly questionable procedure) the ionization threshold is predicted to be approximately $10^{14} W/cm^2$, independent of wavelength (static field limit). In fact ionization thresholds of $\approx 3 \times 10^{13} W/cm^2$ are observed at wavelengths of <u>both</u> 1 μm and 10 μm, so the static limit may be close at hand.[5,6] Since these two wavelengths correspond to absorption of 11 photons or 104 photons, respectively, it is unlikely that perturbation theory would predict this equivalence.

From the considerations above, it appears that ATI occurs in a regime that, if not beyond perturbation theory, is certainly approaching its limits. The theoretical treatment of the process must therefore include effects which are usually neglected in weak field calculations. One of these is the energy shift of the atomic states. Strong oscillating fields can change the energy of an atomic bound state by inducing polarization. I^n scaling assumes that the ionization potential (I.P.) of the atom remains constant. In fact, at intensities where multiphoton ionization occurs, the work required to ionize an atom can increase by several photons due to dynamic Stark effects. These have been observed in xenon experiments discussed below.

The most striking new feature of photoionization at high intensities is above-threshold ionization (ATI), which is the absorption of excess photons beyond the minimum necessary to ionize and atom or molecule (see Fig. 1).[7,8] ATI is widely observed in photoionization at high intensities, and its predominance strongly suggests that minimum-order perturbation theory must be modifed at high intensities. In some cases the minimum-order process is so greatly suppressed, that ATI accounts for nearly all of the photoelectrons. (See Fig. 1.)

1.2 Experimental Limitations

There are several practical limitations to all ATI experiments. First, experiments must use a tight focus. This means that the ionization intensity is difficult to determine. The situation is complicated by depletion of ground state neutral atoms, so that the ionization occurs in parts of the laser focus well below the peak intensity.

In addition, electron dynamics at high intensities are dominated by large accelerations due to the periodic electromagnetic forces. These "ponderomotive" effects alter the electron energy spectra and angular distributions.[9] Finally, to prevent ionization at low intensity as the laser beam turns on, ATI experiments must involve high order nonresonant multiphoton ionization. The most widely studied system is xenon (I.P. = 12.127 eV) photoionized by Nd:YAG radiation ($\hbar\omega$ = 1.165 eV).

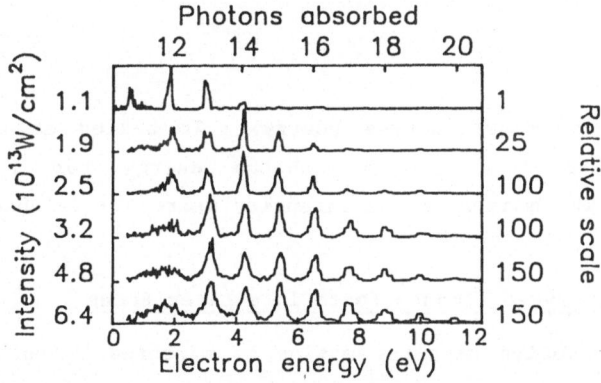

Fig. 1 Photoelectron spectra showing above-threshold ionization in xenon by 1.165 eV photons. The laser was linearly polarized and 100 psec in duration, focused to a 12 μm gaussian waist, with peak intensity shown on the left axis. The electrons were detected along the laser polarization direction, and energies were analyzed by time-of-flight. (From Ref. 9.)

2. THE CLASSICAL WIGGLING ELECTRON

In this section we examine more closely the effect of a periodic electric field of the laser on the dynamics of free and bound electrons.

2.1 Ponderomotive Energy and Wiggling

Thomson scattering is the reradiation caused by the wiggling motion of free electrons in a light field. The wiggle amplitude itself is usually far too small to produce observable deviations in an electron's motion through the light. With high enough photon flux, however, things can be quite different: Thomson scattering in an ATI experiment is utterly negligible, despite the high intensity, because of the short time the electron spends in the focus. For example, a 1 eV electron traversing a tightly focused Nd:YAG laser beam ($5 \times 10^{13} \text{W/cm}^2$, 10 μm beam waist, 100 psec duration) is likely to Thomson scatter only about one photon out of the 10^{17} or so in the beam. The recoil changes the electron's velocity by less than a tenth of a percent. On the other hand, the wiggle motion has an enormous effect on the electron's path. If the intensity is above 10^{13}W/cm^2, the electron can even backscatter from the light! The momentum is transferred between the electron and the light by _stimulated_ scattering, the photon exchange process responsible for wiggling.

A periodic electric field with amplitude E_0 and angular frequency ω induces a wiggle energy of[10]

$$U_p = \frac{e^2 E_0^2}{4 m_e \omega^2} . \tag{2}$$

U_p is called the ponderomotive energy. It scales as the wavelength squared, and is equal to the photon energy for 1 μm light at $I \approx 10^{13} \text{W/cm}^2$, which is also the intensity where ATI is observed in xenon and krypton.

2.2 The Time-averaged Lorentz Force in a Laser Focus

The ponderomotive wiggling motion is affected by both electric and magnetic fields, and is rather complicated, particularly when spatial and temporal inhomogeneity of the focused radiation field are taken into account. When spatial and temporal variations are slow, however, it is simple to show that an electron in a spatially inhomogeneous laser focus experiences a net force in the direction of lower intensity. For example, consider a focused laser beam propagating along the $\hat{z}$ direction, polarized along $\hat{x}$, as in Fig. 2.

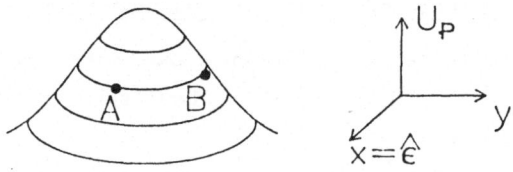

Fig. 2   Cross section of a laser focus.   Lines are contours of constant
intensity.

At point A, an electron wiggles up and down with a displacement
about the mean position $\mathbf{x}_A$

$$\delta\mathbf{x} = -\hat{x}\,\frac{eE(\mathbf{x}_A)}{m_e\omega^2}\,\cos\omega t. \qquad (3)$$

The motion is actually anharmonic, since the field intensity varies over
the wiggle amplitude.   For small motion, the Lorentz force is

$$m_e\ddot{\mathbf{x}} = \hat{x}\big(eE(\mathbf{x}_A) + e\,\frac{\partial}{\partial x}\,E(\mathbf{x}_A)\delta x\big)\cos\omega t. \qquad (4)$$

Substituting Eq. 2 for δx, the time averaged force is

$$\langle m_e\ddot{\mathbf{x}}\rangle = -\hat{x}\,\frac{\partial}{\partial x}\,\frac{e^2E(\mathbf{x}_A)^2}{4m_e\omega^2} = -\nabla_x U_P, \qquad (5)$$

and is directed out of the focus.   An electron at B, where the intensity
gradient is along the $\hat{y}$ direction, experiences a Lorentz force

$$m_e\ddot{\mathbf{x}} = e\mathbf{E} + e\,\frac{\mathbf{v}}{c}\times\mathbf{B}$$

$$\approx \hat{x}eE(\mathbf{x}_B)\cos\omega t - \frac{e^2E(\mathbf{x}_B)}{m_e\omega c}\,\sin\omega t(\hat{x}\times\mathbf{B}). \qquad (6)$$

The magnetic field at point B is

$$\mathbf{B}(\mathbf{x}_B) \approx \hat{y}E(\mathbf{x}_B)\cos\omega t - \hat{z}\,\frac{c}{\omega}\,\frac{\partial}{\partial y}\,E(\mathbf{x}_B)\sin\omega t, \qquad (7)$$

so the time-averaged force is

$$\langle m_e\ddot{\mathbf{x}}\rangle = -\hat{y}\,\frac{\partial}{\partial y}\,\frac{e^2E(\mathbf{x}_B)^2}{4m_e\omega^2} = -\nabla_y U_P, \qquad (8)$$

again, out of the focus.

2.3   The Classical Hamiltonian and the Ponderomotive Potential

Ponderomotive force calculations are most easily handled with Hamiltonian mechanics. The classical Hamiltonian for a spinless free electron may be written

$$H = \frac{[\mathbf{p} - \frac{e}{c}\,\mathbf{A}(\mathbf{x},t)]^2}{2m_e} \; . \tag{9}$$

$\mathbf{A}$ is the vector potential for the laser radiation field, with amplitude A_0. The time-averaged Hamiltonian is

$$\langle H \rangle = \frac{\mathbf{p}^2}{2m_e} + \frac{e^2 A_0^2(\mathbf{x})}{4m_e c^2} \; . \tag{10}$$

The second term is the ponderomotive energy U_p in Eq. 1. Although its origin is the wiggle kinetic energy of the electron, it appears in the time-averged Hamiltonian as an effective potential energy, the "ponderomotive potential." For a nearly monochromatic source such as a laser, U_p simply follows the laser intensity distribution. An electron traversing the focus moves <u>on time average</u> as though it were in the presence of a repulsive potential. Since the electron's total energy is conserved, the net effect is to exchange momentum with the laser beam.

Actually, U_p is a conservative potential only if it is constant. If the intensity changes in time as the electron moves through the light, the electron's energy is no longer conserved, since the Hamiltonian is explicitly time-dependent. The incremental energy change during a time interval δt is simply

$$\delta E = \frac{\partial U_p}{\partial t}\,\delta t \; . \tag{11}$$

An electron overtaken by a light pulse gains ponderomotive potential energy, which can be converted to kinetic energy by accelerating in the ponderomotive potential energy, which can be converted to kinetic energy by accelerating in the ponderomotive potential gradients. In this way, the light does work on the electron just as a water wave can do work on a surfer. The amount of energy exhanged between the electron and the light pulse can be smaller than the light quantum $\hbar\omega$, since the energy change is not caused by photon absorption, but rather is due to stimulated <u>exchange</u> of photons between different <u>frequency</u> modes of the laser pulse.[10]

2.4 Electrons Scattering from a Laser Beam

Electron scattering from a laser beam was demonstrated in a recent experiment employing a pulsed beam of electrons and an intense focused Nd:YAG laser.[11] A circularly polarized 100 psec laser pulse with a peak ponderomotive potential of 8 eV, was focused between the electron source and the detector, so that the ponderomotive potential could block the electron pulse. A pulsed point source of monochromatic electrons at severl discrete energies below 5 eV was provided by photoionizing xenon using 532 nm 100 psec light pulses. The relative timing could be varied to show that the scattering was due to the presence of the light only. (See Fig. 3.)

Electron "surfing," that is, inelastic scattering, occured on the rising or falling temporal edge of the 100 picosecond laser pulse. Figure 3 shows how the .54 eV electron peak was shifted to higher energy when the electrons gained ponderomotive potential energy by passing through the rising leading edge of the laser pulse. Conversely, when the electrons arrived late in the pulse, they decelerated due to the spatial gradients, and then remained at lower energy as the laser pulse passed by.

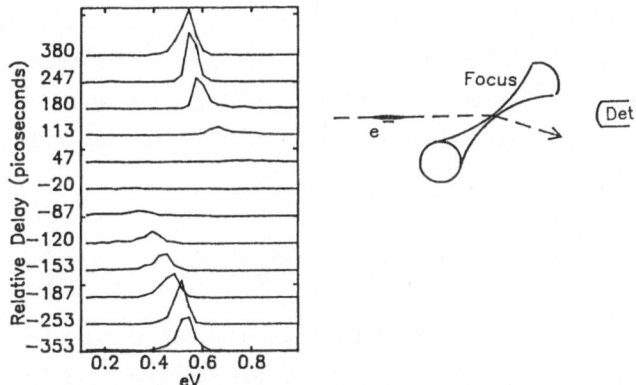

Fig. 3 Scattering of electrons by light. 0.54 eV electrons were directed through the focus of a 100 psec laser pulse with peak ponderomotive potential of 8 eV. Those arriving early or late were undeflected. Electrons passing through the leading and trailing edges were accelerated or decelerated, respectively, due to the phenomenon of "surfing," described in the text. Electrons encountering the peak of the pulse were scattering away from the detector. (From Ref. 11.)

2.5 Ponderomotive Effects in Atoms

The energy shift of a bound electron in a laser field depends on the **A·p** term as well as the A^2 term in the Hamiltonian. For a single electron moving under the combined influence of a laser field and a central potential, the Hamiltonian is

$$H = \frac{[\mathbf{p} - \frac{e}{c}\mathbf{A}(\mathbf{x},t)]^2}{2m_e} + V(r). \tag{12}$$

The classical analog of the Stark shift of a bound state is the light field-induced polarization energy, which depends on the driving force, the presence of resonances, and the binding strength. The energy acquired by a classical dipole excited well below resonance is inversely proportional to the restoring force constant. By analogy, the ground state of a tightly bound atom such as xenon might be relatively unaffected by ponderomotive forces in a Nd:YAG focus, where the driving frequency is nearly an order of magnitude below any resonances. Loosely bound excited states near the ionization limit, however, certainly shift by the full ponderomotive potential, just like free electrons. Since the ionization limit increases, while the ground state remains nearly unchanged, <u>more work is required to ionize an atom in an intense field.</u>

Atomic Stark shift calculations in ATI were recently explored by Pan, et al.[12] They pointed out that since the A^2 term does not depend on the momentum or position of the electron, it shifts all states by the same amount as a free electron state. The **A·p** term, however, evidently nearly cancels the A^2 term for tightly bound ground states. The second-order perturbation theory expression for the **A·p** contribution to the dynamic Stark shift of the ground state is:

$$U_{Stark} = \left(\frac{eA_0}{2m_e c}\right)^2 \sum_i \left[|\langle i|\hat{\varepsilon}\cdot\mathbf{p}|\phi\rangle|^2 \left(\frac{1}{E_\phi - E_i - \hbar\omega} + \frac{1}{E_\phi - E_i + \hbar\omega}\right)\right]. \tag{13}$$

If the photon energy is much less than the separation of any excited state from the ground state, U_{Stark} may be expressed as an expansion in $\hbar\omega/(E_\phi - E_i)$:

$$U_{Stark} = \left(\frac{eA_0}{2m_e c}\right)^2 \sum_i \left[|\langle i|\hat{\varepsilon}\cdot\mathbf{p}|\phi\rangle|^2 \frac{2}{(E_\phi - E_i)}\left[1 + \left(\frac{\hbar\omega}{E_\phi - E_i}\right)^2\right]\right]. \tag{14}$$

The first term in this expansion may be summed using the dipole oscillator strength sum rule.[12] Then we obtain

$$U_{Stark} \approx -\left(\frac{e^2 A_0^2}{4m_e c^2}\right) \tag{15}$$

thus cancelling the ponderomotive potential. The next term in the series is the _static_ field Stark effect $U_{D.C.Stark} = \alpha E_0^2/2$, where α is the polarizability of the ground state. This is much less than the ponderomotive potential in the experiments considered here; but for intensities above 10^{15}W/cm^2, the ground state shift may not be negligible, and the perturbation approach may break down altogether.

2.6 Energy Spectra and Angular Distributions in Intense Fields

As the I.P. increases, the photoelectron kinetic energy must decrease. This effect is hidden, however, since the free electron's ponderomotive potential energy is nearly equal to the I.P. shift. The ponderomotive energy is converted into kinetic energy when the electron leaves the light; so ATI spectra show peaks that are unshifted by the ponderomotive potential.

ATI electrons tend to be emitted along the laser polarization direction for linearly polarized light. However, if the electron velocity is reduced by the ponderomotive increase in the I.P., then the emission angle will change due to ponderomotive scattering as the electron escapes the laser focus. Figure 4 shows angular distributions from the lowest ATI peaks in a xenon photoionization experiment by R. Freeman and coworkers.[13] At low intensities, the electrons are sharply directed along the laser polarization, while at large intensities, the angular distributions become nearly isotropic in the azimuthal plane (normal to **k**). Electron trajectory simulations on a model laser focus indicate energy shifts for these ATI peaks of 1 to 3 eV.

An interesting situation arises when the ponderomotive potential increases by more than $\hbar\omega$. Then an extra photon is required to ionize the atom, and the lowest spectral peak should disappear. This effect is smeared out by the spatial and temporal inhomogeneity of a focused laser pulse; but a strong suppression of the threshold peak is evidence for ionization at intensities where the threshold channel is cut off (see Fig. 1).

3. QUANTUM WIGGLING ELECTRONS

3.1 The Volkov State: A Quantum-mechanical Wiggling Electron

Although ponderomotive scattering and surfing are essentially classical effects, ATI is a quantum phenomenon. We must therefore consider the solutions to Schrödinger's equation for an electron in an external radiation field. If there are no static potentials, the equation may be written:

$$\frac{(\mathbf{p} - \frac{e}{c}\mathbf{A})^2}{2m_e}\,\psi(\vec{x},t) = i\hbar\,\frac{\partial}{\partial t}\,\psi(\mathbf{x},t). \tag{16}$$

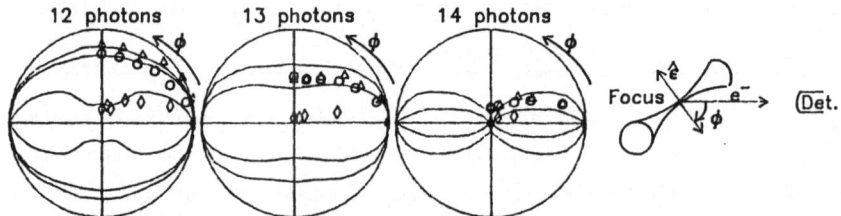

Fig. 4   Intensity-dependent ATI angular distributions in xenon, for absorption of 12, 13 and 14, 1.165 eV photons. The peak laser intensity was 1.5, 3.0 and 6.0 × 10^{13}W/cm^2 for diamonds, circles and triangles, respectively. At low intensities the distributions are sharply peaked along the laser polarization direction. Emission becomes more isotropic at higher intensities due to ponderomotive scattering of the electrons. (From Ref. 13.)

The plane-wave and long wavelength approximation enable direct integration of Eq. 8. The solutions are known as Volkov states.

$$\psi_{Volkov}(\mathbf{x},t) = \exp\Big(\frac{i}{\hbar}\,\mathbf{p}\cdot\mathbf{x} - \frac{i}{\hbar}\,\frac{p^2}{2m_e}\,t$$

$$- \frac{i}{\hbar}\,\frac{e^2 A_0^2}{4\omega m_e c^2}\,t + \frac{i}{\hbar}\,\frac{e^2 A_0^2}{8\omega m_e c^2}\,\sin(2\omega t) - \frac{i}{\hbar}\,\hat{\epsilon}\cdot\mathbf{p}\,\frac{e A_0}{2\omega m_e c}\,\cos(\omega t)\Big). \tag{17}$$

Like plane waves, they are eigenstates of the <u>canonical</u> momentum **p**, but with three additional time-dependent phases due to the oscillating field. The new terms represent the instantaneous kinetic energy of the electron, including a time-averaged ponderomotive term and the periodic fluctuations about this value with frequency ω and 2ω.

3.2   Photoionization in Intense Fields

An ATI photoelectron becomes a Volkov state at large distances from the atom, but it is modifed at short distances by the Coulomb potential. Likewise, the initial state is similar to a neutral atomic ground state, but it is also modified, but the radiation field. The full ATI problem, including both the atomic potential and the radiation field, has yet to be solved analytically, even for the simplest case of a hydrogen atom in a uniform laser beam.

A reasonable first approximation to the full problem is to simply project the initial state wave function for the atom onto Volkov final states. This idea was originally proposed by Keldysh, and has been developed by Reiss, Faisal, and a number of others.[14] It reproduces several experimental features, providing physical insight into the ionizaton process in strong fields. The method projects the initial state wave functon onto Volkov states. Formally, Reiss has shown that this projection is the lowest order term in an S-matrix expansion using the ion potential as a perturbation.

In this approach, the _canonical_ momentum distribution of the final state is the same as the momentum distribution in the initial bound state. In essence, the Coulomb potential that formed the ground state is ignored, so that the bound state just evolves outward with time. The _energy_ of the final state, however, is quite different from the initial state; after all, the electron has absorved all those photons in order to ionize! The energy structure of the Volkov wave function contains information about how many photons were absorbed. The energy spectrum is obtained with a "Floquet" decomposition, which is simply a projection fo the Volkov state on the complete set of plane waves:

$$\psi_{Volkov}(\mathbf{x},t) = e^{\frac{i}{\hbar}(\mathbf{p}\cdot\mathbf{x} - \frac{p^2 t}{2m_e} - U_p t)}$$

$$\times \sum_{n=-\infty}^{\infty} \sum_{m=-\infty}^{\infty} J_m(\frac{e^2 A_0^2}{8\hbar\omega m_e c^2}) J_{n-2m}(\hat{\epsilon}\cdot\mathbf{p}\,\frac{eA_0}{2\omega m_e c}) e^{in\omega t}. \tag{18}$$

Here J_m's are cylindrical Bessel functions, whose arguments are the coefficients of the oscillating terms in the Volkov phase. The energy spectrum is not continuous, but consists of an infinite series of eigenstates separated by $n\hbar\omega$, where n is an integer. Each state contributes to the total with an amplitude proportional to the Bessel function sum. The interpretation is obvious: the nth component in the Floquet decomposition corresponds to the amplitude for absorption of n photons. Thus ATI is simply the projection of the Volkov final state onto electron states of definite energy, $n\hbar\omega$ above the initial state energy.

Of course, neglecting the Coulomb field leads to some serious quantitative disagreements between laser experiments and theoretical predictions. For example, any enhancements in the ionization rate due to intermediate resonances are totally ignored. However, the model naturally accounts for some observations such as channel cutoff, and it has been quite successful in predicting the number and relative magnitude of ATI peaks in Nd:YAG experiments.[15]

The Floquet decomposition shows the absorption of discrete numbers of photons, despite the fact that the electromagnetic field in this problem has not been quantized. The discretization is the result of wave-mechanical interference. Whereas a classical particle exhibits a continuously varying energy as it oscillates up and down in the periodic field, the Volkov wave function is an energy superposition state with an oscillating phase at frequencies ω and 2ω. It's quite easy to show that only energy eigenstates with eigenvalues of $n\hbar\omega$ can be included in this superposition.

Interference phenomena also affect the <u>directions</u> of electron propagation. As the electron emission angle changes with respect to the laser polarization, the Volkov phase factor containing $\mathbf{A} \cdot \mathbf{p}$ also changes. This affects the values of the coefficients in the Floquet expansion. The Bessel function coefficients may even pass through zero, indicating that ATI electrons with a given energy may not propagate in those directions. Angle-dependent propagation of ATI electrons may be compared to Bragg scattering in crystal. Whereas the Bragg peak is caused by constructive interference of x-ray waves scattering from a spatially periodic potential, the ATI electrons are matter waves scattering from a <u>temporally</u> periodic potential. In both cases, wave interference strongly modulates the scattering distributions.

These electron phase-matching effects were recently demonstrated in xenon and krypton ATI experiments utilizing <u>elliptically</u> polarized 1.06 μm light.[16] By varying the ellipticity of the light, it's possible to control the variation in $\mathbf{A} \cdot \mathbf{p}$ with angle to optimize the interference effects. In this case, the light was slightly off circularly polzarized. Angular distributions in the azimuthal plane are shown in Fig. 5. The predictions based on the Floquet expansion are also shown. The best match between theory and experiment occurs for a slightly more eccentric retardation in the calculation. The source of this discrepancy is not known, but may e related to Coulomb effects neglected in the Keldysh approximation. Nonetheless, the general agreement is quite striking.

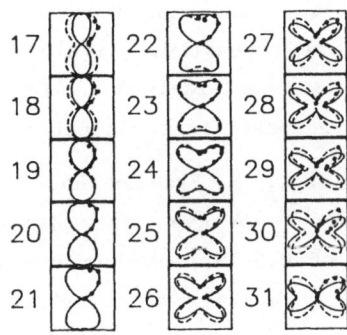

Fig. 5 Krypton photoelectron azimuthal angular distributions for ATI peaks corresponding to absorption of 17-31 photons, for nearly circularly polarized light (retardation $\xi = 80°$). Solid lines are the theoretical calculations with the $\xi = 70°$, and $U_p = 2$ eV. (From Ref. 16.)

4. ATOMIC STRUCTURE IN INTENSE FIELDS

The previous analysis ignores atomic structure almost entirely. The atom confines the electron prior to the expeirment, and then conveniently disappears the moment the laser turns on. Furthermore, the nonresonant approximation ignores all structure other than the ground state and the ionization potential. This leads to a simple intuitive picture of ionization into wiggling electron states. But in fact, the high-lying bound states may have an important influence on the ionization rate. This is because the loosely bound highly excited states must have aponderomotive shift comparable to free electrons. For shifts as large as $\hbar\omega$, these states must come into multi-photon resonance with the laser field at some intensity. If parity and angular momentum rules permit, an enhancement in the ionization rate may occur.

Unfortunately, the final state ponderomotive accelerations in the gradients of the laser focus tend to cancel the effect of the level shifts on the energy of the electrons. The resonances therefore do not produce multiple energy peaks, although they may still alter the overall ATI rates and angular distributions. However, under certain circumstances, the final state ponderomotive effects may be effectively turned off. If the laser pulse duration is very short, the ponderomotive force does not act for a long enough time to alter the electron's kinetic energy or momentum. For pulses of about one picosecond or less, the electrons just "kick out" of the ponderomotive potential, giving up their ponderomotive potential energy, but retaining the initial velocity imparted during ionization. Since the ionization potential of the atom increases linearly with the intensity, the electron's energy is a record of the intensity when it was ionized. In this way, intermediate resonances might be seen as peaks in the electron energy distribution.

In a recent experiment, R. Freeman and co-workers have probably observed this phenomenon.[17] A dye laser (wavelength 616 nm) with a variable pulse duration of 0.5-13 picoseconds was used to ionize xenon. For longer pulsewidths, typical ATI peaks were observed at the appropriate energies. However, when the pulsewidth was reduced below 1 picosecond, the peaks shifted and split into a number of sharp features. Figure 6 shows the short pulse ATI spectrum superimposed on the even parity single-electron excitation spectrum for xenon. Most of the peaks coincide with allowed six-photon resonances, provided the resonant intermediate states have shifted by the same amount as the ionization limit.

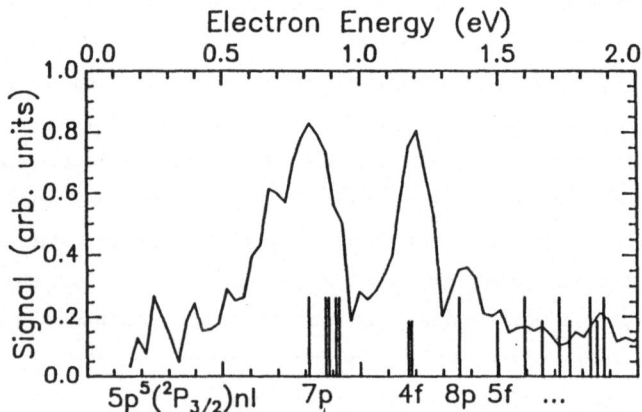

Fig. 6 Photoelectron spectrum in xenon for 7 photon ionization by sub-picosecond 616 nm pulses. The peaks line up well with the expected positions of six-photon resonant enhancements, provided that the resonant intermediate states shift by the full ponderomotive potential of a free electron in the laser field. (From Ref. 17.)

5. CONCLUSIONS

ATI experiments have greatly increased our understanding of the interaction of intense fields with electrons. Ponderomotive forces and final state effects are now fairly well understood. Much has also been learned about the quantum nature of the wiggling electron continuum. The newest results suggest that atomic structure may be strongly altered by intense fields. Ultra-short pulsed lasers will be extremely valuable tools in exploring these questions in the future.

It is a pleasure to acknowledge useful discussions with M. Bashkansky and R. R. Freeman.

REFERENCES

1. Parts of this paper have been adapted from P. Bucksbaum, "Above-threshold ionization and the quantum mechanics of wiggling electrons," in Proceedings of the NATO Advanced Research Workshop on Atomic and Molecular Processes with Short Intense Laser Pulses, ed. by A. Bandrauk (Plenum Press, New York, 1987).

2. L. A. Lompre, G. Mainfray and C. Manus, J. Phys. B 13, 85 (1980).

3. See, for example, the feature issue on multielectron excitations in atoms, Jour. Opt. Soc. B 4, ed. by W. E. Cooke and T. J. McIlrath, 701-862 (1987).

4. H. A. Bethe and E. Salpeter, Quantum Mechanics of One- and Two-Electron Atoms (Springer Verlag, Berlin, 1957) Sect. 54.

5. F. Yergeau, G. Petite and P. Agostini, J. Phys. B 19, L663 (1986).

6. S. L. Chin, F. Yergeau and P. Lavigne, J. Phys. B 18, L213 (1985).

7. P. Agostini, F. Fabre, G. Mainfray, G. Petite and N. Rahman, Phys. Rev. Lett. 42, 1127 (1979).

8. P. Kruit, J. Kimman and M. van der Wiel, J. Phys. B 14, L597 (1981).

9. T. J. McIlrath, P. H. Bucksbaum, R. R. Freeman and M. Bashkansky, Phys. Rev. A 35, 4611 (1987).

10. T.W.B. Kibble, Phys. Rev. 150, 1060 (1966); J. H. Eberly, Progress in Optics VII, ed. by E. Wolf (North-Holland, Amsterdam, 1969) p. 361.

11. P. H. Bucksbaum, M. Bashkansky and T. J. McIlrath, Phys. Rev. Lett. 58, 2590 (1987).

12. L. Pan, L. Armstrong, Jr. and J. H. Eberly, Jour. Opt. Soc. B 3, 1319 (1986); M. Mittleman, Introduction to the Theory of Laser-Atom Interactions (Plenum Press, New York, 1982).

13. R. R. Freeman, T. J. McIlrath, P. H. Bucksbaum and M. Bashkansky, Phys. Rev. Lett. 57, 3156 (1986).

14. L. V. Keldysh, Z. Eksp. Teor. Fiz. 47, 1945 (1964) [Sov. Phys. JETP 20, 1307 (1964); H. R. Reiss, Phys. Rev. A 22, 1786 (1980); F.H.M. Faisal, J. Phys. B 6, L89 (1973).

15. H. R. Reiss, J. Phys. B 20, L79 (1987).

16. M. Bashkansky, P. H. Bucksbaum and D. W. Schumacher, Phys. Rev. Lett. 59, 274 (1987).

17. R. R. Freeman, P. H. Bucksbaum, H. Milchberg, S. Darack, D. Schumacher and M. E. Geusic, Phys. Rev. Lett. 59, 1092 (1987).

ELECTRON ENERGY SPECTRA IN HIGH INTENSITY MULTIPHOTON IONIZATION

Guillaume Petite and Pierre Agostini

Service de Physique des Atomes et des Surfaces
CEN Saclay-91191-Gif sur Yvette, France

INTRODUCTION

When an atom is placed in an intense electromagnetic field, it can be ionized even if the photon energy is less than its ionization threshold. This process, know as Multiphoton Ionization (MPI), has been studied now for almost twenty years[1] and becomes the dominant process when very high intensities are used. It has been observed in a wide range of experimental conditions, for laser intensities between 10^7 Wcm^{-2} and 10^{15} Wcm^{-2}, and wavelengths ranging from 193 nm to 10600 nm (in the latter case, several hundreds of photons can be absorbed in the MPI process).[2] Not only single, but also multiple ionization has been observed in such experiments in the case of very high intensities.[3,4] MPI has also been observed in the radio frequency domain, as discussed in another chapter of this book.

Of course, ionization is not all that happens to the atom submitted to an intense e.m. field. Strong modifications of the atomic spectrum occur, such as a.c. Stark shifts, which have to be taken into account if one wants to understand the MPI process. There is no better example of this than the case of resonant MPI where a.c. Stark shifts of the ground and resonant states strongly modify the dynamics of the ionization process at high intensities.[5] As we will see, the influence of the field on the final state of the MPI process is also at the center of the understanding of many of the recent experiments.

The use of Electron Energy Spectroscopy (EES) has been introduced only lately in the study of MPI. It immediately revealed new aspects of MPI, and particularly demonstrated the existence of Above Threshold

Ionization (ATI), a process in which more than the minimum necessary number of electrons are absorbed.[6] This process has been extensively studied these past few years, but is still only partially understood in the case of very high intensities. EES proved to be a very powerful tool in bringing new information about the final state of the MPI process, of both the electron and the ion.

The aim of this paper is twofold. First we want to show, using recent experiments, how EES helps obtaining a good description of the final state of the MPI process, even in the case where strong e.m. fields are used. Second, in a case where the experimental result can be compared to an "ab initio" calculation, we show MPI experiments can bring new information on the atomic structure. Of course, one has in this case to limit oneself to the case of "intermediate" intensities, where the MPI process itself is well understood, even taking into account some of the high intensity effects. We first briefly describe the general experimental conditions of a high intensity MPI experiment, and what can be expected as its result. We then consider the case of 11 photon ionization of xenon at the Nd:YAG fundamental frequency, and show on this example how a physical description of the final state of the electron in high intensity MPI can be obtained from EES. Finally, MPI of strontium in the visible optical range will serve as an example showing the kind of spectroscopic information which can be obtained in the interplay between theory and experiment.

GENERAL EXPERIMENTAL CONDITIONS

Laser intensity: with up-to-date giant laser systems, one can achieve intensities up to 10^{17} Wcm^{-2}, but it is essential to realise that there is a limitation to the intensity that the atom is really going to experience due to the atom itself, and to the fact that laser pulses have a finite rise-time. As the field is gradually turned-on, ionization starts in the leading edge of the pule and may very well be complete before the maximum intensity is reached in the laser pulse, so that the atom only sees a fraction of the peak intensity. If one now considers a series of experiments in which a laser pulse with a given duration is shined on an atom with gradually increasing peak intensities, the peak intensity for which ionization is total after the passing of one laser pulse is generally referred to as the "saturation intensity."[1] An increase of the peak intensity beyond this value cannot produce more ionization, except by extension of the interaction volume to the wings of the laser spatial distribution. The saturation intensity depends on the specific experiment (atom, wavelength and pulse duration), and sets an upper limit to what can effectively be achieved in terms of "strong field." In the typical high intensity experiment discussed below, the saturation intensity is of the order of 10^{13} Wcm^{-2}.

The electric field corresponding to such an intensity is of the order of 10^8 Vcm^{-1}. It is still only a fraction of the field seen by an electron in the ground state, but one does·not have to go to highly excited states to see the atomic field drop below such values (n > 4 would be enough in the case of hydrogen). If one now consider ionization of an electron in the ground state, it only has to travel a few Angstroms before it sees an inversion in the order of the fields which, for a (typical) 1 eV energy in the final state, takes about 10^{-15} s, less than one laser period. All this is enough to show that MPI at such intensities is indeed a strong field problem, and to anticipate complications in the theory due to the fact that, during the ionization process, there will be a change in the respective orders of the e.m. and Coulomb fields.

The electron energy spectrum: what can be expected as electron energy spectrum can be simply drived from energy conservation:

$$E_e = (N + S) E_p - E_i \tag{1}$$

where E_e, E_p, and E_i are respectively the electron, photon and ionization energies. N photons have to be absorbed to reach the ionization threshold, but S more photons can be absorbed in the ionization continuum. Figure 1 shows the relevant atomic energy levels and the different ionization processes to be considered in a case discussed below: MPI of strontium at wavelengths between 557 and 575 nm. N = 3 photons allow in this case to reach the first ionization threshold and leave the ion in its ground state. This is the minimum order process, and therefore it is expected to state. This is the minimum order process, and therefore it is expected to be dominant. However S extra photons can be absorbed in the ionization continuum, always leaving the ion in the ground state. This process is known as Above Threshold Ionizaiton (ATI)[6,18] and yields faster electrons, whose energy spectrum consists of a series of lines evenly spaced by the photon energy. By apsorption of more photons (4 are enough in the case of Fig. 1) one can also reach higher ionization limits, leaving the ion in one of its excited states. The EES spectrum now displays several sets of lines corresponding to the different ionization limits. Finally, as shown on Fig. 1, ionization of the ion can occur, and the corrpesponding EES spectra will then superimpose to the one obtained from simple ionization. This shows that EES gives detailed information on the final state of the ionization process, concerning both the electron and the ion. Also when different ionization channels contribute to the process, EES allows to study these different channels separately, as we will see below.

Experimental techniques: we do not intend to present the details of an MPI experiment (they can be found in most of the references above). We will just mention that the lasers have to be either Q-switched or Mode-locked, owing to the intensities considered. The latter are generally preferred because they deliver Fourier-limited pulses, which are better accounted for by theory. Any electron spectroscopy technique can be used, but the pulsed character of the experiment almost automatically selects time-of-flight spectroscopy, which also applies to mass-spectroscopy of the ions. The most important experimental problem besides that of the determination of an absolute value of the laser intensity (not generally obtained to better than a factor of two), is that of space charge which can strongly effect the low energy electrons obtained in such experiments. Therefore one has to work at a very low signal level, which in turn commands the use of counting techniques and also requires very low background pressures ($<10^{-8}$ Torr).

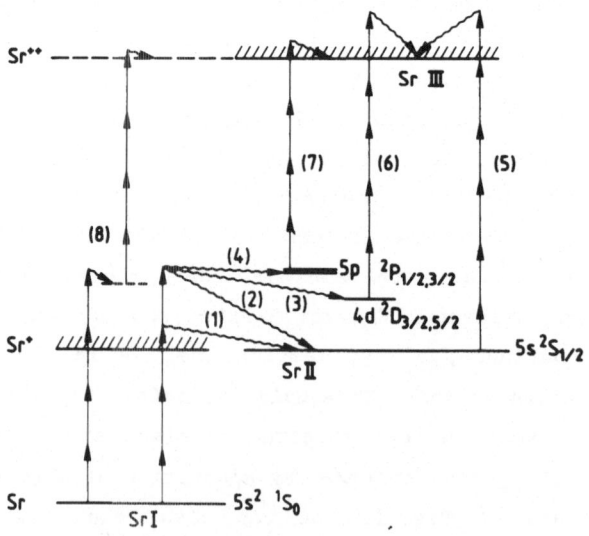

Figure 1. MPI of stontium at wavelengths around 565 nm, illustration of the different possible ionization channels.

Above Threshold Ionization has been observed in so many different conditions that one can say that it is an intrinsic part of the MPI process. However, its importance greatly depends on the experimental parameters, mainly the laser wavelength and intensity. At moderate intensities (I < 10^{12} Wcm^{-2} and/or for wavelengths in the near U.V. range, its magnitude is small and it can be considered merely as a correction to the lowest order process. Even if calculations are by no means simple, ATI offers in this case no major conceptual difficulties and, in some cases quantitative agreement can be obtained between the results of an ab initio calculation of ATI probabilities and that of the experiment.[16] The ATI dynamics is in this case well described by Lowest Order Perturbation Theory (LOPT in this chapter), which predicts a I^{N+S} power law dependence for the probability of a process involving absorption of S excess photons in the ionization continuum, S being generally small in this (S = 1 or 2). The situation is quite different when the intensity is increased, or the photon frequency reduced, which is typically the case in MPI of rare gases by near IR light. A large number of excess photons can then be absorbed, and the probability for such processes becomes eventually much larger than that of the lowest order one. LOPT can obviously not account for such a situation and many theoretical attempts are currently in progress to solve this problem, none of them having succeeded so far in accounting for the whole of the high-intensity ATI phenomenology. This matter is by far too complicated to be reviewed in this chapter, and the interested reader is referred to Ref. 19 and papers quoted therein. Most of the difficulty lies in the fact that the final state of the process cannot be considered in this case as a weakly perturbed Coulomb state. This was clearly shown in recent ATI experiments that we now discuss.

Figure 2 shows a typical result obtained in MPI of xenon by 1064 nm light of a Nd:YAG laser and contains most of the high intensity ATI phenomenology. Spectrum (a) serves as a reference spectrum and has been obtained for a laser intensity of 2.8×10^{12} Wcm^{-2} and a pulse duration of 136 ps. Spectrum (b) is obtained at 7.5×10^{12} Wcm^{-2} and for the same pulse duration. Spectrum (c) is obtained at the latter intensity but for a pulse duration of 50 ps. The vertical lines show the energies at which the ATI peaks are expected using Eq. (1) and the ionization potential taken from Moore's tables.

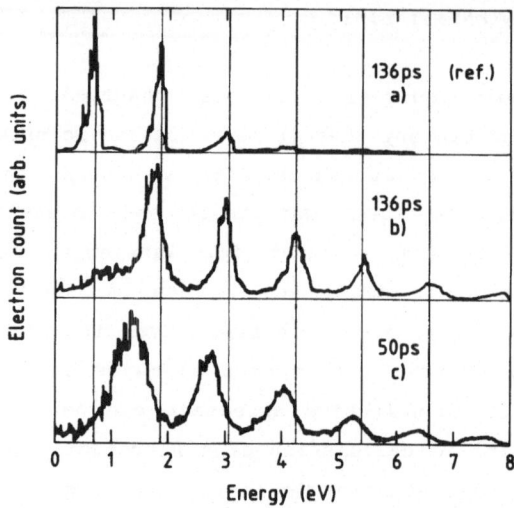

Figure 2. Different ATI spectra obtained in Xenon at 1064 nm, for different laser pulse durations and intensities.

Besides the increase of the number of ATI peaks resulting from the increase in the intensity between spectra (a) and (b), we would like to concentrate on two aspects of the high intensity spectra:

(i)   On spectrum (b), the first AIT peak is almost absent.   This effect has been repeatedly observed in such experiments and has received the name of "peak suppression."   The different ATI peaks otherwise appear at their expected energies.

(ii)   The overall aspect of spectrum (c) is quite similar to that of spectrum (b) except that the low energy peaks are red shifted.   It was shown that this shift depends linearly on the laser intensity and decreases with the electron energy[8] (the high energy peaks of spectrum (c) are unshifted).

These two phenomena are direct consequences of the strong perturbation of the final state by the laser field.   When released from the atom by the MPI process, the electron still feels the effect of the e.m. field.   It thus undergoes a rapid oscillatory movement, at the laser frequency, whose average classical energy is given by:

$$\Delta = e^2 E^2 / 4m\omega^2 \qquad (2)$$

where E is the laser field amplitude and ω the laser frequency. Thus, energy conservation as expressed by Eq. (1) writes:

$$E_e = E_{dr} + \Delta = (N + S) E_p - E_i \qquad (3)$$

where E_{dr} is the elctron average classical drift kinetic energy. We have included in the ionization energy E_i an eventual a.c. Stark shift of the atom's ground state. All quantities in (3) have to be positive. By definition of N, when S = 0, the right hand side of (3) is less than the photon energy. It readily follows than if Δ is of the order of E_p, Eq. (3) might be impossible to satisfy with S = 0, or even S = 1,2... if the laser intensity, and thus Δ is increased. The peak suppression effect can thus be expected from energy conservation. In the above mechanism, the oscillation energy of the electron in the laser field, often call "ponderomotive potential," acts as an effective shift of the ionization potential. Its magnitude of course depends on the laser intensity and frequency. In the experiment presented above, at a wavelength of 1064 nm, it is of about 1 eV for an intensity of 10^{13} Wcm^{-2}, which is enough to explain the disappearance of the first peak of spectrum (b), whose initial energy is only 0.65 eV.

The red-shift of the low energy peaks observed on spectrum (c) is another indication of the up-shift of the final state by Δ. Given Eq. (3), and the fact that the quantity measured in an experiment is the drift energy of the electron, one would expect a shift of the whole spectrum by an amount Δ, not observed on spectra like (b). However, one has to remember that the electron energy is measured **outside** the laser field. This energy E_{out} can then be calculated from the energy **inside** the field ($E_{in} = E_e$ as given by Eq. (3)) in the following way:

$$E_{out} = E_{in} + \int_{in}^{out} \mathbf{F} \cdot d\mathbf{x} \qquad (4)$$

where the integral on the right-hand side represents the work of the **total** electromagnetic force. The dynamical variables of the electron and the e.m. force are then separated in a high frequency part (oscillation: x_h, v_h, F_h) and a low frequency one (drift: x_l, v_l, F_l), and the variations of the low frequency components over a few periods are neglected. Cross terms in the integral in Eq. (4) then average out so that:

$$E_{out} = E_{in} + \int_{in}^{out} (\mathbf{F_h} \cdot \mathbf{v_h} + \mathbf{F_l} \cdot \mathbf{v_l})\, dt \qquad (5)$$

381

where

$$F_h = -e \; E(\mathbf{x}_1, \; t) = -e \; \varepsilon(\mathbf{x}_1, \; t) \; \cos \omega t \qquad (6)$$

and

$$F_1 = -(e^2/4m\omega^2) \; \nabla \; (\varepsilon^2(\mathbf{x}_1, t)) \qquad (7)$$

is the usual ponderomotive force.[20,21] One can then show[8,20] that is one assumes a guassian beam profile of beam waist Ω, and a temporal envelope for the field of the kind $\varepsilon(t) = \varepsilon_0 \exp(-|t|/t_0)$, one has for an electron emitted at time $t = 0$ in the center of the focal volume, and with its initial drift velocity aligned along the laser polarization:

$$E_{out} - E_{in} = \Delta \theta \qquad (8)$$

$$\theta = \sqrt{\pi}\beta \; \exp(\beta^2) \cdot erf_c(\beta) \qquad (9)$$

where

$$\beta = \Omega/2Vt_0 = 1.2 \times 10^{-12} \; (\Omega/t_0) \; E^{-1/2}$$

with V being the electron drift velocity. β has a simple meaning: it is the ratio of the time (Ω/V) it takes the electron to exit the laser focus to the pulse duration $(2t_0)$. Given the asymptotic behavior of the complementary error function, it is easy to study the two extreme cases:

 - $\beta \to 0$: infinite pulse duration ("beam" experiment). The electron has the time to experience the total spatial gradient. In this case, $\theta = 0$. The total energy of the electron is conserved and oscillation energy is converted to drift energy when the electron exits the laser focus. The energy peaks are therefore found at the position expected from the spectroscopic tables: **The quantity measured in this case is the total electron energy after ionization.**

 - $\beta \gg 1$: short pulse situation. The laser field is turned off before the electron has the time to leave the laser focus. It then feels the gradient essentially along the laser propagation direction, that is perpendicularly to its velocity. In this case $\theta = 1$, the oscillation energy is lost in the process and the electron peak is red-shifted by an amount Δ: **the quantity measured in this case is just the electron drift energy after ionization.**

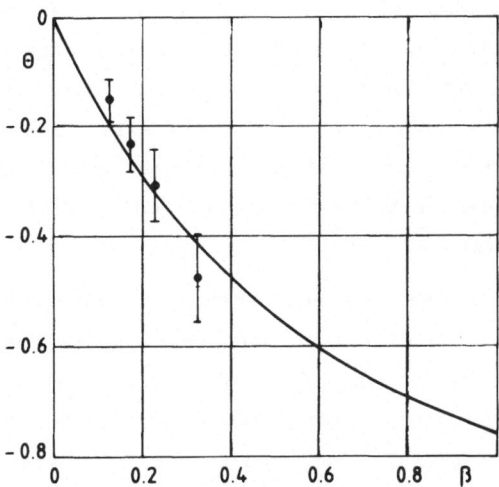

Figure 3. Energy red shift (in units of Δ) of the different peaks of spectrum 2(c) vs. $\beta = t_{ex}/2t_o$ (see text). Full line: theory (Eqs. 8 and 9); bars: experiment.

Intermediate cases lead to an only partial transfer of oscillation energy to drift energy. This is precisely the case of spectrum (c). With a 50 ps pulse duration and a 20 μm beam waist, only the faster electrons have the time to escape before the laser is turned off. The peaks are therefore red-shifted by an amount depending on their initial drift energy, as can be computed from Eqs. (8) and (9). The result of such a calculation is compared on figure 3 to the shifts measured on the different peaks of spectrum (c), and the agreement is quite satisfactory.

Spectrum b is an example of the long pulse situation, where no shifts is measured. With a 50 ps pulse, this situation can also be obtained by reducing the beam waist and thus the escape time of the electron. In the above experiment, the beam waist was therefore reduced from 20 μm to 10 μm and this resulted in a disappearance of the shifts with a 50 ps pulse. The short pulse situation was realised in an experiment with ultra-short laser pulses (100 fs) at a wavelength of 620 nm. In this experiment,[23] even though high energy ATI peaks were produced, no electron was fast enough to escape the laser focus before the laser was turned off. This resulted in an overall shift of the spectrum by an amount δ.

To summarize, observation of energy shifts of the ATI electron spectra give a direct evidence of a stong perturbation of the final state of the MPI process when high laser intensities are used. A consequence of this perturbation is the peak suppression effect observed in all situations (short or long pulses). These results now have to be accounted for in a theory of high intensity ATI, because a precise understanding of the dynamics of this process will only be obtained when the coupling mechanism between the (perturbed) ground state and the final state will be understood.

No such problems are met at intermediate intensities. A satisfying theoretical description of ATI is obtained in this case in the framework of the usual perturbative theories, eventually including selected higher order corrections to account for distortions of the atomic spectrum. It is then possible to sue the results of EES experiments to extract information on the atomic structure which were not otherwise accessible. An example of this is given in the next section, in the case of strontium.

MPI/ATI AT INTERMEDIATE INTENSITIES: THE STRONTIUM CASE

A partial atomic spectrum of SrI and SrII has been shown on figure 1, and we jsut want to recall that, because the stontium ion has low-lying excited states, it is possible with absorption of only one excess visible photon to reach higher ionization limits. Such a process will now compete with first order ATI, so that the EES spectrum is expected to present a quite different aspect.

Using a tunable laser it is possible to study intermediate resonances on states (not shown on the figure) which can be either below the first ionization limit, or above. In the latter case, these states are intermediate states for the higher order processess yielding an excited ion only (they act as autoionizing states in the case of the three photon process). They are two-excited electrons states, and two photon resonances on such states below the first ionization limit can also be reached in the wavelength range accessible to a Rhodamine 6G dye laser. We have studied MPI of strontium using such a laser, which delivered Fourier limited 20 ps pulses, for intensities ranging between 10^{11} Wcm^{-2} and 3×10^{12} Wcm^{-2}. Both single and double MPI were observed. Intermediate resonances could be studied on both the electron or the ion signal, the latter being by far the most comfortable and was used when channel selectivity was not necessary (two photon resonances).

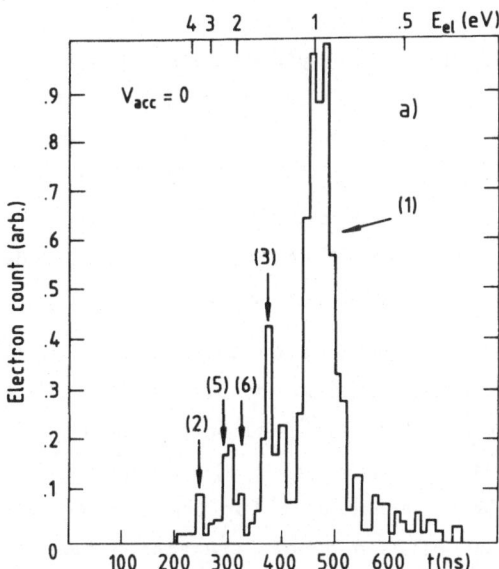

Figure 4. Electron energy spectrum obtained in MPI of strontium at 562 nm and for a laser intensity of 3×10^{12} Wcm^{-2}.

Figure 4 shows an EES spectrum obtained at a wavelength of 562 nm, and an intensity of 3×10^{12} Wcm^{-2}. All the processes shown on Fig. 1 yield electron peaks which are labeled according to the channel numbering used in Fig. 1. Absorption of three photons only allows to leave the ion in the ground state, and a 0.9 eV electron is released in the process. The corresponding peak (1) is the larges one in the spectrum. After absorption of one excess photon, the ion can be left in three different states: (i) the ground state; this is an ATI process as discussed in the preceeding section and it yields the peak (2) which, because of the lower intensity is now small compared to the lowest order peak. (ii) the 4d state; an electron with an energy of 1.25 eV is released in the process, yielding peak (3) of the spectrum. (iii) the 5p state; the electron is emitted with a 0.1 eV energy only, so that the corresponding peak is not seen on the spectrum. However, using a small acceleration (1 V) for the electron, this peak is easily detected and can be studied in the same conditions as peak (3).

Another feature is clearly visible on Fig. 4, in the 1.8 – 2.0 eV region. This peak was shown[24] to be due to electrons released in the sequential double ionization processes using the 5s and 4d ion states as intermediate states.

Returning to the four-photon processes, several remarks can be made. First considering ATI leaving the ion in the ground state, its order of magnitude is the same as what was observed in cesium at comparable intensities. The two other four photon processes seem on the contrary more probable (they yield much larger peaks), though they involve excitation of the core electron. This is due to the presence, in the energy region reached by absorption of three photons, of autoionizing states of the type 4d,nl or 5p,nl. The influence of these states can be studied in a channel selective way, by measuring separately the wavelength dependence of the different peaks amplitudes. The result of such an experiment is shown on figure 5.

Several comments can be made on this figure. Resonances are clearly seen in both linear and circular polarization. They had already been observed in the MPI ion signal,[25] and only some of them could be assigned to states known from U.V. absorption spectroscopy.[26] A result like that of figure 5 could be of great help for theat matter. in the independent electron picture, and since a dipole transition operator only acts on one electron (namely the ejected electron), an intermediate three photon resonant state must have a core correspond g to the ion state obtained in the four photon ionization process on which the resonance appears. In this picture, a resonance appearing in channel (3) of figure 1 (ion state 4d) should have no effect on channel (4) (ion state 5p). Of course all the resonances can appear on channel (1), autoionization being not subject to the dipole selection rule. A glance at figure 5 is enough to see that this picture is by far too simple, which is no surprise for a heavy atom like strontium. All resonances but one (labelled V on figure 5) appear in a quite similar fashion in both channels (3) and (4), because of the strong configuration mixing between the different three-photon resonant states. How much mixing is there is precisely one of the possible outcomes of such an experiment, provided a good control of resonant MPI theory, including autoionization.[27] Before presenting the theoretical interpretation of this data, we would like to recall a last experimental result concerning a.c. Stark shifts.

The above mentioned resonances appear to be unshifted by the laser field, which is not the case of an other group of resonances observed at slightly longer wavelengths. These are two photon resonances on states $5p^2$ and $5s5d$. The $5p^2$ can be either a $^1P_{0,1,2}$, whereas $5s5d$ can be either a $^1D_2$ or a $^3D_{1,2,3}$, only states with an even J satisfying the two photon selection rules. The a.c. Stark shifts of the two photon resonances were studied as a function of the laser intensity. Only states $5s5d$ ($^1D_2$, $^3D_2$) and $5p^2$ ($^3P_{0,2}$) gave rise to detectable resonances. Only resonances on states $5p^2$ $^3P_0$ and $5s5d$ $^3D_2$ were measureably shifted. The shifts show the usual linear intensity dependence, which can be characterized by the value of the slope which is $-2.0(+1.1/-0.5)$ for the tp^2 state and $0.5(+0.3/-0/2)$ for the 5s5d state (in $cm^{-1}/GW.cm^{-2}$).

These results were compared to the result of an ab-initio calculation using Multiconfiguration Hartree-Fock wavefunctions.[28] Such wavefunctions can reproduce the sate energies with a few percent accuracy only, which is not enough for MPI calculations which are very sensitive in some cases to such energies. Therefore the actual energies were used in the calculations. Very soon it became clear that Stark shifts, in particular, could not be reproduced without taking into account singlet-triplet mixing, which was then calculated using angular distributions of the emitted electron obtained in another laboratory,[29] so that it is not, in this calculation, adjustable. The importance of singlet-triplet mixing is clearly examplified in Table I, where the experimental vlaues of the a.c. Stark shift coefficients measured above are compared to the results of three calculations. In calculation I and II, coupling of the $5p^2$ states with the 5pns series was introduced. Singlet-triplet was taken into account in calculation II, but not in calculation I. Calculation III includes, in addition to calculation II, states 5p5d and 4d4f which lie close to one photon from the $5p^2$, and can therefore be anticipated to be important in this calculation. In the case of the 5s5d state, only one calculation was made, including states 5p5d and 5s5f, and singlet-triplet mixing was taken into account.

Table I. Comparison between experimental and theoretical values of the a.c. Stark shift coefficients (in $cm^{-1}/GWcm^{-2}$). See text for precisions on the different calculations.

state	experiment	calc. I	calc. II	calc. III
$5p^2\ ^3P_0$	$-2.0(+1.1/-0.5)$	-4.26	-2.2	-2.41
$5p^2\ ^3P_2$	$-0.03(+/-0.1)$	-0.91	-0.021	-0.0225
$5s5d\ ^3D_2$	$0.5(+0.3/-0.2)$		1.0	

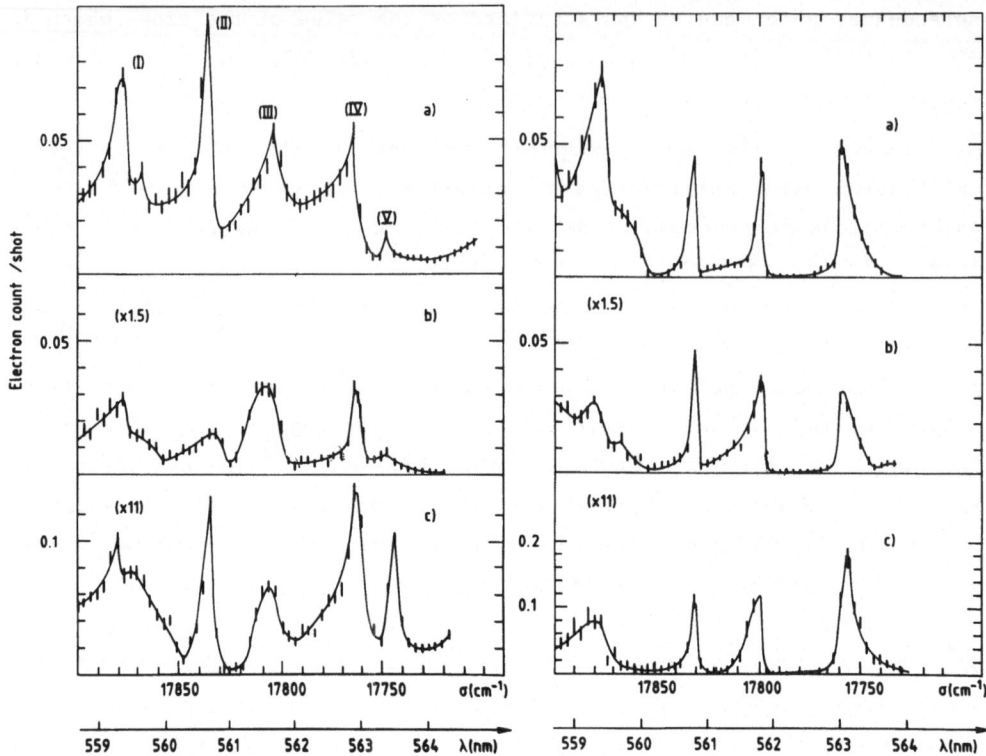

Figure 5.   Wavelength dependence of the amplitudes of the electron peaks
(fig. 4) corresponding to MPI channels: (a)- 1, (b)- 3, (c)- 4
(on fig. 1).   Laser polarization left: linear, right: circular.

Comparing the different results in the $5p^2$ case immediately shows
the importance of singlet-triplet mixing.   Depending whether it is
included or not makes a difference of almost a factor of two, and brings
the calculated value within the experimental error bar, whereas including
additional states in the calculation only makes a 10% difference in the
coefficient.

The amount of singlet-triplet mixing these calculations conclude to
is quite important.   In the case of the 5p6s state for instance, one
finds:

$$5p6s \; (^3P_1) = a \; (^3P_1) + b \; (^1P_1)$$

$$5p6s \; (^1P_1) = -b \; (^3P_1) + a \; (^1P_1)$$

with $a^2 = 0.536$ and $b^2 = 0.464$, which is certainly an important effect.

Another outcome of such calculations concerns the identification of
the three photon resonances of Fig. 5.   Some of them were already known

from older spectroscopic work, but some others were not, as usual in multiphoton spectroscopy experiments. The different data presented here, through a somewhat complicated process, led to propose the following assignment:

I: $4d4f\ ^3D_3$ II: $5p^2\ ^3P_2$ III: $5p5d\ ^3D_3$

IV: $4d4f\ ^1F_3$ V: $5p5d\ ^3D_1$ VI: $5p6s\ ^1P_1$

where roman numbers correspond to the numbering of resonances of Fig. 5, VI designates the feature appearing on some spectra on the long wavelength side of resonance I and which is not numbered on the figure. Additional data by Feldman et al.[30] was used in the process. Note that the 5p5d states are seen here for the first time.

Of course, the above assignments are a rough approximation for a more complicated situation because they are based on an independent electron picture, which is clearly inadequate here. This is well demonstrated by calculating the configuration interaction coefficients, another outcome of the interplay between theory and the above data. One finds that the states presented above as the $5p^2$ states in fact have almost 50% of $4d^2$ in them, a result recently confirmed by R-matrix calculations.[31] This strong mixing explains why 4dnf states are so easily reached through a three photon absorption process starting from $5s^2$ ground state. Another example of the strong configuration mixing affecting the autoionizing states is obtained by considering the state named $4d4f(^3D_2)$ in the above assignment, and whose wavefunction really is:

$$"|4d4f\rangle" = -\ 0.61\ |4d4f\rangle + 0.67\ |4d5f\rangle - 0.23\ |5p5d\rangle + 0.22\ |4d7p\rangle + \ldots$$

where only states with a coefficient larger than 0.1 have been retained. Such a decomposition clearly shows that the independent electron picture does not provide the best possible basis set, at lease inside this particular group of states of strontium.

As a conclusion to this section, we have demonstrated how application of EES to MPI can, in a case where theory does not encounter major conceptual difficulties, yield a large amount of information about the atomic structure, even at intensities large enough to seriously modify this structure.

CONCLUSION

We have presented two rather different situations met in the study of atoms submitted to intense light fields. In high intensity multiphoton ionization, the strong perturbation of the final state by the laser field is the dominant effect. It gives rise to deep modifications of the electron spectra and to spectacular energy shifts observable with ultrashort pulses. This situation can obviously not be handled by standard perturbation theory. One of the reasons for this is the inversion, in the course of the multiphoton ionization process, of the respective orders of magnitude of the two electric fields involved here, the Coulomb and electromagnetic fields.

In multiphoton ionization of alkaline earth atoms, Electron Energy Spectroscopy provides direct information about the different ionization channels. Multiphoton ionization itself is a very interesting tool because the states involved in multiphoton transitions considerably differ from those reached in traditional spectroscopy. The perturbations of the atomic spectrum induced by the laser field (a.c. Stark shifts) have proved to be valuable checks of the complex computations required by theory. Specific measurements in strontium have revealed quantitatively the importance of configuration mixing and singlet-triplet mixing.

REFERENCES

1. J. Morellec, D. Normand and G. Petite, in Adv. At. Mol. Phys., 18, 97, (1982)

2. F. Yergeau, S. L. Chin, and P. Lavigne, J. Phys. B. At. Mol. Phys., 20, 723 (1987).

3. A. L'Huillier, L. A. Lompre, G. Mainfray and C. Manus, Phys. Rev. Lett. 48, 1814 (1982).

4. T. S. Luk, H. Pummer, K. Boyer, M. Shadidi, M. Egger and C. K. Rhodes, Phys. Rev. Lett., 51, 110 (1983).

5. J. Morellec, D. Normand and G. Petite, Phys. Rev. A, 14, 300 (1976).

6. P. Agostini, F. Fabre, G. Mainfray, G. Petite and N. K. Rahman, Phys. Rev. Lett., 42, 1127 (1979).

7. P. Agostini, M. Clement, F. Fabre and G. Petite, J. Phys. B. At. Mol. Phys., 14, L491 (1981).

8. P. Agostini, J. Kupersztych, L. Lompre, G. Petite and F. Yergeau, Phys. Rev. A. (Rap. Comm.), 36, in press (1987).

9. P. H. Bucksbaum, M. Bashansky, R. R. Freeman, T. J. McIlrath and L. F. Dimauro, Phys. Rev. Lett., 56, 2590 (1986).

10. R. R. Freeman, P. H. Bucksbeam, H. Milchberg, S. Darack, D. Schumacher and M. E. Geusic, to be published (1987).

11. H. J. Humpert, R. Hippler, H. Schwier and O. Lutz, in "Fundamental Processes in Atomic Collision Physics," ed. H. Kleinpoppen, J. S. Briggs and H. O. Lutz, New York: Plenum (1985).

12. P. Kruit, J. Kimman an W. J. Van der Wiel, J. Phys. B. At. Mol. Phys., 14, L597 (1981).

13. P. Kruit, J. Kimman, H. G. Muller and M. J. Van der Wiel, Phys. Rev. A., 28, 248 (1983).

14. L. A. Lompre, A. L'Huillier, G. Mainfray and C. Manus, J. Opt. Soc. Am., B2, 1906 (1985)

15. L. A. Lompre, G. Mainfray, C. Manus and J. Kupersztych, J. Phys. B. At. Mol. Phys., 20, 1009 (1987).

16. G. Petite, F. Fabre, P. Agostini, M. Crance and M. Aymar, Phys. Rev. A, 29, 2677 (1984)

17. G. Petite, P. Agostini and F. Yergeau, J. Opt. Soc. Am., B4, 765 (1987)

18. F. Yergeau, G. Petite, and P. Agostini, J. Phys. B. At. Mol. Phys. 19, L663 (1987)

19. Proceedings of the International Conference on Multiphoton Processes IV (Boulder, July, 1987). S. J. Smith and P. Knight, eds., Cambridge University Press, to be published.

20. L. D. Landau and E. M. Lifshits in Mechanics, Oxford: Pergamon, p. 94, (1960).

21. T.W.B. Kibble, Phys. Rev., 150, 1060 (1966).

22. J. Kupersztych, to be published (1987).

23. H. G. Muller, H. B. Van Linden van den Heuvell, P. Agostini, G. Petite A. Antonetti, M. Franco and A. Migus to be published.

24. P. Agostini and G. Petite, Phys. Rev. A (rap. Comm.), 32, 3800 (1985).

25. P. Agostini and G. Petite, J. Phys. B. AT. Mol. Phys., 18, L281 (1985).

26. W.R.S. Garton, G. L. Grasdalen, W. H. Parkinson and E. M. Reeves, J. Phys. B. At. Mol. Phys., 1, 144 (1968).

27. Young Soon Kim and P. Lambropoulos, Phys. Rev. A., 29, 3159 (1984).

28. X. Tang, P. Lambropoulos, A. L'Huillier, P. Agostini and G. Petite, to be published.

29. D. Feldmann and K. H. Welge, private comm.

30. D. Feldmann, J. Krautwald, S. L. Chin, A. Van Hellfeld and
    K. H. Welge, J. Phys. B. At. Mol. Phys., $\underline{15}$, 1663 (1982).
31. M. Aymar, E. Luc Koenig and S. Watanabe, J. Phys. B. At. Mol. Phys.,
    in press (1987). M. Aymar, J. Phys. B. At. Mol. Phys., in press.

H_2 IN INTENSE LASER RADIATION – MULTIPHOTON TWO ELECTRON DIRECT PROCESSES

Munir H. Nayfeh, J. Mazumder and D. C. Humm

Department of Physics
Department of Mechanical Engineering
University of Illinois at Urbana-Champaign
1110 W. Green Street
Urbana, IL 61801

ABSTRACT

We have observed ions with energies of up to 12.5 eV from a sample of molecular hydorgen subjected to an intense 28 ns, 308 nm laser beam. We show that these ions are strong evidence for a two-electron many-photon direct collective process, and discuss the nature of this process. The rapid dissociation occurs on a femtosecond timescale from an initial internuclear separation as short as 0.58 Å, well within the classically forbidden region.

The H_2 molecule has recently been studied by a number of groups using resonantly enhanced multiphoton ionization by multicolor narrow-band tunable radiation.[1-4] In these studies, the total H_2^+ ion current is monitored as well as the energy spectrum of the ejected electrons as a function of the wavelength of the radiation. Such powerful complementary techniques can be used to study the structure of the resonant intermediate state, to determine the branching ratios into various electronic, vibrational, and rotational levels of the product molecular ion (and achieve selectivity), and to directly investigate the dynamics of the complete multiphoton ionization process. One study compared measured photoelectron energy spectrum peaks to predictions from rotational selection rules and a calculation of partial waves of the ionized electron to analyze the excited molecular state.[3]

Here a somewhat different and complementary study of H_2 is done, of processes with nonresonant intermediate states in a more intense field. Early studies of this type were performed with a Nd (1.06 μm) laser[5]; we use a 308 nm, 25 ns beam. In addition, we measure the kinetic energy of the H+ ions produced. This repulsive dissociation or "Coulomb explosion" method has been a useful tool in studying diatomic molecules[6,7]. It provides a tool to measure femtosecond dynamics independent of the laser pulse duration[7], and gives an accurate measurement of the internuclear separation at which a process is occurring.

The current study may also be relevant to experiments on the nonresonant multiphoton ionization of atoms in intense fields. H_2, a two-electron system, may bear similarities to He. Controversy has arisen in the process of interpreting recent results on atoms, especially Xe.[8-10] Collective emission occurring on a time scale of 1 fs was originally postulated, but later theoretical study concluded that successive stripping of single electrons was the dominant mechanism.[11] The current experiment bears upon this question, because the Coulomb explosion technique we utilize provides the femtosecond "clock" necessary to distinguish between direct and sequential stripping of electrons. In addition, the H_2 molecule is a relatively simple system, amenable to detailed analysis and understanding.

We now give a brief description of the experimental setup (see Fig. 1). Hydrogen molecules are observed in a molecular beam formed when they diffuse through a multicollimator assembly from a region of H_2 gas at a pressure of several mtorr. This loosely collimated beam has a density of about $10^{12}/cm^3$. It passes between the lower two of three parallel electrodes in a diffusion pumped cell. The lower two electrodes are horizontal, flat metal plates separated by 8 mm, with a 3x5 mm slot covered by fine mesh (the analyzing grid) in the middle electrode to allow the passage of ions to the detector. They are biased at values between 50 volts retarding and 100 volts accelerating in order to analyze the energy of the produced ions. The upper electrode is a wire mesh with an extraction voltage that is kept at −5 kV. It is connected to a metal tube leading directly to an eighteen-stage Venetian-blind electron multiplier capable of single ion detection.

The laser radiation used in the experiments is the 308 nm unpolarized fundamental of an XeCl excimer laser which gives 400 mJ per 25 ns pulse. The beam is focused to a fine line halfway between the analyzing electrodes and directly beneath the analyzing grid. An estimate of the average power density based on the advertised beam divergence is 10^{10} W/cm^2. This should be considered a minimum; we expect the center of the focal spot to be much more intense, especially because the highly divergent part of the beam departs over a 4 meter path it

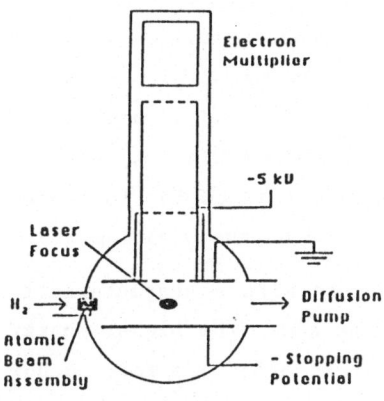

Figure 1.

Experimental Apparatus. The
ions experienced first a
stopping potential and then
an extraction voltage.

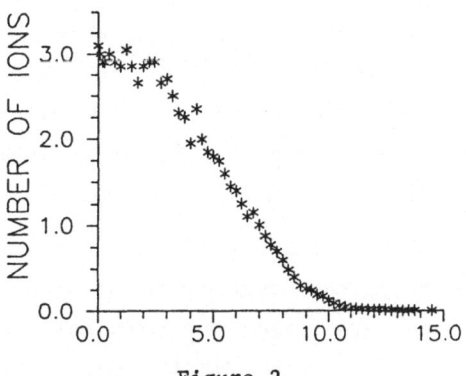

Figure 2.

The ion yield at a laser
intensity of 2.9 I_0. These
high-energy ions are
described by the processes
shown in Figs. 7 and 8.

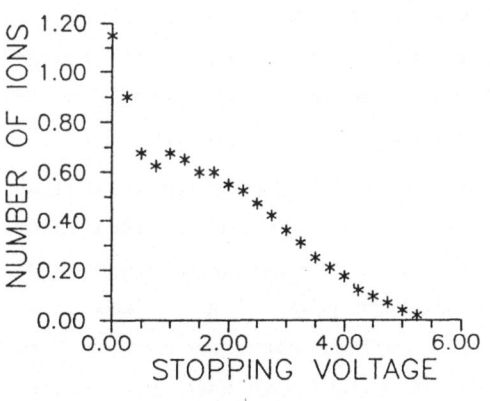

Figure 3.

The ion yield at a laser
intensity of 1.7 I_0. These 5
eV ions are described by the
processes shown in Fig. 6.

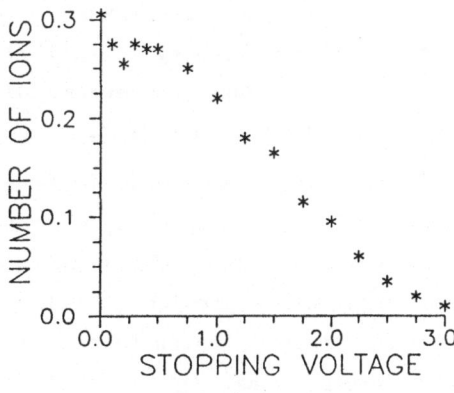

Figure 4.

The ion yield at a laser
intensity of I_0. These 3 eV
ions are described by the
processes shown in Fig. 6.

travels before focusing. The relative intensity of each pulse is monitored by a photodiode, which measures the light reflected by the window of the cell.

The energy of the ions is determined by measuring the voltage necessary to stop them. Unfortunately, this procedure provides analysis only of the velocity component parallel to the stopping field. Ions are produced with velocities in other directions as well. As a result, for an in with energy E_0, a distribution of energies from 0 to E_0 will be observed, because a lesser stopping potential will suffice for ions with only a component of their velocity in the direction of the stopping field.

When there are several species of ions, each with a characteristic energy, the lower energy ions cannot be distinguished from higher energy ions with large transverse velocities. Since this effect depends on how many ions with significant transverse velocities are accepted by the detector, it can be alleviated by reducing the acceptance solid angle of the detector (by changing the analyzing grid area or the distance of the focal point from it). In spite of the large solid angle we were able to acheive some resolution of the lower energy ions by utilizing the power dependence of the processes involved. We quenched the higher energy ions by lowering the intensity of the laser radiation, but we realize they will be quenched only if the intensity dependence of the ion yield varies monotonically with the ion energy. We may be missing low-energy ions produced by high order transitions, and such will be studied in further detail in the future. Finally, we do not believe that space-charge effects can change the energy distribution, because our pulse time is long, and the ions are fast, so they should not accumulate.

Figs. 2-4 give the experimental results: the number of ions reaching the detector as a function of stopping potential for three selected laser intensities of increasing magnitude. In Fig. 5 we show an enlargement of the high energy portion of Fig. 2. At highest power (Figs. 2 and 5) we measure stopping potentials of up to 12.5 volts, corresponding to ions with energies as high as 12.5 eV. It is these ions that give strong evidence for direct processes. At the next lower power (Fig. 3), the fast ions are gone, but we see ions with energies of 5 eV. At the lowest power (Fig. 4), the 5 eV ions disappear, and ions of 3 eV are observed. Data was also taken for high power at approximately twice the pressure, and no change was observed in the energy of the fastest ions, showing the absence of space-charge effects, as explained above.

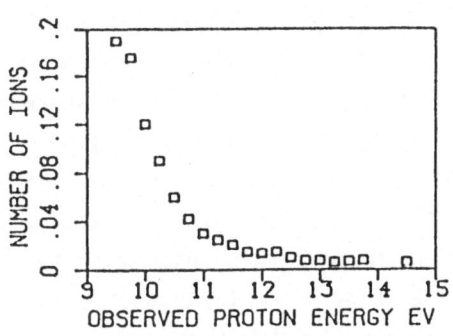

Figure 5.

An enlargement of the high-
energy portion of Fig. 4.

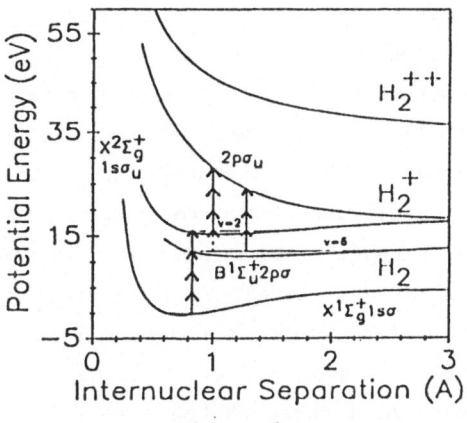

Figure 6.

Sequential excitations which
explain observations of low-
energy ions.

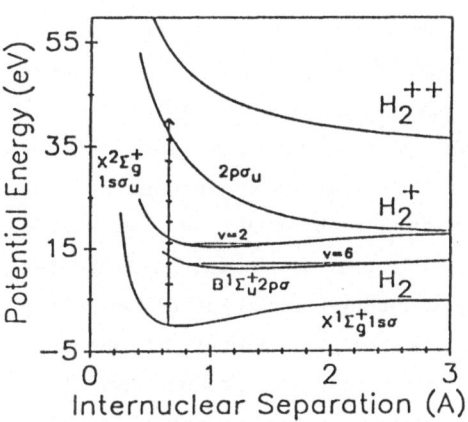

Figure 7.

Direct dissociative
excitation/ionization from
inner turning point of ground
state, giving rise to ions of
kinetic energy 9.8 eV each.

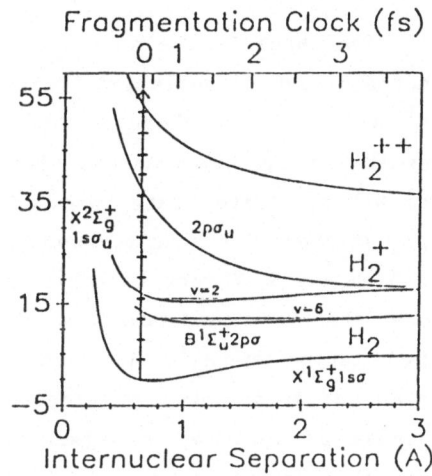

Figure 8.

Direct two-electron
ionization to the two-
electron continuum, starting
from the inner turning point
of the ground state, and
giving rise to ions of
kinetic energy 11.1 eV each.

We now analyze the data and discuss the various mechanisms responsible for the energy distribution of the ions we observed (see Figs. 2-5). We will start with the low energy ions, and then proceed on to the more new and exciting high-energy ones. As shown in Fig. 6, the excitation scheme we used populates the v=6 vibrational level of the B state through a near three photon resonance of detuning -.01 eV (which is about the laser bandwidth), as well as the H_2^+ (v=2) bound ground state by near four photon resonance of detuning +0.17 eV (given to the ejected electron). The oscillator strength for the transition to the v=6 vibrational level of the B state is 2.7×10^{-2}. Note that all excitations must take place at constant internuclear separation, in keeping with the Franck-Condon effect. We believe that the 5 eV ions in Fig. 3 arise through a three photon resonant excitation from the H_2^+ (v=2) state (at internuclear separation 1.00 Å) to the repulsive $2p\sigma_u$ ground state of the ion. At this internuclear separation, the energy of the $2p\sigma_u$ state is 28 eV: as a result the dissociation of the H_2^+ into H(1s) and H^+ fragments liberates 10 eV of kinetic energy shared by the products (5 eV each). Selection rules inhibit a similar process using two photons to produce 3 eV ions.

The three (and also five) eV ions can be explained by three (four) photon direct dissociative ionization of the B state of H_2 to the $2p\sigma_u$ of H_2^+ at an internuclear separation of 1.28 Å (1.00 Å). This process results in the simultaneous ejection of one electron and dissociation of the H_2^+ repulsive state into H(1s) and H^+ fragments. The energy of the $2p\sigma_u$ at these internuclear separations is 24 eV and 28 eV, producing ions of 3 eV and 5 eV respectively, agreeing with our observations. We should note that the dissociative ionization of the B state, unlike the dissociation of H_2^+, does not require resonance interaction with the laser radiation and can proceed by an even or odd number of photons, because an electron is ejected. This will tend to produce red (low energy) wings to the 5 and 3 eV energies without altering the blue (high energy) wings.

Now for the direct evidence of direct, simultaneous, excitation of two electrons by many photons via nonresonant intermediate states through one of two processes, the central thesis of our paper. This evidence arises from analysis of the high energy ions observed in Fig. 5. It is clear from Figs. 7 and 8 that it is not possible to produce such ions by any process starting from the B (v=6) state nor from the H_2^+ (v=2) ground state because their inner turning points, at 0.78 Å and 0.77 Å respectively, are too large. These ions, however, can be accounted for if the excitation proceeds directly from the ground state. This is because the inner turning point of the ground state occurs at an internuclear distance of 0.65 Å, the smallest distance of any turning point in the system. In order to avoid relaxation of the internuclear separation as the excitation proceeds, the process should not involve resonance interaction with any intermediate vibrational highly excited state of the system. Using a numerical tabulation of the energy level diagram[12,13], we first examine direct dissociative ionization of the ground state (at 0.65 Å) via the repulsive ground state of H_2^+, as shown in Fig. 7. This is a two-electron excitation, with one electron ionized and one excited to the $2p\sigma_u$ state of H_2^+. At 0.65 Å, the dissociate ionization threshold occurs at 37.7 eV which is 19.6 eV above the dissociation limit, thus giving ions of 9.8 eV each. It is a 10 photon excitation with the ejected electron acquiring 2.5 eV kinetic energy. The ion energy of 9.8 eV, however, is less than the maximum of the energies in the experimental results of Fig. 5.

The alternative mechanism for the production of high energy ions is direct two electron excitation to the two electron continuum, illustrated in Fig. 8. In this process, two electrons are stripped simultaneously, followed by Coulomb explosion of the two protons. As an example, with excitation from near the smallest turning point (at 0.65 Å), the two electron ionization threshold occurs at 53.8 eV. Using 31.64 eV for the H^+-H^+ system at infinite distance gives 11.1 eV for the kinetic energy of each of the resulting protons. At this turning point the process proceeds via the absorption of 14 photons, and so the ejected electrons share a kinetic energy of ~2 eV. Similar calculations at other R values also show that the two electron ionization processes yield ions of energies higher than the dissociative ionization case. In fact, this is due to the electronic energy of the p orbital of the H_2^+ repulsive state. This energy is shown in Fig. 9 along with the energy of other orbitals.

Fig. 8 also illustrates the time scale of the process. On the upper horizontal axis, the time taken by the protons to Coulomb explode from an initial position of 0.6 Å is given (the dissociative ionization time scale is similar). It takes about 1 fs for the protons to move from an initial internuclear separation of 0.6 Å to one of 1.0 Å, so we are not only exciting from the forbidden region, but also examining very short time scales. Since 1 fs is a typical electronic time scale, this process is direct or collective as far as the electrons are concerned. Note that the fs time scale is a clear and direct result of our experimental observations. In a sequential stripping process, the internuclear separation would have time to relax between strippings, and the resulting Coulomb explosion energy would be from the ground state internuclear separation of H_2^+, less than half the energy of the H_2 Coulomb explosion. These arguments apply equally well to the dissociative direct excitation/ionization. Since sequential stripping (or sequential excitation of one and stripping of the other) would drastically decrease the kinetic energy of the ions coming from the Coulomb explosion, and we have observed high energy ions, far higher than can be explained by sequential stripping, we have no choice but to conclude that we have observed a multiphoton multielectron direct or collective excitation. The direct or collective, within a femtosecond, processes shown in Figs. 7 and 8 are the only way to explain the kind of high ion energies we have seen.

Now that our central result has been established, there remain a couple of interesting additional questions we can study further with a more detailed analysis of the data and some simple theoretical arguments. The first is the observation of excitation from the forbidden region of the vibrational wavefunction, showing the sensitivity of the Coulomb explosion technique not only to very short times, but also very short distances (10 pm). The second is the question of which direct multiphoton multielectron process we are observing. We have good reason to believe that it is the direct two-electron ionization rather than the direct dissociative excitation/ionization.

Because our measurements, as seen in Fig. 5, show ions of energies larger than 11.1 eV, namely ions in the range 11–12.5 eV, excitation from the classically forbidden region of the ground state (R < 0.65 Å) must have taken place. Although the vibrational wavefunction in this region is exponentially decaying, it is not vanishingly small. In fact, at the edge of the forbidden region, it is at approximately 57% of its peak value.[14] We calculated the internuclear separations at which the two electron ionization and the dissociative ionization processes produce ions of a given energy (i. e. the internuclear potential curve), and illustrate it in Fig. 11. The lower horizontal axis gives the energy and we use the upper horizontal axis to illustrate the corresponding R values for each of the two processes. This gives a correspondence between R and E, different for each of the two processes, and so we can write functions of R, such as the ground state vibrational wavefunction, as functions of E, the measured energy, making such functions amenable to measurement. This assumes that the Franck–Condon effect forces all excitations to take

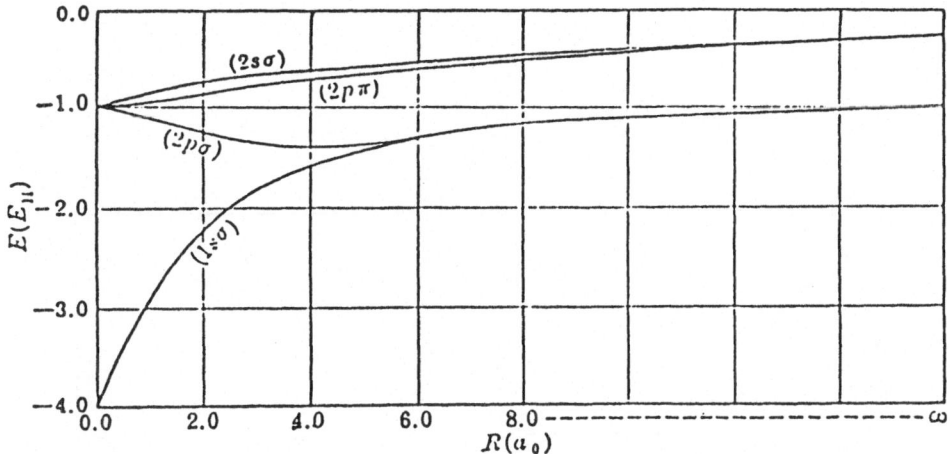

Figure 9.   Electronic energy of the $2p\sigma_u$ state and others.

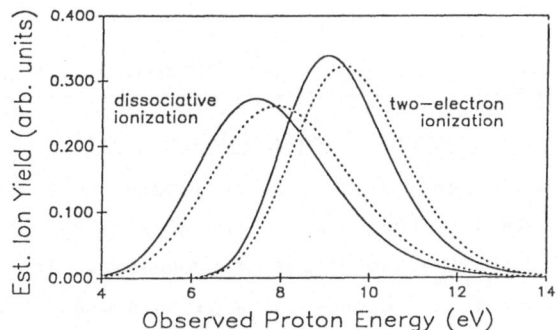

Figure 10.   The probability density of the ground state vibrational
wavefunction of H_2 as a function, not of internuclear
separation, but instead of the detected energy of the ions
coming from each of the two processes, dissociative
ionization and two-electron ionization. This estimate of
the ion yield (ignoring effects due to detuning, electron
wavefunctions, or # of photons) is compared to the estimates
(dotted lines) one gets from calculating (numerically) the
actual wavefunction overlaps that result in the
Franck-Condon effect.

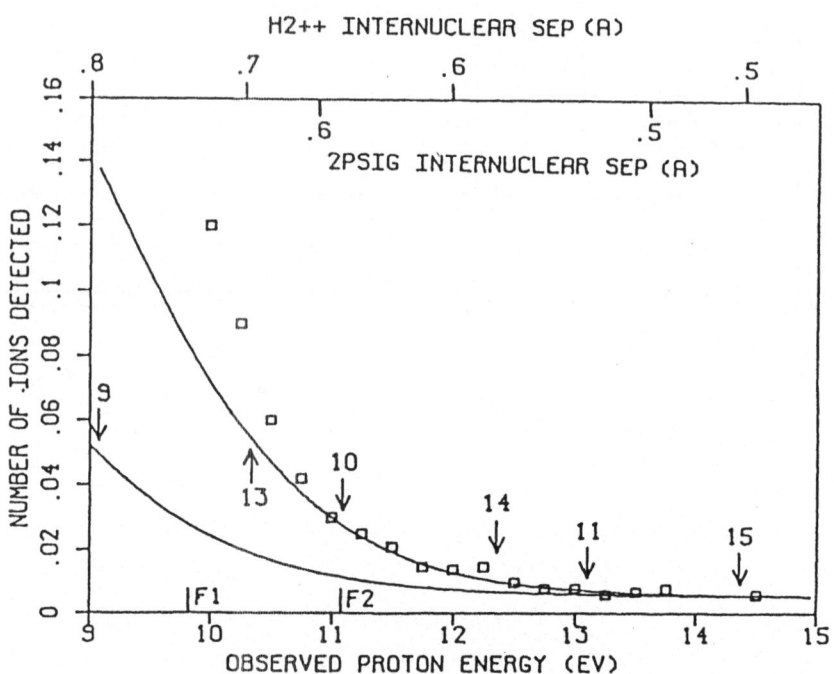

Figure 11. The two upper scales give the internuclear separations that would lead to the observed energies given on the bottom scale, for the two different processes. The arrows give the observed proton energy for a given number of photons absorbed if the electron(s) do not carry off any energy (threshold of ionization). There are 9, 10, or 11 photons for the dissociative ionization process and 13, 14, or 5 photons for the two-electron ionization process. F1 and F2 give the observed proton energy for which the protons are initially at the edge of the forbidden region of the ground state vibrational wavefunction (0.65 Å) in each process.

place at constant internuclear separation. Of course, the particular function of E obtained depends on the process giving rise to the measured energy, as the relationship between R and E depends on the process. As an example, we plot in Fig. 10 the probability density of the vibrational ground state of H_2 as a function of the observed ion energy for each process (same wavefunction, different processes). We compare with the exact calculation of the wavefunction overlaps that produce the Franck-Condon effect, showing fairly close agreement. All other things being equal (no dependence on the detuning or the number of photons), one would expect this to give the ion yield of each process as a function of energy, a measurable quantity.

If one knows the process, one can measure distances very accurately. From Fig. 11, an uncertainty of 1/4 eV in the ion kinetic energy measurement corresponds to a position uncertainty of only ~.01 Å. More accurate energy measurements would make this even smaller. It appears that the Coulomb explosion technique may allow the accurate measurement of vibrational wavefunctions.

To experimentally distinguish the dissociative from the two-electron ionization process, one could use a coincidence measurement, since the two-electron ionization produces two ions, which are both detected, while the other process produces one neutral, which remains undetected. The two processes require a different number of photons (10 vs. 14), so one might expect the power dependence to distinguish them. However, the exponent of the power dependence is an inaccurate measure of the number of photons, often underestimating it in an unpredictable way. For example, an exponent of 8 has been observed for a 14 photon process[15] and a 9 photon process[16]. We observed an exponent of approximately 8 at 9 V stopping potential and approximately 3 at 2 V, confirming at least that the higher energy ions come from a much higher-order process.

Although the measurements don't distinguish the two processes, some simple theoretical arguments strongly favor the two-electron ionization. Fig. 10 shows that the dissociative ionization process is less probable, all other things being equal, at high energies. To see the importance of this effect, we integrate the curves in Fig. 10 and plot them in Fig. 11. Because the curves represent the yield of ions with a given energy, they must be integrated for comparison with data obtained from a stopping potential experiment, since all ions with an energy greater than the stopping potential are detected and therefore summed. This procedure does not give an exact representation of our detector function, but it is a good first approximation. The curves clearly show that in the region of interest, around 12.5 eV, the dissociative ionization process is less probable than the two electron ionization by a factor of about 3.5. In the same figure we present the experimental data, normalized to the probability density at a two electron ionization energy of 11 eV. The agreement is quite excellent at the high energy side, but deviations are seen at energies less than 9 eV. Of course, we got similar conclusions when we normalized the results to the probability density evaluated at dissociative ionization energies. Thus, we have further evidence matching our data with the two direct processes, but still no clear means of discerning between them except a probability factor strongly favoring the two-electron ionization, which would be part of a branching ratio.

Detection efficiency also favors the two-electron ionization. The detection efficiency is less for dissociative ionization because events in which the neutral atoms are ejected towards the detector (relevant to the case of dissociative ionization) will be missed unless the laser manages to ionize the atoms with 100 percent efficiency before they leave the interaction region. Under the conditions of our experiment we expect to ionize only a few percent of the hydrogen atoms. This conclusion is based on some measurements we made in which the H_2 beam was converted to an H beam by a DC discharge. Our measurements indicated that 10 percent or less of the H atoms are ionized by the laser beam. This percentage drops further in the actual experiment because the H atoms produced in the dissociative ionization process are not thermal; they are produced with a kinetic energy of 12 eV. At this energy they have velocities of 4.8×10^6 cm/s, and cross the beam in ~2 ns. Since the laser pulse duration is 25 ns, we expect the rate of ionization to drop by a factor of ~6 from that of the thermal atom. Thus, virtually all of the neutrals will be missed, and only half as many events will be detected from dissociative ionization as from two-electron ionization.

Thus, as a result of consideration of vibrational wavefunctions and detection efficiencies we project that the two-electron ionization process is favored by a factor of 7 over the dissociative ionization process. Moreover, there are other factors which may influence the branching ratios of the two processes. These include: the order of the multiphoton process, the photoionization cross section, and the polarization of the laser radiation.

The two processes each involve a "large" number of photons (12-14) and no resonant intermediate states. In this case, the parameter $4\pi r_0 F/\delta^2$ is useful, where r_0 is the classical radius of the electron, and k is an effective average detuning of all the intermediate states involved.[17] Using simple arguments based on an earlier analysis of atoms by Bebb and Gold[17], one finds that the ionization cross section is proportional to

$$(4\pi r_0 F/\delta^2)^N \int \phi_f{}^* e^{iN\underline{k}\cdot\underline{r}} \phi_i d^3\underline{r}$$

where N is the number of photons, and m_i, m_f are the initial and final state wavefunctions respectively. If $\delta^2 = 4\pi r_0 F$, the first factor, coming from the number of photons, will be the same for both processes. At the powers we are using, we find that the average detuning required for the condition to be achieved is 0.17 eV which is not totally unrealistic. The second factor favors the two-electron ionization process. Since $N\underline{k}\cdot\underline{r} \approx .03$, the dipole approximation is valid, and the integral reduces to an overlap integral between the initial and final two-electron wavefunctions, which can be broken down into a sum of products of one-electron overlap integrals. In the case of dissociative ionization, every term in the sum contains a factor of the one-electron overlap integral between the $1s\sigma g$ ground state and the $2p\sigma_u$ excited state, which is zero from parity considerations. This does not happen in the two-electron ionization, where all the overlap integrals are between ground state electrons and free electrons. Thus, we expect the dissociative ionization to be suppressed. The suppression, of course, may not be as strong if the number of photons is not high enough.

Because of the large number of photons involved, we expect no polarization effects when the laser radiation is unpolarized as in our experiment. However, we would expect some effects for a completely polarized beam.

The locations of various multiphoton processes are displayed on the energy axis of Fig. 10. It is interesting to note that there is evidence in the data of the 14 photon final state resonance in the two electron ionization process. This resonance occurs in excitation from the forbidden region of R. Also one can see evidence for the 13 photon resonance of the same process. Both of these resonances cause deviation of the data from the predictions based on the R dependence of the wavefunction of the ground and final state. On the other hand, it is not clear if we have any evidence for the 11 photon resonance of the dissociative ionization process, which would account for ion energies of 13 eV. There might be some evidence for the 10th order resonance of this process occuring at 11.1 eV; however, it is not as strong as the rise in the yield at the 13 photon resonance of the two electron ionization process.

In conclusion, we have observed multiphoton direct two-electron excitation via nonresonant intermediate states, and we believe that we have evidence for the simultaneous ionization of two electrons by many photons. It is remarkable that these processes can be investigated, and such short time scales examined, with a laser of such low power and long pulse duration. In fact, the long pulse and the great sensitivity of the Coulomb explosion method allow us to completely rule out sequential excitation as an explanation of the high-energy ions. We have also observed excitation from the forbidden region of the vibrational wavefunction of a molecule, and opened the door to the possibility of actually measuring such wavefunctions accurately. Future work will involve detailed studies of the branching ratios, power dependencies and polarization effects, and experiments on heavier diatomic molecules.

References

1. P. M. Johnson and C. E. Otis, Ann. Rev. Phys. Chem. 32:139 (1981).
2. S. L. Anderson, G. D. Kubiak, and R. N. Zare, Chem. Phys. Lett. 105:22 (1984).
3. S. T. Pratt, P. M. Dehmer, J. L. Dehmer, J. Chem. Phys. 78:4315 (1983).
4. J. Hessler and W. Glab, in press.
5. M. Lu Van, G. Mainfray, C. Manus, and I. Tugov, Phys. Rev. Lett. 29:1134 (1972).
6. A. K. Edwards and R. M. Wood, J. Chem. Phys. 76:2938 (1982), with R. L. Ezell, Phys. Rev. A 34:4411 (1986).
7. L. J. Fransinski, K. Codling, P. Hatherly, J. Barr, I. N. Ross, and W. T. Toner, Phys. Rev. Lett. 58:2424 (1987).
8. G. Kyrala, presented at the Int. Conf. on Mltph. Prcss. IV, Boulder, CO (1987).
9. A. L'Hullier, L-A. Lompre, G. Mainfray, and C. Manus, J. Physique 44:1247 (1983).
10. U. Johann, T. S. Luk, H. Egger, and C. K. Rhodes, Phys. Rev. A 34:1084 (1986).
11. P. Lambropoulos, Phys. Today 40(1):S25 (1987).
12. T. E. Sharp, J. Atomic Data 2:119 (1973).
13. D. R. Bates and R. H. G. Reid, Adv. Atomic and Mol. Phys. 4:13 (1968).

14. G. Herzberg, Spectra of Diatomic Molecules, p. 77, Van Nostrand Reinhold (1950).

15. P. Agostini, G. Barjot, J. F. Bonnal, G. Mainfray, C. Manus, J. Morellec, IEEE J. Quantum Elec. 4:667 (1968).

16. G. S. Voronov, G. A. Delone, N. B. Delone, Sov. Phys. JETP Lett. 3:313 (1967), Sov. Phys. JETP 24:1122 (1967).

17. A. Gold and H.B. Bebb, Phys. Rev. Lett. 14:60 (1965).

MULTIPHOTON IONIZATION OF ATOMIC HYDROGEN BY EXTRA PHOTONS

Erna Karule

Institute of Physics, Latvian SSR Academy of Sciences
229021 Riga, Salaspils, USSR

INTRODUCTION

Multiphoton ionization (MPI) of atoms by extra photons or so-called above threshold ionization (ATI) is the one-step ionization process when atomic electron absorbs more photons than energetically necessary for ionization. We investigate ATI theoretically and our approach is based first on the method of the Sturmian expansion of the transition matrix elements developed by us earlier for MPI of H by minimum number of photons.[1] Secondly, as Sturmian expansion diverges in the case of ATI, the analytical continuation ofthe divergent part of the radial transition matrix elements is carried out using technique similar to that proposed for the case of two-photon ATI.[2-4]

TRANSITION MATRIX ELEMENTS FOR ATI

If we do the summation over all angular variables of the intermediate states step by step, then the transition rate for (N+1) – photon ionization of H by one extra photon in the case of linearly polarized light may be written in the form

$$Q^{(N+1)}/I^N = 4\pi^2 \alpha \, a_o^2 \omega I_o^{-N} \sum_{L_{N+1}} |T^{(N+1)}(n_o 1_o, E_{N+1} L_{N+1} | \omega)|^2 \tag{1}$$

where $Q^{(N+1)}$ is the total cross section, I is intensity of light in W/cm^2, a_o is Bohr's radius, α is the fine structure constant, $I_o = ce^2/2\pi a_o^4 = 14.038 \times 10^{16} \, W/cm^2$, ω is the energy of a photon.

In the dipole approximation using Sturmian expansion for Coulomb Green's function the transition matrix elements for (N+1)-photon ATI of H are obtained in the form

$$T^{(N+1)} = -C_{N+1} \sum_{L_N = L_{N+1} \pm 1} \sum_{L_{N-1} = L_N \pm 1} B(L_{N+1}, L_N) B(L_N, L_{N-1}) \times$$

$$\sum_{m_{N-1}=1}^{\infty} Y(m_{N-1}, L_{N+1}, L_N, L_{N-1}) X_{N-1}(m_{N-1}, L_{N-1})$$

(2)

where $B(L_a, L_b) = (4 - 1/L^2(max))^{-1/2} \, 2C/(2E)^{L/2}$ is the normalization factor of the Coulomb wavefunction. X_j can be calculated recursively down to X_{N-1}

$$X_j(m_j, L_j) = q_j^{-2L_j - 1} (m_j)_{2L_j + 1} / (m_j + L_j - q) \sum_{L_{j-1} = L_j \pm 1} B(L_j, L_{j-1}) \times$$

$$f(m_j + L_j, L_j, q_j; m_{j-1} + L_{j-1}, L_{j-1}, q_{j-1}) X_{j-1}(m_{j-1}, L_{j-1})$$

(3)

By definition we have

$$X_o(m_o, L_o) = X_o(n_o - 1_o, 1_o) = n_o^{-2-1_o} ((n_o - 1_o)_{21_o+1})^{1/2}$$

If before integration over radial variables we use recurrence relations for the confluent hypergeometric functions then for f we get

$$f(m, L, p; n, 1, q) = f(n, 1, q; m, L, p) = (-1)^{L-m} 2^{21+2} p^{L-1} \times$$

$$(pq/(q - p))^{21+4}((21+1)!)^{-1} z^{-n-m} \times$$

(4)

$$\sum_{s=0}^{2} C_2^s ((L - m + 1)_s (s - m - L - 2)_{2-s} \delta_{L, 1-1} - \delta_{L, 1+1}) \times$$

$$(1 + m - s - np/q) z^{s-2} \, {}_2F_1(-m + 1 - s, -n + 1 - 1, 21 + 2; 1 - z^2)$$

$$C_2^s = 2!/s!/(2 - s)! \qquad\qquad z = (q + p)/(q - p)$$

The summation over intermediate states with positive energies is included in the expression for Y, which diverges if the Sturmian expansion is used directly. To get convergent expression for Y we must replace the sum over intermediate states by two sums. One is a power series, which may be rewritten as a polynomial. Another formally corresponds to Appell's hypergeometric series. These double series have the expansion in terms of Gaussian hypergeometric functions. For ATI these double series can be analytically continuated in a form of the expansion in terms of other Gaussian hypergeometric functions. If $N\omega > 1/(2n_o^2)$ convergent expression for Y may be written in form

$$Y = (2\omega)^{-2}(2L_{N-1} + 2)_{L_N-L_{N-1}+2}((2L_N+1)!)^{-1}(T_1 + T_2) \qquad (5)$$

where T_1 is a polynomial

$$T_1 = 2^{-2}\omega^{-1}q_N q_{N-1}^{L_{N-1}+L_{N+1}}(2L_{N+1}+2)_u/gz_3^{-q_{N+1}-L_{N+1}}(1 - z_1)^w(1 + z_3)^t \times$$

$$\sum_{s=0}^{m_{N-1}-1} (1 - m_{N-1})_s(w + 2)_s(I - z_1)^s/(2L_{N-1} + 2)_s/s! \times$$

$$\sum_{n=1}^{d} (-d)_n(g)_n(1+z_1^{-1})^n/(2L_N+2)_n/n! \sum_{k=0}^{n-1} (2L_N+2)_k/(g+1)_k(1+1/z_1)^{-k} \times$$

$$\sum_{j=0}^{k} (-k)_j(t + 6)_j(2L_N+2)_j/j!/(1+q_{N-1}/q_{N+1})^j(1-q_{N-1}/q_{N+1})^j \times$$

$$\times \ _2F_1(-q_{N+1} + L_{N+1} + 1, -u - j; 2L_{N+1} + 2; 1 - z_3) \qquad (6)$$

where

$$w = L_{N-1} + L_N + 2, \quad t = L_N + L_{N+1} - 2, \quad u = L_N - L_{N+1} + 2$$

$$d = s + L_{N-1} - L_N + 2, \quad g = q_N + L_N + 1, \quad z_2 = (q_{N+1} + q_N)/(q_{N+1} - q_N)$$

$$z_1 = (q_{N-1} + q_N)/(q_{N-1} - q_N), \qquad z_3 = (q_{N+1} + q_{N-1})/(q_{N+1} - q_{N-1})$$

As a polynomial expression (6) is valid both below and above the N-photon ionization threshold. For T_2 expression convergent above the N-photon ionization threshold may be written

$$T_2 = q_{N-1}^{L_{N-1}-L_N} q_{N+1}^{t-5}(1 - z_1)^{L_{N-1}-L_N} (L_{N+1} - q_{N+1} + 1)_u \times$$

$$h(m_{N-1}, L_{N-1}, L_N, q_N) (T_A - DT_B)$$

where

$$h(m, L, 1, q) = \sum_{n=0}^{m-1} (1 - m)_n(L + 1 + 4)_n(1 - z_1)^n/(2L + 2)_n/n! \times$$

$$_2F_1(1 - L - n - 2, q + 1 + 1; 21 + 2; 1 + 1/z_1),$$

$$T_B = z_2^{1-c} z_1^a(1 + q_{N+1}/q_N)^{k-1}(1/q_{N-1} + 1/q_N)^{-k+1} \times$$

$$(1 - z_2^{-2})^k \ \Gamma(a) \ \Gamma(b)/\Gamma(c) \sum_{s=0}^{i} (-i)_s(e)_s/(c)_s/z_2^s/s! \times$$

$$\sum_{n=0}^{j} (-j)_n(b - 4 + s)_n/(c + s)_n/z_2^n/n! \ _2F_1(a, b + 2L_N + s + n - 3,$$

$$c + s + n; z_2^{-2})$$

where $a = L_N - q_N + 1$, $b = q_{N+1} - L_N + 2$, $c = q_{N+1} - q_N - 1$

$i = L_N - L_{N+1} + 2$, $j = 4 - i$, $k = 2L_N + 1$, $e = q_{N+1} + L_{N+1} + 1$

Expression (8) for T_B is convergent for all frequencies where the ionization by one additional photon is possible. To get the whole expression for T_2 convergent in this region it is necessary to use two different expressions for D and T_A. One in the vicinity of the N-photon ionization threshold (on the threshold $|z_1| = |z_2| = 1$), similar to that used in the case of two-photon ionization and another in the region where $|1/z_2| < 0.38$. The latest may be written

$$D = (-1)^{L_N+1} \exp(-\pi(2E_N)^{-1/2}) \, \Gamma(1 - c)/\Gamma(5 - b) \, \Gamma(c)/\Gamma(b)$$

$$T_A = -(\omega q_N/2)^{-2L_N-1}/q_{N+1} \, (-z_1)^{3-q_N} z_2^{-L_N-1} \sum_{n=0}^{j} (-j)_n/(-z_1)^n/n! \times$$

$$\sum_{s=0}^{i} (-i)_s(e)_s/(b - 4)_s \, (-z_1)^s/s! \sum_{k=0}^{\infty} (L_N + v + 2)_k/\beta/(z_1 z_2)^k/k! \times$$

$$_2F_1(L_N - v, \, \beta; \, q_N - v - k; \, z_1/z_2)$$

where $v = q_{N+1} + s + n - 3$ $\beta = q_N - q_{N+1} + 2 - s - n - k$.

Above the N-photon ionization threshold $|z_1| < 1$, therefore the convergence of the double series in expression for T_A depends mainly on z_2. The regions for both analytical continuations for D and T_A overlap, which is used to test the accuracy of the transition rate calculations. The structure of expressions (1) - (9) allows to considerably diminish the computation time and therefore is suitable to evaluate the transition rates when the number of photons is large.

Table 1. Transition rates for two-photon, three-photon, four-photon, and five-photon ionization of the ground state hydrogen by linearly polarized light.

λ (Å)	N	$Q^{(N)}/I^{N-1}10^{15-N}$ $(W^{1-N}\,cm^{2N})$	λ (Å)	N	$Q^{(N)}/I^{N-1}10^{15N}$ $(W^{1-N}\,cm^{2N})$	λ (Å)	N	$Q^{(N)}/I^{N-1}10^{15N}$ $(W^{1-N}\,cm^{2N})$
1823	2	1.11(−2)	2735	3	6.11(−2)	3646	4	1.73(+4)
	3	5.03(−3)		4	2.34(−1		5	8.09(+4)
1800	2	1.14(−2)	2650	3	7.98(−2)	3630	4	6.37(+1)
	3	4.74(−3)		4	2.49(−1)		5	2.99(+2)
1740	2	1.07(−2)	2550	3	1.47(−1)	3500	4	8.79(−1)
	3	3.67(−3)		4	3.78(−1)		5	4.44(+0)
1600	2	9.14(−3)	2480	3	5.15(−1)	3420	4	4.78(−1)
	3	2.02(−3)		4	1.18(+0)		5	2.36(+0)
1500	2	8.37(−3)	2460	3	1.26(+0)	3300	4	3.69(−1)
	3	1.35(−3)		4	2.81(+0)		5	1.55(+0)
1400	2	8.45(−3)	2433	3	3.10(+2)	3250	4	3.82(−1)
	3	1.01(−3)		4	6.58(+2)		5	1.46(+0)
1300	2	1.28(−2)	2430	3	4.29(+2)	3150	4	6.11(−1)
	3	1.18(−3)		4	9.05(+2)		5	1.91(+0)
1240	2	5.87(−2)	2425	3	1.90(+1)	3080	4	7.18(+1)
	3	5.05(−2)		4	3.97(+1)		5	2.06(+2)
1220	2	1.24(+0)	2420	3	5.72(+0)	3079	4	1.67(+2)
	3	1.09(−1)		4	1.18(+1)		5	4.81(+2)
1216	2	2.11(+2)	2405	3	9.44(−1)	3078	4	8.16(+2)
	3	1.89(+1)		4	1.88(+0)		5	2.35(+3)
1200	2	5.80(−2)	2360	3	9.58(−2)	3077	4	1.23(+4)
	3	5.55(−3)		4	1.61(−1)		5	3.36(+4)
1100	2	4.05(−4)	2260	3	1.82(−2)	3076	4	3.32(+2)
	3	3.42(−5)		4	1.32(−2)		5	9.62(+2)
1050	2	1.60(−3)	2200	3	1.59(−2)	3060	4	7.77(−1)
	3	6.73(−4)		4	7.61(−3)		5	2.47(+0)
1030	2	2.08(−2)	2160	3	1.69(−2)	3026	4	7.39(−2)
	3	1.02(−3)		4	7.03(−3)		5	2.61(−1)
1027	2	1.94(−1)	2053	3	4.04(+0)	2920	4	1.90(+1)
	3	1.03(−2)		4	2.54(+0)		5	4.58(+1)
1026	2	1.36(+0)	2050	3	3.65(+0)	2919	4	5.32(+1)
	3	7.35(−2)		4	2.50(+0)		5	1.30(+2)
1000	2	1.97(−4)	1944	3	5.87(−1)	2916	4	2.84(+1)
	3	1.56(−5)		4	2.80(−1)		5	7.20(+1)
975	2	6.79(−3)	1900	3	8.33(−1)	2848	4	1.15(+1)
	3	2.66(−4)		4	2.78(−1)		5	2.78(+1)
971	2	9.05(−3)	1890	3	2.77(−3)	2814	4	2.65(+1)
	3	5.03(−4)		4	1.09(−3)		5	5.73(+1)
949	2	7.89(−3)	1877	3	5.27(−2)	2794	4	1.64(+0)
	3	4.11(−4)		4	1.34(−2)		5	3.18(+0)

Table 2. Transition rates for six-photon, seven-photon, eight-photon, and nine-photon ionization of the ground state hydrogen by linearly polarized light.

λ (A)	N	$Q^{(N)}/I^{N-1}10^{15-N}$ (W^{1-N} cm^{2N})	λ (A)	N	$Q^{(N)}/I^{N-1}10^{15N}$ (W^{1-N} cm^{2N})	λ (A)	N	$Q^{(N)}/I^{N-1}10^{15N}$ (W^{1-N} cm^{2N})
5468	6	2.55(+2)	6381	7	1.35(+4)	7290	8	8.96(+8)
	7	6.34(+4)		8	4.79(+5)		9	6.89(+10)
5400	6	3.14(+2)	6280	7	4.47(+4)	7270	8	2.86(+7)
	7	7.05(+3)		8	1.40(+6)		9	2.19(+9)
5320	6	4.57(+2)	6220	7	1.95(+5)	7250	8	1.00(+7)
	7	9.45(+3)		8	5.63(+6)		9	7.58(+8)
5200	6	1.85(+3)	6170	7	4.83(+6)	7210	8	5.92(+6)
	7	3.59(+4)		8	1.29(+8)		9	4.31(+8)
5160	6	7.59(+3)	6158	7	1.03(+8)	7190	8	1.73(+7)
	7	1.47(+5)		8	2.72(+9)		9	1.22(+9)
5140	6	5.10(+4)	6150	7	7.64(+7)	7185	8	5.17(+7)
	7	9.84(+5)		8	1.99(+9)		9	3.59(+9)
5130	6	3.45(+6)	6140	7	8.33(+6)	7175	8	2.26(+7)
	7	6.66(+7)		8	2.14(+8)		9	1.51(+9)
5125	6	4.38(+5)	6120	7	2.29(+6)	7150	8	1.02(+5)
	7	8.45(+6)		8	5.72(+7)		9	6.34(+6)
5000	6	8.95(+1)	6079	7	8.86(+8)	7080	8	1.83(+5)
	7	1.57(+3)		8	2.27(+10)		9	1.20(+7)
4880	6	1.62(+5)	6070	7	2.39(+6)	6960	8	2.43(+5)
	7	2.62(+6)		8	6.40(+7)		9	1.30(+7)
4870	6	4.32(+6)	6050	7	4.91(+4)	6900	8	3.53(+5)
	7	7.04(+7)		8	1.87(+6)		9	1.66(+7)
4864	6	4.14(+9)	6010	7	5.26(+3)	6820	8	5.07(+6)
	7	6.81(+10)		8	2.41(+5)		9	1.73(+8)
4861	6	1.07(+9)	5880	7	2.74(+4)	6805	8	6.00(+7)
	7	1.76(+10)		8	4.58(+5)		9	1.80(+9)
4840	6	1.70(+4)	5840	7	6.67(+5)	6775	8	1.20(+5)
	7	3.16(+5)		8	9.22(+6)		9	3.08(+6)
4805	6	2.01(+2)	5838	7	1.83(+6)	6650	8	6.10(+7)
	7	5.06(+3)		8	2.49(+7)		9	2.32(+9)
4760	6	1.93(+3)	5700	7	1.24(+6)	6648	8	2.20(+9)
	7	2.74(+4)		8	1.68(+7)		9	8.20(+10)
4750	6	8.73(+4)	5698	7	8.35(+6)	6564	8	1.27(+8)
	7	1.27(+6)		8	1.11(+8)		9	4.92(+9)
4746	6	1.26(+4)	5680	7	3.21(+2)	6516	8	6.50(+7)
	7	1.89(+5)		8	1.27(+4)		9	2.58(+9)
4725	6	3.40(+1)	5627	7	5.79(+7)	6515	8	2.94(+8)
	7	9.17(+2)		8	8.00(+8)		9	1.15(+10)
4690	6	2.13(+4)	5600	7	5.25(+3)	6500	8	4.61(+4)
	7	2.75(+5)		8	8.77(+4)		9	3.02(+6)

RESULTS AND CONCLUSIONS

Transition rates for ATI of the ground state H by linearly polarized light are computed for the total number of photons from three up to nine in the frequency range where the ionization by one extra photon is possible. Simultaneously transition rates are calculated for the multiphoton ionization MPI by minimum number of photons. Some results are presented in Tables 1 and 2. There is no sharp wavelength dependence due to the set of intermediate states in continuum. Therefore, the lowest order ATI dispersion curves are simular to those for ionization by minimum number of photons.

Recently, the first ATI experiment for H is reported by Muller et al.[5] They measured the ratio of five- and four-photon ionization at the 3p resonance and calculated ATI process through 3p state. Results of this paper and that of theirs[5] are presented in Table 3.

Table 3. The ratio $Q^{(5)}/Q^{(4)}$ at the 3p resonance.

3076 (Å)	ATI (our results)	Experiment[5]	ATI[5]
$Q^{(5)}/Q^{(4)} \times 10^6$ $(GW/cm^2)^{-1}$	2.90	3.12	3.305

The ratio calculated by us changes insignificantly in the region close to the 3p resonance (Fig. 1). Experimental measurements agree well with the two theoretical estimates.

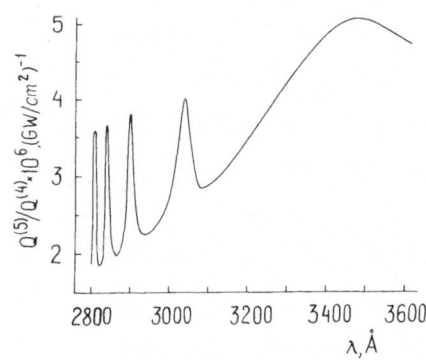

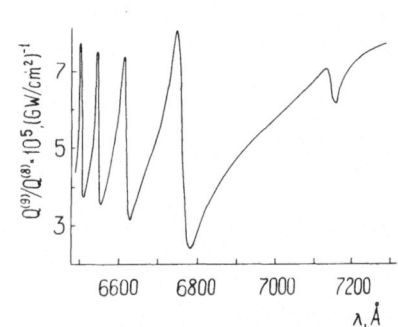

Fig. 1 The ratio $Q^{(5)}/Q^{(4)}$, $Q^{(5)}$ corresponds to ATI.

Fig. 2 The ratio $Q^{(9)}/Q^{(8)}$, $Q^{(9)}$ corresponds to ATI.

The ratio of N-photon ATI and (N-1)-photon "normal" MPI is greater near the (N-1)-photon ionization threshold and is growing with N (Figs. 1 and 2). Minima on the curves (Figs. 1 and 2) correspond to the regions near the resonances onthe MPI and ATI dispersion curves. Maxima on the curves (Figs. 1 and 2) are due to the deeper minima for the (N-1)-photon dispersion curves as for the N-photon ATI dispersion curves.

With respect to the order of magnitude, ATI processes differ insignificantly from "normal" MPI processes where the same number of photons are participating.

The above method may be applied to the ATI of excited states as it was done for two-photon ATI.[3,4]

REFERENCES

1. E. Karule, Multiphoton ionization of atomic hydrogen in the ground state, in: "Atomic Processes" (in Russian), eds. R. K. Peterkop, Zinatne, Riga, pp. 5-24 (1975).

2. E. Karule, Two-photon ionisation of atomic hydrogen simultaneously with one-photon ionisation, J. Phys. B: At. Mol. Phys. 11, 441-447 (1978).

3. E. Karule, Two-photon above-threshold ionisation of atomic hydrogen in the excited states, in: "Nonlinear Processes in Two-Electron Atoms," (in Russian), eds. N. B. Delone, Academy of Sciences, USSR, Moscow, pp. 209-235 (1984).

4. E. Karule, Two-photon ionisation of excited atomic hydrogen at the photoelectric threshold and above, J. Phys. B: At. Mol. Phys. 18, 2207-2218 (1985).

5. H. G. Muller, H. B. van Linden de Heuvell, M. J. van der Wiel, Experiments on above-threshold ionization of atomic hydrogen, Phys. Rev. A34, 236-243 1986.

STRONG-FIELD EXCITATION OF CONTINUUM RESONANCES

Z. A. Melhem and P. L. Knight

Blackett Laboratory
Imperial College
London SW7 2BZ, United Kingdom

ABSTRACT

Recent experiments on multiphoton ionization in the presence of strong laser fields have indicated the existence of strong continuum-continuum transitions referred to as Above Threshold Ionization (ATI), and observed as distinct peaks in the photoelectron spectra, spaced by the photon energy. The peaks exhibit strong saturation; lowest-order peaks are suppressed at higher intensities. Existing theoretical models assume the electron continuum is structureless and unchanging with energy. We have investigated the relaxation of the flat continuum via the essential state representation, employed in strong-field discrete-discrete transitions, to discrete-continuum transitions. We present electronic spectra produced within a structured continuum (for example representing symmetric autoionizing resonances). We show that the prescribed structure will lead to a dramatic change in the electronic spectra, allowing not only saturation to change peak heights but also to split lines through AC Stark effects. We will report on generalizations to arbitrary numbers of extra continuum-continuum transitions.

INTRODUCTION

The rapid advances in the use of intense lasers (with powers larger than $10^{13}W/cm^2$) to study atoms has led to a considerable interestin multiphoton ionization of atoms. Many experiments have demonstrated the existence of strong continuum-continuum-(cc-) transitions or Above Threshold Ionization (ATI). These experiments[1] have shown how the saturation of cc-transitions leads to a distribution of photoelectrons over a range of continuum states. Kruit et al.[2] have presented a detailed analysis of ATI in the high intensity regime (10^{13} W/cm^{-2}) on 11-photon ionization of Xe by 1064 nm photons. They have observed the vanishing of the low-order peaks, a total order of nonlinearity close to that appropriate for the minimum number of photons necessary for ionization and the shifting of the energy at which the envelope of the photoelectron spectrum maximizes as the intensity increases accompanied with a considerable broadening of the peak.

Theoretically, ATI has been investigated by many groups (e.g. Deng and Eberly,[3] Edwards et al.,[4] Bialynicka-Birula,[5] and Shore and Knight[6]). In these approaches the N-photon ground state-to continuum step is replaced by some effective coupling. In the Den and Eberly approach the cc matrix elements are assumed to be constants and cc-transitions are connecting an infinite set of flat continua with the first continuum connected to the ground state by the effective coupling.

Structure within a continuum (e.g. autoionizing resonances) will significantly modify the continuum excitation dynamics. Transitions between high-q symmetric autoionizing resonances will behave much as bound-bound transitions between states embedded in a flat continuum and interacting with it to provide widths characteristic of these resonances. These embedded states behave for most purposes as if they were smeared-out bound states in the continuum which will be shifted and broadened.

In this work we investigate the relaxation of the flat continuum via the dressed state approach. The "essential state" representation is used to study the resonant interaction of an intense field with an atomic or molecular transition (e.g. Cohen-Tannoudji,[7] Feneuille,[8], and Knight and Milonni[9]). For an atom-plus-field system, the dressed states are linear combinations of the unperturbed degenerate states, whose degeneracy is lifted by the atom-field interaction. In what follows we will concentrate on discussing continuum resonance excitation with N-embedded states.

Formal Theory

The model to be considered is represented by a total Hamiltonian H constructed from H_0, describing the unperturbed atom and V(t) the interaction of the atom with the laser field, given by

$$\hat{V} = -\underline{\hat{d}} \cdot \underline{\varepsilon} \qquad (1)$$

where $\underline{\hat{d}}$ is the electric dipole operator ($e\underline{\hat{r}}$) of the atom or molecule and ε is the electric field of the radiation. The atomic states are given a width phenomenologically (for simplicity, although it is easy to justify from formal decay theory[10]) by introducing into the equations of motion a damping term and by regarding the eigen-energies in the atom as being "dressed" by the laser field.

It is possible to expand the wavefunction of the driven atomic system in terms of complete sets of uncoupled atomic states $|k\rangle \equiv \psi_k$, i.e.

$$|\Psi(k) > \sum_k a_k e^{-i\omega_k t}|k\rangle \tag{2}$$

where $\omega_k = E_k/\hbar$ and $a_k(t)$ are the time-dependent coefficients that describe how the coupling V mixes the unperturbed states. The amplitudes $a_k(t)$ are determined by the Schrodinger equation for the time development of $|\Psi(t)\rangle$ and the fulfillment of the orthonormality condition. The wavefunction of the coupled system is normalized such that

$$\langle\Psi|\Psi\rangle = 1 \tag{3}$$

The exact equations of motion for the probablity amplitudes are obtained by employing the wavefunction expansion for $|\Psi(t)\rangle$ in the full time dependent Schrodinger equation, i.e.

$$i\hbar \frac{\partial}{\partial t} |\Psi(t)\rangle = (\hat{H}_o + \hat{V})| \Psi(t)\rangle \tag{4}$$

from this we obtain

$$i\hbar\dot{a}_1(t) = \sum_{k \neq 1} a_k(t) e^{i\omega_{1k}t} \langle 1|\hat{V}|k\rangle = - i \frac{\gamma}{2} a_k(t) \tag{5}$$

where ω_{1k} ($= \omega_1 - \omega_k$) $\equiv$ transition frequency, and $\langle 1|\hat{V}(t)|k\rangle \equiv V_{1k}(t)$ and excites transition between all accessible states $|k\rangle$ to the state $|1\rangle$.

The atom, in our model, is initially in the ground state $|i\rangle$, at $t = 0$ we have $a_1(0) = 1$ and $a_f(0) = 0$. Equation 5 can be rewritten as:

$$\dot{a}_1(t) = -i \sum_{k \neq 1} a_k(t) e^{i\omega_{1k}t} V_{1k}(t) - i \frac{\gamma}{2} a_k(t) \tag{6}$$

where we have put $\hbar = c = 1$, and a_1 are amplitudes for $|1\rangle$ in the interaction representation.

Description of the Model and the Equations of Motion

This atomic system has an initially populated ground state (with energy E_0) coupled by the laser field to a ladder of resonance states $|n\rangle$ embedded in a flat continuum $|E\rangle$ (Fig. 1). There are $(N - 1)$ atomic levels characterized by the atomic energies $E_N(N = 1, 2, ...)$ which are in (or near) resonance with the laser energy ω_L. Provided that all the off-resonance energies are not too large, we will assume that only adjacent states are coupled by the radiative interaction $V_{1k}(t)$. The embedding of the state E_N in the flat continuum can be incorporated by assigning a decay width γ_N.

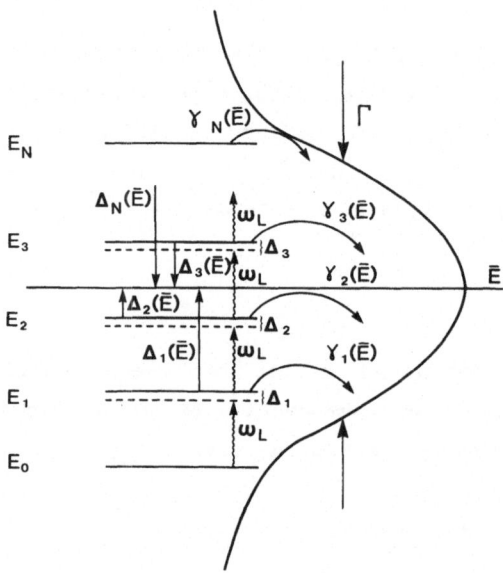

Fig. 1 Schematic diagram showing the embedding of N-discrete states into the continuum.

Employing the Rotating Wave Approximation (RWA) leads to the following equations of motion for the multilevel system:

$$\dot{C}_0(t) = -i\,V_{01}\,C_1(t) - i\,\varepsilon_0\,C_0(t)$$
$$\dot{C}_1(t) = -i\,V_{10}\,C_0(t) - iV_{12}\,C_2(t) - i\,\varepsilon_1\,C_1(t)$$
$$\dot{C}_2(t) = -i\,V_{21}\,C_1(t) - iV_{23}\,C_3(t) - i\,\varepsilon_2\,C_2(t)$$

$$\vdots$$

$$\dot{C}_{N-1}(t) = -i\,V_{N-1,N-2}\,C_{N-2}(t) - iV_{N-1,N}\,C_N(t) - i\,\varepsilon_{N-1}\,C_{N-1}(t)$$
$$\dot{C}N(t) = -i\,V_{N,N-1}\,C_{N-1}(t) - i\,\varepsilon_N\,C_N(t)$$

where

$$\varepsilon_0 = E_0 + n\omega_L$$
$$\varepsilon_N = \varepsilon_0 + \Delta_N - i\,\frac{\gamma}{2}\,N$$
$$\Delta_N = N\,\Delta \tag{7}$$

with $\Delta(= \omega - \omega_L)$ the detuning and ω is the energy spacing between the equally spaced levels (for simplicity we have considered an equally-spaced ladder of levels; it is simple to relax this restriction).

We solve this problem for any value of N by using a matrix method. The equations of motion can be represented by

$$i\,\frac{d}{dt}\,C_N(t) = \underline{M}\,C_N(t) \tag{8}$$

where $\underline{M}$ is a $(N \times N)$ tridiagonal complex symmetric matrix given by:

$$i\,\frac{d}{dt}
\begin{pmatrix}
c_0 \\
c_1 \\
c_2 \\
\vdots \\
\vdots \\
c_{N-1} \\
c_N
\end{pmatrix}
=
\begin{pmatrix}
\varepsilon_0 & V_{01} & & & & \\
V_{10} & \varepsilon_1 & V_{12} & & & \\
& V_{21} & \varepsilon_2 & V_{23} & & \\
& & \vdots & \vdots & & \\
& & \vdots & \vdots & & \\
& & V_{N-1,N-2} & \varepsilon_{N-1} & V_{N-1,N} \\
& & & V_{N,N-1} & \varepsilon_N
\end{pmatrix}
\begin{pmatrix}
c_0 \\
c_1 \\
c_2 \\
\vdots \\
\vdots \\
c_{N-1} \\
c_N
\end{pmatrix}$$

We calculate the eigenvalues and the eigenvectors of the matrix $\underline{M}$, (the effective Hamiltonian for this model). By assuming the existence of a similarity transformation such that,

$$\underline{S}^{-1}\,\underline{M}\,\underline{S} = \text{Diagonal }(\lambda_1,\ \lambda_2,\ \cdots\ \lambda_n) \tag{9}$$

Equation 9 can be written as

$$i\,\frac{d}{dt}\,\underline{\tilde{C}}\,(t) = A\,\underline{\tilde{C}}\,(t) \tag{10}$$

where
$$\underline{\tilde{C}} = \underline{S}^{-1}\,\underline{C}\,(t) \tag{11}$$

$$\underline{A} = \underline{S}^{-1}\,\underline{M}\,\underline{S}$$

The initial condition is: $\underline{C}\,(t = 0) = \underline{C}(0) = \begin{matrix} 1 \\ 0 \\ 0 \\ \cdot \\ \cdot \\ \cdot \\ 0 \end{matrix}$ (12)

From Eqs. (10-12) we get

$$\underline{\tilde{C}}(0) = \underline{S}^{-1} \, \underline{C}(0)$$

and the amplitudes of the embedded states $C_p(t)$ will be written as

$$C_p(t) = \sum_{q=1}^{N} S_{pq} \, \tilde{C}_q(0) \, e^{-i\lambda qt} \tag{13}$$

It is worth noting here that for a one level system (N = 1), we simply reproduce the Fermi Golden Rule, while for a two-level system (N = 2) we will have

$$M = \begin{matrix} \varepsilon_0 & V_{01} \\ V_{10} & \varepsilon_1 \end{matrix} \tag{14}$$

which is the two-level Rabi problem (8,9) incorporating an additional decay channel.

The amplitudes of the embedded states can be used to calculate the amplitudes of the continuum and the continuum spectrum $|C_\varepsilon(t)|^2$, i.e.

$$i \frac{d}{dt} C_E(t) = E \, C_E(t) + \sum_{p=1}^{N} V_{PE} \, C_P(t) \tag{15}$$

This first-order differential equation can be solved in terms of the resonance amplitudes $C_P(t)$, to give

$$C_E(t) = \sum_{q=1}^{N} \frac{T_q}{\lambda_q - E_P} \left(e^{-i\lambda_q t} - e^{-iE_P t} \right) \tag{16}$$

where

$$T_q = S_{q1}^{-1} \sum_{p=2}^{N} V_{PE} \, S_{pq} \tag{17}$$

Results

We present photoelectron spectra produced by multiphoton ionization within a structured continuum. Here we confine ourselves with calculations for N = 3 [(n = N − 1) transitions] atomic system. We first choose for simplicity the continuum loss γ_N to be constant ($\gamma_N = \gamma$) for all the embedded states. The value of the perturbation which mixes the embedded state with the continuum (V_{1k}) is chosen to be constant ($V_{1k} = V$). The contribution to a series of levels coupled to a broad background resonance

$$\gamma_N(\overline{E}) = \frac{\Gamma/\pi \, |\overline{V}_N|^2}{(E_N - \overline{E})^2 + \Gamma^2/4} \tag{18}$$

When the broad background width Γ is very large compared with the level spacing and the Rabi frequencies, the decay is very fast and the excitation will behave as if the continuum were flat. No Rabi oscillations can occur between the initial state and higher lying states and the electron spectra are a series of single lines of the resonance positions, spaced by the photon energy. But when the widths are smaller, Rabi oscillations can occur to structure and AC-Stark split the photoelectron spectra. In Fig. 2 we show the effect of a narrow resonance on the electron spectrum, when the Rabi oscillations significantly split the line and change the relative heights of the electon peaks. We report elsewhere a detailed study of these effects on continuum–continuum excitation.

This work was supported in part by the UK Science and Engineering Research Council.

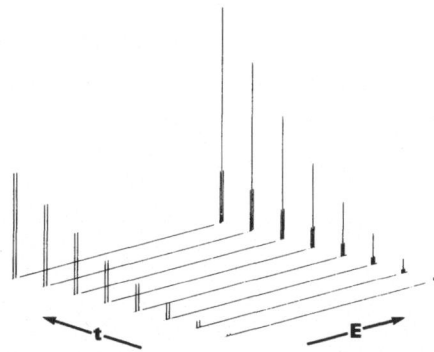

Fig. 2 The time-development of the photoelectron spectrum in the presence of a strong laser field, where $V = \overline{V} = 0.005$; $\omega = \omega_L = 1$; $\gamma_1(\overline{E}) = \gamma_2(\overline{E}) = 0.15758 \times 10^{-5}$; $\Delta_1(\overline{E}) = -0.5$, $\Delta_2(\overline{E}) = 0.5$, $\Gamma = 10$. All units are in a.u.

REFERENCES

1. G. Petite, F. Fabre and P. Agostini, <u>Collisions and Half-Collisions with Lasers</u>, eds. N. K. Rahman and C. Guidotti (Harwood Acad. Pub., 1984) p.203; M. J. Van der Wiel, ibiden, p. 217.

2. P. Kruit, J. Kimman, H. G. Muller and M. J. Van der Wiel, Phys. Rev. A <u>28</u>, 249 (1983).

3. Z. Deng and J. H. Eberly, J. Opt. Soc. Am. B <u>2</u>, 486 (1985).

4. M. Edwards, L. Pan and L. Armstrong, Jr., J. Phys. B <u>17</u>, L515 (1984).

5. Z. Bialynicka-Birula, J. Phys. B <u>17</u>, 3019 (1984).

6. B. W. Shore and P. L. Knight, J. Phys. B <u>20</u>, 413 (1987).

7. C. Cohen-Tannoudji, <u>Frontiers in Laser Spectroscopy</u>, eds. R. Balian, S. Haroche and S. Liberman, Vol. 1 (North-Holland, Amsterdam, 1977) p. 3-102.

8. S. Feneuille, Rep. Prog. Phys. <u>40</u> 1257 (1977).

9. P. L. Knight and P. W. Milonni, Phys. Rep. <u>66</u>, 21 (1980).

10. e.g. R. Loudon, <u>The Quantum Theory of Light</u>, Second Edition (Oxford University Press, Oxford, 1983).

LOW-ENERGY RADIATIVE SCATTERING IN A LASER FIELD

Farhad H. M. Faisal

Fakultät für Physik
Universität Bielefeld
Federal Republic of Germany

We point out some important recent developments in the radiative e-atom scattering problem and briefly report on the recently developed radiative close-coupling method with application to low energy e + H^+ scattering in a laser field.

Study of electron-atom scattering in a laser field has made significant progress recently. Already, evidence of absorption and emission of a large number of CO_2-laser protons by the scattering electron has been found in the experiments of Weingartshofer and collaborators.[1,2] Also, inelastic scattering of electrons at the energies of ordinary resonance scattering but in the presence of a CO_2-laser, have been observed in experiments by Andrick and Langhans.[3]

Very recently Mason and Newell[4] found the first experimental evidence of the process of "simulutaneous electron-photon excitation" of atoms. Possible atomic transition by simultaneous action of a photon and an electron was first envisaged by Maria Göppert-Mayer[5] in 1931. Buimistrov[6] in 1969 suggested investigating the process in an external laser field. Quantitative calculations[7-10] of various properties of the process began in 1976 using the second order perturbation theory within the Bethe-Born picture. In recent years the second order calculations have been improved[11-13] to go beyond the Bethe approximation, and the presence of a new interference structure in the differential cross section has been predicted. These calculations are essentially concerned with high energy scattering where the plane wave approximation of the free electron may be used. The above experiments have been generally performed for rather low scattering energy, of the order of the binding energy of the atomic electron or so. In the experiment of Mason and Newell,[4] for example, the final electron energy is in fact extremely low, being a fraction of an electron volt and clearly requires analysis based on low-energy scattering theory.

A very recent progress on the theoretical side has been the development of the radiative close-coupling method.[14,15,20] It opens up the possibility of ab initio analysis of radiative scattering experiments, particularly at low energy. As in the ordinary electron-atom scattering processes it is expected[16] that low-energy radiative scattering would exhibit resonance phenomena some of which could be new in the sense that they would occur only in the presence of the field. For example, semi-empirical and non-perturbative model analysis[17-19] of radiative e-atom scattering at low energy has already revealed, among other things, a new scattering resonance-mechanism due to the temporary formation of a true negative ion and its subsequent break-up by photon absorption. Here, I intend to briefly discuss the newly developed radiative close-coupling method and the results of its first application to the basic problem of $e + H^+$ scattering in an excimer laser field.[14,15] The calculations have provided numerical evidence for a whole class of "capture-escape"-Rydberg resonances, in the field modified Rutherford scattering as well as in the inverse Bremsstrahlung cross sections.

One of the basic problems in radiative scattering is to understand how the $e + H^+$ scattering cross sections, described exactly by the well-known Rutherford formula, changes in the presence of a strong laser field. Besides its intrinsic value in scattering physics, a knowledge of the cross sections of such processes is of interest in large scale laser-plasma interaction. Furthermore this problem is a prototype of a large class of e-ion interaction problems in the presence of the laser.

Up to now this apparently simple scattering problem has defied exact solution. This is primarily due to the analytical difficulties associated with the long-range Coulomb potential and with the increased dimension of the Schrödinger equation in the presence of the radiation field. The latter fact show up in both semiclassical and quantum field formulations. In the former representation the time dimension enters explicitly; in the latter case the extra dimension of the fock-space needs to be dealt with.

In view of the lack of any analytical solution we are led to attack the problem by direct numerical means. To this end we extend the powerful close-coupling method[21] of solution of the ordinary electron scattering problems and incorporate the interaction of the radiation field via the Floquet representation[22] of the scattering equations.

The Schrödinger equation of the system is ($e = \hbar = m = 1$)

$$i \frac{\partial}{\partial t} \Psi(\vec{r},t) = [-\frac{1}{2} \nabla^2 - \frac{\vec{A}(t)}{c} \cdot \hat{p} + \frac{1}{2c^2} \vec{A}^2(t) - \frac{Z}{r}]\Psi(\vec{r},t) \tag{1}$$

where we have chosen a circularly polarized radiation field

$$\vec{A}(t) = A_0[\hat{\varepsilon}_x \cos(\omega t + \delta) - \hat{\varepsilon}_y \sin(\omega t + \delta)] \tag{2}$$

$\hat{\varepsilon}_x$ and $\hat{\varepsilon}_y$ are the orthogonal unit vector, A_0 is the peak vector potential, ω is the frequency and σ is a constant phase; $\hat{p} = -i\vec{\nabla}$, $z = 1$ for H^+. In the present form the interaction Hamiltonian in (1) does not fall off with increasing distance and is thus not convenient for direct application of the close-coupling asymtotics. We therefore change the representation by making the unitary transformation[23,24]

$$\Psi(\vec{r},t) = \exp(-i \int^t dt[-\frac{\vec{A}(t)}{c} \cdot \hat{p} + \frac{1}{2c^2} \vec{A}(t)^2])\Phi(\vec{r},t) \tag{3}$$

in (1) to obtain

$$i \frac{\partial}{\partial t} \Phi(\vec{r},t) = [-\frac{1}{2} \nabla^2 - \frac{Z}{|\vec{r} - \vec{\alpha}_0(t)|}]\Phi(\vec{r},t) \tag{4}$$

where $\vec{\alpha}_0(t) = \frac{1}{c} \int^t dt\, \vec{A}(t)$.

The interaction Hamiltonian in (4) has a Coulomb asymptotic behavior which can be conveniently treated within the close-coupling method. We make a Floquet plus partial wave expansion of $\Phi(\vec{r},t)$,

$$\Phi(\vec{r},t) = \sum_{n=-\infty}^{\infty} \sum_{lm} e^{-iEt+in(\omega t + \delta)} \frac{1}{r} F_{nlm}(r) Y_{lm}(\hat{r}) \tag{5}$$

and substitute it in (4), equate equal coefficients of $\exp(in(\omega t + \sigma))$ and project onto $Y^*_{lm}(r)$ to obtain the Floquet representation of the radial close-coupling equations for the channel functions $F_{nlm}(r)$ in the channel $i = (nlm)$:

$$[\frac{d^2}{dr^2} + k_n^2 - \frac{l(l+1)}{r^2} + \frac{2Z}{\sqrt{r^2 + \alpha_0^2}}]F_{nlm}(r) = \tag{6}$$

$$= -2Z \sum_{\lambda=1, p=0}^{\infty, \lambda} \sum_{l'm'} \langle Y_{lm}|V_{\lambda_p}(\theta,\phi)|Y_{l'm'}\rangle F_{n-\lambda_p,l',m'}(r)$$

where the multipole-coupling potentials are

$$V_{\lambda_p}(\theta,\phi) = \frac{(2\lambda - 1)!!}{2^\lambda \lambda!} \binom{\lambda}{p} \left(\frac{r\alpha_0 \sin\theta}{r^2 + \alpha_0^2}\right)^\lambda \frac{1}{\sqrt{r^2 + \alpha_0^2}} e^{i\lambda_p \phi} \tag{7}$$

with $\lambda_p = \lambda - 2p$ and $K_n = [2(E - n\omega)]^{1/2}$. The asymptotic behavior of the jth solution in the ith open channel for the scattering problem is

$$F_i^{(j)}(r)_{r\to\infty} = \frac{1}{\sqrt{k_i}} \left[\delta_{ij}\sin(\theta_i) + K_{ij}\cos(\theta_i)\right] \quad i,j=1,2,\ldots,n_{op} \tag{8}$$

where $\theta_i = k_i r - \frac{1}{2} l_i \pi + \frac{Z}{k_i} \ln 2k_i r + \sigma_{l_i}$, σ_{l_i} is the Coulomb phase shift and n_{op} is the number of open channels included. $K = (K_{ij})$ is the real K-matrix which is related to the (complex) S-matrix by $S_{ij} = \left[\frac{1 + iK}{1 - iK}\right]_{ij}$. Exponentially decreasing boundary conditions are made to satisfy for the closed channels. The radiative scattering amplitude in which the momentum of the incident electron, $\vec{k}_0$, is in the direction $\hat{k}_0$ and the final momentum, $\vec{k}_N$, after exchange of N photons, is in the direction $\hat{k}_N$, is found to be related to the open channel S-matrix elements by

$$f_{0\to N}(\hat{\kappa}_0,\hat{\kappa}_N) = f_{Coulomb}(\theta)\delta_{N,0} \tag{9}$$

$$+ \sum_{l_0 m_0 l m} \frac{2\pi i}{\sqrt{(k_0 k_N)}} i^{l_0 - l} e^{i(\sigma_{l_0} + \sigma_l)} Y^*_{l_0 m_0}(\hat{k}_0) Y_{lm}(\hat{k}_N) \left[\delta_{N,0}\delta_{l l_0}\delta_{m m_0} - S^{n l m}_{0 1 l_0 m_0}\right]$$

σ_l are the lth partial wave Coulomb phase shifts and $f_{Coulomb}(\theta)$ is the ordinary Rutherford scattering amplitude where θ is the angle between the incident and the scattered directions, for N = 0. Hence from (9) the elastic cross section modified by the field is

$$\frac{d\sigma^{(0)}}{d\Omega}(\theta) = \frac{d\sigma_{Coulomb}(\theta)}{d\Omega} + |f^{(0)}_{rad}(\theta)|^2 + 2\text{Re}[f^*_{Coulomb}(\theta)f^{(0)}_{rad}(\theta)]. \tag{10}$$

where $f^{(0)}_{rad}(\theta)$ is the term with N = 0 in the second line of (9). Similarly for the inelastic processes of the stimulated absorption (N < 0) and emission (N > 0) the differential cross sections are

$$\frac{d\sigma^{(N)}}{d\Omega}(\theta) = \frac{k_N}{k_0} |f^{(N)}_{rad}(\hat{k}_0,\hat{k}_N)|^2 \tag{11}$$

where $f^{(N)}_{rad}(\hat{k}_0,\hat{k}_N)$ is given by the second line of (9) for $N \neq 0$. Hence, from (11) the angle integrated total cross sections of N photon absorption or emission are given by

$$\sigma^{(N)}(\hat{k}_0) = \frac{4\pi^2}{k_0^2} \sum_{lm} \left| \sum_{l_0 m_0} i^{l_0} e^{i\sigma_{l_0}} S^{N l m}_{0 1 l_0 m_0} Y^*_{l_0 m_0}(\hat{k}_0) \right|^2 \tag{12}$$

The numerical solution of the above extended close-coupling equations for the e + H$^+$ radiative scattering problem has revealed a new class of resonances which dominate the cross sections of the elastic (Rutherford) as well as the inelastic (inverse Bremsstrahlung) processes in the presence of the field. These resonances will be referrerd to as "capture-escape Rydberg resonances" due to their physical origin.

Figure 1 provides the first numerical evidence of such resonances corresponding to n = 2 to 5 principal quantum numbers of the intermediate neutral H-atom. They are due to the "reaction" which can be schematized as

$$\hbar\omega + e + H^+ \rightarrow H(n) + 2\hbar\omega \rightarrow \hbar\omega + e + H^+.$$

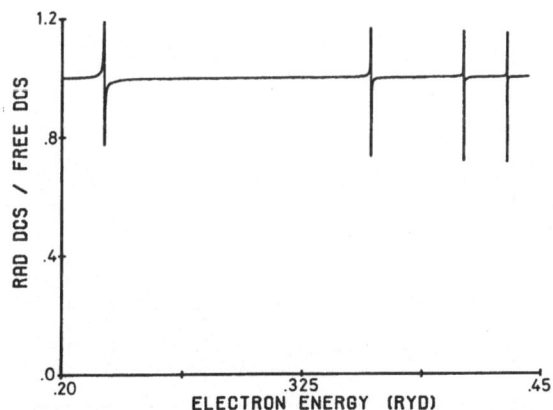

Fig. 1 Ratio of the field modified elastic e + H$^+$ to the ordinary Rutherford scattering as a function of incident electron energy. Field strength F_0 = .005 a.u. $\hbar\omega$ = 6.419 eV. Incident electron along Ω_0 = (90°,0°) and final direction Ω = (90°,90°). Z-axis is along laser propagation direction of the circularly polarized field. "Capture escape" resonances for n = 2 to 5 are to be noted.

The most prominent features of this result are (a) the existence of a series of very clear resonance structures and (b) away from the resonances the field modified elastic cross sections are rather closely given by the unmodified Rutherford cross section. The resonance structures are indications for the fact that the scattering electrons become captured by the proton in the bound states of the H-atom and then reemitted in the same incident energy channel. This is a resonant induced phenomenon. Thus at a given laser frequency which matches the energy difference between the incident electron energy (positive) and a

Rydberg level energy (negative) the laser field can force the electron to emit a photon and cause it to be captured temporarily in the Rydberg-state till the subsequent absorption of a photon permits the electron to escape from this state into the continuum again. The delay introduced by this capture-escape episode, shows up as a resonance in the scattering signal.

In Fig. 2 we present the corresponding inelastic cross section for the absorption of a photon. This process can be thought of as an inverse Bremsstrahlung process in a strong field at low electron energies. In the absence of the field no such channel is possible and hence the background to this signal is intrinsically zero. Here too we obtain numerical evidence of a very prominent sequence of the "capture-escape-Rydberg resonances."

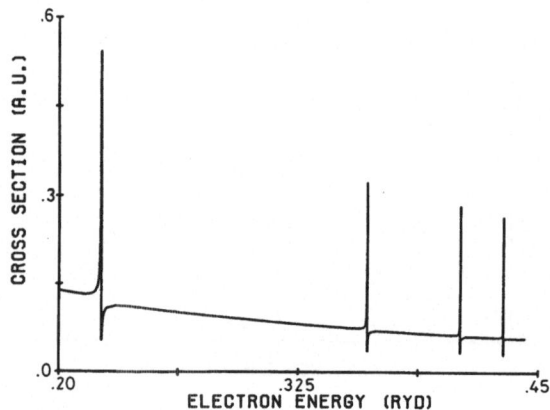

Fig. 2 Angle-integrated one-photon inverse Bremsstrahlung cross section as a function of incident electron energy. Field strength F_0 = .005 a.u., $\hbar\omega$ = 6.419 eV. Incident electron along Ω_0 = (90°,0°). Z-axis along the propagation direction of the circularly polarized field. "Capture-escape" Rydberg resonances for n = 2 to 5 are present against a zero background signal.

In Fig. 3 we show the n = 2 resonance in magnification. It has a width of ≈ 5 meV. We note that with increasing field strength this and the other resonances for higher n tend to broaden. In particular the higher Rydberg resonances (which are weak at a given intensity) begin to appear significantly with increasing field strength. We also remark that the width of such a resonance with a given n can be thought of as a measure of the strong field photoionization rate of that particular Rydberg state of the neutral atom.

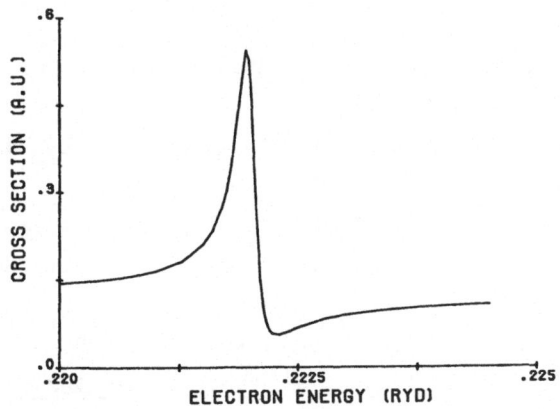

Fig. 3 Magnification of the n = 2 "capture-escape" Rydberg resonance for the angle-integrated one-photon inverse-Bremsstrahlung cross-section. Incident electron along $\Omega_0 = (90°,0°)$. Field strength F_0 = .005 a.u., $\hbar\omega$ = 6.419 eV. Z-axis along the propagation direction of the circularly polarized field. The width at this field strength ≈ 5 meV.

Finally we note that these resonances can be "tuned" either by varying the photon frequency at a fixed electron energy or by varying the electron energy at a given photon frequency. This flexibility in tuning the resonances combined with the fact that their widths can be manipulated by changing the field intensity may prove to be useful in observing them experimentally.

REFERENCES

1. A. Weingartshofer, J. K. Holmes, G. Caudle, E. M. Clarke and H. Krüger, Phys. Rev. Lett. 39, 269 (1977).

2. A. Weingartshofer, J. K. Holmes, J. Sabbagh and S. L. Chin, J. Phys. B: At. Mol. Phys. 16 (1983).

3. D. Andrick and L. Langhans, J. Phys. B: At. Mol. Phys. 11, 2355 (1978).

4. N. J. Mason and W. R. Newell, J. Phys. B: At. Mol. Phys. 20, L323 (1987).

5. M. Göppert-Mayer, Aun. der Physik 9, 27 (1931).

6. V. M. Buimistrov, Phys. Lett. 30A, 136 (1969).

7. N. K. Rahman and F.H.M. Faisal, J. Phys. B: At. Mol. Phys. 9, L275 (1976).

8. N. K. Rahman and F.H.M. Faisal, Phys. Lett. 57A, 426 (1976).

9. N. K. Rahman and F.H.M. Faisal, J. Phys. B: At. Mol. Phys. $\underline{11}$, 2003 (1978).

10. S. Jetzke, F.H.M. Faisal, R. Hippler and H. O. Lutz, Z. für Physik, $\underline{A315}$, 271 (1984).

11. A. Lami and N. K. Rahman, J. Phys. B: At. Mol. Phys. $\underline{16}$, L201 (1983).

12. A. Dubois, A. Maquet and S. Jetzke, Phys. Rev. A $\underline{34}$, 1888 (1986).

13. S. Jetzke, J. Broad and A. Maquet, J. Phys. B: At. Mol. Phys. $\underline{20}$, 2887 (1987).

14. L. Dimou and F.H.M. Faisal, Phys. Rev. Lett. $\underline{59}$, 872 (1987).

15. L. Dimou and F.H.M. Faisal, in Photons and Continuum States of Atoms and Molecules, ed. N. K. Rahman, C. Guidotti and M. Allegrini (Springer-Verlag, Heidelberg, 1987) p. 240.

16. F.H.M. Faisal, Comments At. Mol. Phys. $\underline{15}$, 119 (1984).

17. F.H.M. Faisal in Photon-Assisted Collisions and Related Topics, ed. N. K. Rahman and C. Guidotti (Harwood Acad. Pub., New York, 1982) p. 287.

18. L. Dimou and F.H.M. Faisal, in Collisions and Half-Collisions with Lasers, ed. N. K. Rahman and C. Guidotti (Harwood Acad. Pub., New York, 1984) p. 121.

19. F.H.M. Faisal, J. Phys. (Paris) Coll. Cl. $\underline{46}$ (1985).

20. F.H.M. Faisal, in Fundamentals of Laser Interactions, ed. F. Ehlotzky (Springer-Verlag, Heidelberg, 1985) p. 16.

21. P. G. Burke and M. J. Seaton, in Methods in Computational Physics: Advances in Research and Applications, ed. B. Aldev et al. (Academic, New York, 1971) p. 1.

22. F.H.M. Faisal, Theory of Multiphoton Processes (Plenum, New York, 1987) Ch. 12.

23. W. Henneberger, Phys. Rev. Lett. $\underline{21}$, 838 (1968).

24. F.H.M. Faisal, J. Phys. B: At. Mol. Phys. $\underline{6}$, L89 (1973).

NEW RESULTS ON MULTIPHOTON FREE-FREE TRANSITIONS

F. Trombetta,[1] B. Piraux,[2] and G. Ferrante[3]

[1]Istitute di Fisica dell'Università, Università
di Palermo, Italy
[2]Optics Section, Blackett Laboratory, Imperial College
London, England
[3]Diptimento di Fisica Teorica, Università di Messina, Italy

INTRODUCTION

Recently some of the present authors have reported on calculations concerning free-free transitions in the presence of very strong laser fields,[1] in which a number of peculiar features were clearly displayed (among others, well pronounced maxima in the total cross sections and an oscillatory structure). In particular, those calculations concerned total cross sections of direct and inverse multiphoton bremsstrahlung, for several numbers of exchanged photons as functions of the field intensity. Figs. 1 and 2 show a sample of the typical results reported in Ref. 1.

Fig. 1 shows total cross sections for absorption of different numbers of laser photons when electron are scattered by a screened coulomb potential and the field (assumed purely coherent) is polarized along the incoming particle beam direction.

Calculations similar to those reported in Fig. 1 have been previously performed by Brehme[2] and discussed by Geltman[3] and, more recently, also by Daniele et al.,[4] who, in particular, considered also the influence of the laser field spatial inhomogeneity. In the previous analysis, however, calculations have stopped at the onset of the oscillatory behaviour, arguing that possible use of Volkov waves in the limit of very intense fields could be improper. Thus, no effort was made to explore numerically the oscillatory, high intensity domain and to interpret the results in physical terms. As a consequence, the discussion

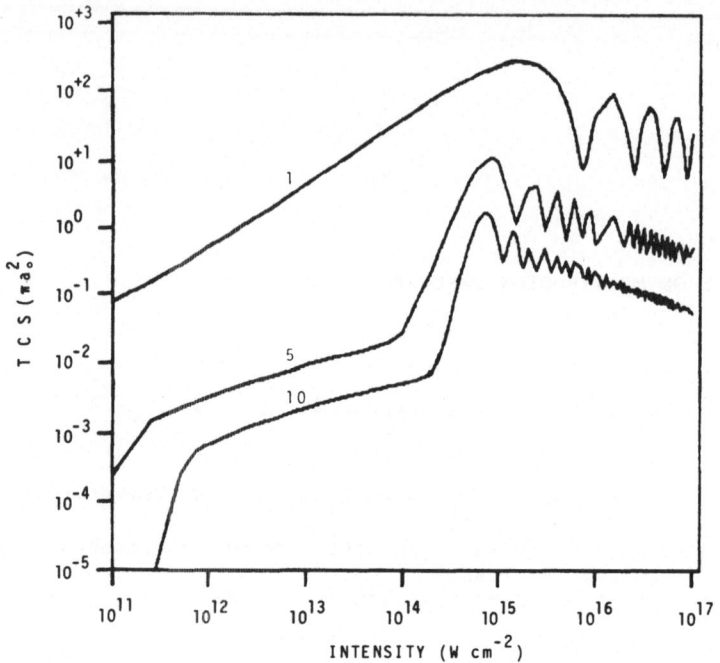

Figure 1. Total cross sections (TCS) vs. the field intensity for one, five and ten photons absorbed. From Ref. 1.

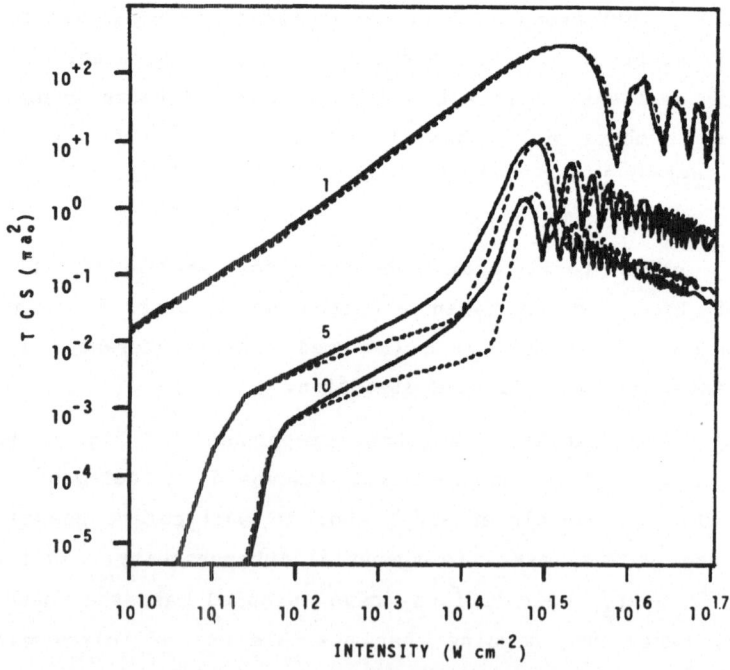

Figure 2. Total cross sections (TCS) vs. the field intensity for absorption (dashed lines) and emission (solid lines) of one, five and ten photons. From Ref. 1.

given by Geltman, being based on incomplete data, missed important points such as the physical conditions leading to the maxima in the cross sections. Now this aspect of the problem, together with the physics lying behind it, has been investigated in detail and, at the present level of analysis, it may be considered understood.[1]

The Fig. 2 shows total cross sections both for absorption and emission of different numbers of laser photons vs. the intensity, for the same geometry as in Fig. 1. In Fig. 2 the most striking feature is the different behaviour of emissions and absorptions just before the first maxima (except the one-photon case).

It is not difficult to recognize that calculations like those of Fig. 2 provide a rather exhaustive answer to the detailed behaviour of the so-called "Marcuse effect."[3,5-7] In other words, the problem concerned with this effect is to understand in detail the absorption vs. emission balance when electrons are scattered by ions in the parallel geometry (the field polarization parallel to the incoming particle beam). Due to its obvious significance, the problem has been considered also previously,[5-7] but never in a physically clear and computationally exhaustive way. The Fig. 2 reports on a significant sample of calculations, showing which and when channels with emission prevail over those with absorption, and helps to understand the physical mechanism responsible for it. Results like those of Fig. 2 help also to fully appreciate the recent analysis by Bivona et al.[8] on the multiquantum structure of the absorption coefficient in the inverse bremsstrahlung.

A peculiarity (and a limitation) of the results shown in Figs. 1 and 2, as well as of almost all the previous quantum mechanical calculations of the same problem, is that the analysis is based on the First Born Approximation (FBA). In spite of the important circumstance that the reported results may be given, as a rule, a transparent physical interpretation, the first question to arise is to which extent the FBA results may be trusted. It calls for similar analysis within treatments going beyond the FBA.

A second question is to determine whether the limits of validity of the FBA in the absence of any field still hold for cases when very strong fields are present. The question is of interest in its own right and in the past has received no attention except for some unproved statements[9,10] and qualitative remarks.[11] For instance, in Ref. 9 it is stated that when the amplitude of the classical oscillatory velocity v_0 ($= E_0/\omega$ in atomic units, E_0 being the electric field amplitude and ω the frequency) equals the initial particle velocity v_i, the FBA cannot be used (in parallel geometry) even if the field-free criterion of validity would be fulfilled; in Ref. 10, instead, it is stated that when $v_o \gg v_i$ the criterion of validity of the FBA is the same as in the field-free case, provided v_i is replaced by v_o. Generally, instead, it is taken that the field-free conditions of validity of the FBA remain unchanged also in the presence of a strong laser field.

Prompted by the aforesaid considerations, we have undertaken the tasks: (a) of deriving the criterion of validity of the FBA directly from the field-assisted formulation of the scattering problem, without relying on the field-free results; and (b) of studying the same field-assisted scattering problem in the same high-intensity domain within treatments which go beyond the FBA (eikonal and second Born approximations). A number of results on this line are reported below. In concluding this introduction, we would like to point out that the lack, up to now, of comprehensive calculations beyond the FBA for physically interesting domains of the direct and inverse multiphoton bremsstrahlung witnesses only of the difficulty of the problem and not of its scarce relevance. As the FBA is the only viable approximation allowing to cover the many aspects brought about by the several parameters of the problem, it should be not underappreciated in the present context the usefulness of this time-honoured first-order approximation.

THE CRITERION OF VALIDITY OF THE FIRST BORN APPROXIMATION TO THE MULTIPHOTON FREE-FREE TRANSITIONS

In this section we derive the criterion of validity of the First Born Approximation (FBA) for laser-assisted potential scattering. Atomic units are used throughout.

Let us start from the Lippmann-Schwinger equation

$$\psi^+_{k_i} = \chi_{k_i} + \int dt' \int d\underline{r}' \; G^+_o (\underline{r},t;\underline{r}',t') \; V(\underline{r}') \; \psi^+_{k_i} (\underline{r}',t') \tag{1}$$

$\psi^+_{k_i}$ and G^+_o being respectively the exact wavefunction and the Green's function in the presence of the radiation field only; χ_{k_i} is the Volkov wave, i.e. the exact wavefuntion of a free electron embedded in a radiation field, taken in the long wavelength approximation.

Similarly to the field-free treatments, we require that

$$|\chi_{k_i}| \gg |\int dt' \int d\underline{r}' \, G_o^+(\underline{r},t;\underline{r}',t') \, V(\underline{r}') \, \chi_{k_i}(\underline{r}',t')| \tag{2}$$

for any $\underline{r}$ and t. Using the expression for the Green's function

$$G_o^+ (\underline{r},t;\underline{r}',t') = (-i/\hbar) \int d\underline{k} \, \chi_k(\underline{r},t) \, \chi_k^*(\underline{r}',t') \, u(t-t') \tag{3}$$

u denoting the step function, and the Volkov wave

$$\chi_k(\underline{r},t) = \exp(i\underline{k}\cdot\underline{r} - i(\hbar k^2/2m) \, t + i\underline{\alpha}_o \cdot \underline{k} \, \sin\omega t) \quad \underline{\alpha}_o = \underline{E}_o/\omega^2 \tag{4}$$

the oscillating exponentials are then expanded in Bessel functions. In (4) the A^2-factor is not written because in the present problem it cancels exactly. A straightforward calculation transforms the condition (2) to

$$(1/8\pi^2) \, |\sum_{-\infty}^{n_o} k_n \int d\Omega_k \, J_n(\underline{\alpha}_o \cdot \underline{k}_n) \, \tilde{v}(\underline{k}_n - \underline{k}_i)| \ll 1 \tag{5}$$

where

$$k_n = k_i \, (1-2n\omega)^{1/2} \qquad n_o = \text{integer part of } (k_i^2/2\omega) \tag{6}$$

and $\tilde{V}(q)$ denotes the Fourier transform of the potential.

The criterion (5) holds for any local potential $V(r)$ and for any angle between the field and the incident momentum. In the absence of the field, Eq. (5) reduces to

$$(4\pi)^{-1} \, |\int d\underline{r}' \, \frac{\exp(ik_i r')}{r'} \, V(\underline{r}') \, \exp(i\underline{k}\cdot\underline{r}')| \ll 1 \tag{7}$$

i.e., to the field-free criterion.

Assuming $\underline{E}_o$ is parallel to $\underline{k}_i$ and a screened coulomb potential

$$V(r) = Z \exp(-r/r_o) \, / \, r \tag{8}$$

the criterion (5) becomes

$$(Z/2k) \cdot |\sum_{-\infty}^{n_o} \int_{-1}^{1} dx \, J_n(\alpha_o k_n x)/(x_n-x)| \ll 1 \tag{9}$$

where

$$x_n = [1 + r_o^2 (k_i^2 + k_n^2)] / 2r_o^2 k_i k_n \qquad (10)$$

The right hand member appearing in the Eq. (9) has the structure of a sum over the multiphoton channels which, at given incident energy and field frequency, are energetically open. The criterion (9) has been used for numerical calculations, a sample of which is shown in Figs. 3-4.

The basic result contained in Figs. 3 and 4 is that the presence of a radiation field in the case of parallel geometry broadens the conditions of validity of the FBA as compared with the field-free case, and that nothing particular occurs in the behaviour of F^1 when $v_o \cong v_i$. Concerning other geometries, it is not clear whether this conclusion still holds, and an analysis is presently in progress.

For weak fields, one has

$$z \; |F_{ff} - F^1| \ll 1 \qquad (11)$$

where

$$F_{ff} = (2k_i)^{-1} \log(1 + 4k_i^2 r_o^2) \qquad (12)$$

reproduces the field-free criterion and

$$F^1 = \begin{array}{ll} 4 \; (v_o/v_i) \; (1 - 2k_i^2 r_o^2) & k_i r_o \ll 1 \\ 2 \; (v_o/v_i) \; \log \; [4/\sqrt{(\omega/\varepsilon)^2 + (2/k_i k_o)^2}] & k_i r_o \gg 1 \end{array} \qquad (13)$$

ε being the incident energy ($= k_i^2/2$). It can generally be shown (i.e., for any value of $k_i r_o$) that $F > 0$, so that a radiation field enlarges the validity of the FBA (at least, for the parallel geometry considered here).

In the weak-field limit, the field enters only through the classical oscillatory velocity v_o; for $k_i r_o \gg 1$, F^1 is very sensitive to v_o, because the argument of the logarithm is very large. Further, when $k_i r_o < \varepsilon/\omega$ the frequency dependence is fully contained in v_o; generally, at low frequency, by increasing the field strength, F^1 increases rapidly improving the conditions for the validity of the FBA.

438

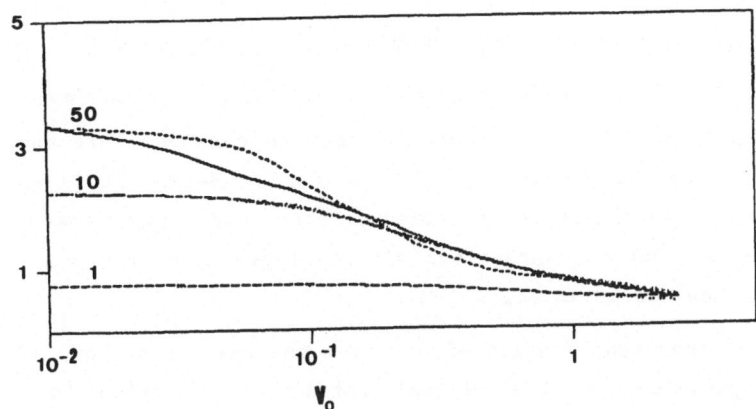

Figure 3. The left hand member of Eq. (9) vs. the oscillatory velocity (in at. un.). The incident momentum is $k_i = 3$. The numbers on the curves denote the range (in Bohr radii) of the Yukawa potential. For the range 50 a.u. the upper curve refers to a field photon energy of 0.0735 a.u. (2 eV), the lower one to 0.0184 a.u. (0.5 eV). Z is 1 a.u.

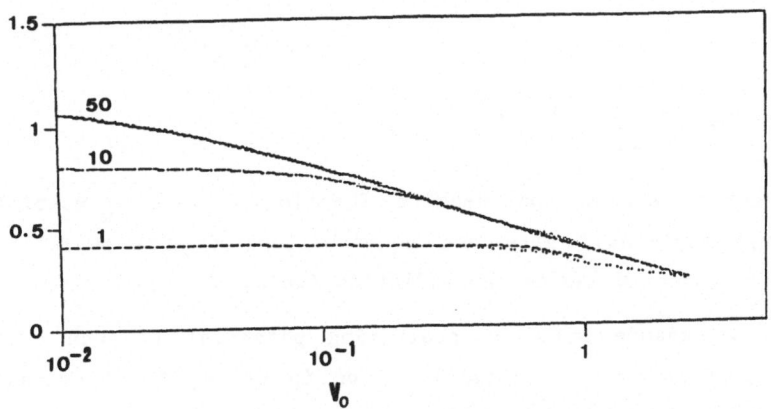

Figure 4. As in Fig. 3, but for an incident momentum $k_i = 6$ a.u.

The examples reported in the Figs. 3-4 show the remarkable fact that the above derived weak-field estimates in the essence extend to higher values of v_o and then of the field as well; in other words, the high intensity numerical results are in full qualitative agreement with the trends anticipated by the anlytical weak field estimates.

AN EIKONAL TREATMENT OF THE MULTIPHOTON FREE-FREE TRANSITIONS

The results reported in Figs. 1-2 are based on the FBA formulas exacly calculated. It is of some interest to consider here the eikonal approximation to the physical problem at hand. Below, it is shown that performing reasonable simplifications, from the exact eikonal cross section one may obtain expressions clearly reproducing the same physical content as those shown in Figs. 1-2.

We may conveniently start either from the results of Ref. 12 or from general expression for the eikonal total cross section in the form derived in Ref. 9, namely

$$\sigma^E = 4 \int d^2b \int (2\pi)^{-1} da \tag{14}$$

$$\sin^2[\chi_0 + \sum_{n\neq 0} J_n(nv_o/v_i) \exp(-ina) \chi_n(b,v_i)]$$

with

$$\chi_n(b,v_i) = (2v_i)^{-1} \int_{-\infty}^{\infty} dz\, V(b,z) \exp(-in\omega z/v_i) \tag{15}$$

and

$$\chi_0 = (2v_i)^{-1} \int_{-\infty}^{+\infty} dz\, V(b,z) \tag{16}$$

the field-free eikonal phase-shift function. Thus χ_n multiplied by $[J_n(nv_o/v_i) \exp(-ina)]$ give the contributions to the overall eikonal phase-shift function due to the radiation field.

Now, we assume that the scattering potential is such that χ_n is weakly dependent on n. Generally, even in strong field contexts, for angle-integrated quantities, only small values of n count, so that the assumption of weak n-dependence is reasonable for finite-range potentials and high relative energies. Thus, we take χ_n out of the infinite sum over n in Eq. (14) either letting n = 0 or, alternatively, taking some average value of n, say $\bar{n}$

$$\underset{n \neq 0}{\Sigma} J_n(nv_o/v_i) \, \exp(-ina) \, \chi_n(b,v_i)$$

$$\cong 2 \, \chi_{\underline{n}}(b,v_i) \, \overset{\infty}{\underset{1}{\Sigma}} \cos(na) \, J_n(nv_o/v_i) \tag{17}$$

For $x = v_o/v_i < 1$, using Eq. 5.7.32.1 of Ref. 13, the sum over n is performed to give

$$\overset{\infty}{\underset{1}{\Sigma}} \cos(na) \, J_n(nv_o/v_i) = (V_o/2V_i) \, [\cos\beta/(1 - \frac{v_o}{v_i} \cos\beta)] \tag{18}$$

with β defined by the equation

$$\beta - (v_o/v_i) \, \sin\beta = a \tag{19}$$

In the a-integration, when $\cos\beta = 1$, (18) becomes very large if, additionally, $v_o = v_i$. In such a case, the contribution to the eikonal phase-shift function due to the field is significantly enhanced, in agreement with our previous findings, and the complete cross section too is expected to be considerable enhanced.

We now make the assumption that the field-free eikonal phase-shift function χ_o is much larger than the contributions due to the field and, accordingly, expand the sine function of Eq. (14). It gives

$$\sigma^E \cong \sigma^E_{ff} + 8v_i^2 \overset{\infty}{\underset{1}{\Sigma}} J_n^2(nv_o/v_i) \int d^2b \, \cos(2\chi_o) \, |\chi_n(b,v_i)|^2 \tag{20}$$

with σ^E_{ff} the field-free cross section and χ_o and χ_n given respectively by the Eqs. (16) and (15). This result is rich in information concerning the behaviour of the cross section and it is in close qualitative agreement with the findings of the FBA ones. For $v_o < v_i$ and letting again $n \to \bar{n}$ in χ_n, the summation in (20) is performed with the aid of Eq. 5.7.31.1 of Ref. 13, yielding

$$\sigma^E \cong \sigma^E_{ff} + 4v_i^2 \, |\chi_{\underline{n}}|^2 \, [(1 - v_o^2/v_i^2)^{1/2} - 1]^{-1/2} \int d^2b \, \cos(2\chi_o) \tag{21}$$

Finally, assuming sufficiently high initial velocities and weak potentials, we approximate the sine by its argument and readily obtain expressions similar to the FBA ones (as it should be).

In conclusion, the eikonal approximation yields expressions for the cross sections having largely the same behaviour as the FBA counter-parts.[1] Presently, a similar analysis is being carried out also within the second Born approximation, and preliminary results lend further support to the FBA result.

CONCLUSIONS

We briefly conclude with the following remarks:

1. We have found that in the important case of parallel geometry (when the radiation field is parallel to the incoming particle beam), the FBA is expected to work much than in the field-free case, especially for relatively long-range potentials and high-intensity lasers. A comprehensive physical interpretation of this result will be presented elsewhere. Here, we anticipate that in the case of parallel geometry this result is probably due to a kind of field-induced channelling of the incoming particle beam along the strong radiation field polarization. In turn, this amounts in an effective way to weaken the scattering potential.

2. In the criterion of validity of the FBA, generally, the characteristic parameters of the incoming particle are compared with those of the scattering potential. In the parallel geometry, for $v_o < v_i$ one expects also a cooling of the particle beam (prevailing of the emission processes). Thus, the broadening of the range of validity of the FBA is fully determined by a weakening of the strength of the scattering potential. For other geometries, particle heating is instead expected, which too should favour a broadening of the validity of the FBA. On the contrary, no channelling is expected and thus no reduction of the scattering potential strength may be anticipated. The limits of validity of the FBA for arbitrary geometries are presently under investigation.

3. A simple analysis of the multiphoton free-free transitions in the parallel geometry performed within the eikonal treatment has fully supported the FBA results. It in turn confirms that the results ofthe Figs. 1-2 have a physical meaning which is independent of the approximation method used to treat the scattering dynamics.

4. The above results are, instead, expected to be dependent on the particular laser model used in the treatment of the problem. We have used here the fundamental model of a purely coherent field, which is likely to be a poor representation of very intense real laser fields. Changing the laser model implies a change in the physics of

the problem, at least as far as the characteristics of the assisting field are concerned; accordingly, the results too may change. An analysis of these aspects as well as of the information provided by a second Born treatment is presently under way.

ACKNOWLEDGEMENTS

Stimulating discussions with R. Daniele and G. Messina are acknowledged. The authors express their thanks to the University of Palermo Computational Center for the computer time generously provided to them. This work is supported in part by the Italian Ministry of Education, the National Group of Structure of Matter and the Sicilian Regional committee for Nuclear and Structure of Matter Researches.

REFERENCES

1. R. Daniele, F. Trombetta, G. Ferrante, P. Cavaliere and F. Morales, Phys. Rev. A36, 000(1987).

2. H. Brehme, Phys. Rev. C3, 837 (1971).

3. S. Geltman, J. Res. NBS 82, 173 (1977).

4. R. Daniele, G. Ferrante, and R. Zangara, Il Nuovo Cimento 2D, 1509 (1983).

5. D. Marcuse, Bell Syst. Tech. J. 41, 1557(1962).

6. F. V. Bunkin, A. E. Kazakov and M. V.Fedorov, Sov. Phys. - Usp. 15, 416 (1973).

7. P. V. Elyutin, Sov. Phys. - JETP 38, 1097 (1974).

8. S. Bivona, R. Zangara and G. Ferrante, Il Nuovo Cimento 7D, 113 (1986).

9. J. I. Gersten and M. H. Mittleman, Phys. Rev. A12, 1840 (1975).

10. a) F. V. Bunkin and R. V. Karapetyan, Izv. BUZ "Radiofizika" 14, 950 (1971), in russian.

    b) R. V. Karapetyan and M. V. Fedorov, Sov. Phys. - JETP 48, 412 (1978).

11. S. Bivona, R. Zangara and G. Ferrante, Phys. Lett. 110A, 375 (1985).

12. G. Ferrante, C. Leone and L. Lo Cascio, J. Phys. B12, 2319 (1979).

13. A. P. Prudnikov, Yu. A. Brychkov and O. I. Marichev, Special Functions, Integrals and Series, Nauka, Moscow (1983) (in russian).

14. B. Wallbank, J. K. Holmes and A. Weingartshofer, "Experimental Differential Cross-Sections for Multiphoton Free-Free Transitions" (to be published).

MULTICHANNEL MULTIPHOTON IONIZATION OF THE

HYDROGEN ATOM BY A CHAOTIC FIELD: ROLE OF THE POLARIZATION

S. Basile,[1] F. Trombetta[1] and G. Ferrante[2]

[1]Intituto di Fisica, Universita di Palermo, Italy
[2]Dipartimento di Fisica Teorica, Univ. di Messina, Italy

This contribution is concerned with the theory of ionization of atoms by strong laser fields, when ionization channels of different photon multiplicity are simultaneously open. Presently, this process is attracting a lot of attention, from both the experimental and theoretical sides.[1]

Here specifically we present a treatment concerning the ionization of hydrogen atoms by a chaotic field. Basic features of our treatment are: (a) it is based on the S-matrix formalism, and exploiting the notion of transition probability per unit time; (b) it is essentially nonperturbative; (c) it is non-Keldysh-type, in the sense that we include in the final continuum electron state the influence of the coulomb interaction with the residual ion; (d) the approximate wavefunction for the ejected electrons is taken as an incoming coulomb function times the same time-dependent phase factor entering the Volkov waves. This 'ansatz' may be shown[2] to be a sort of high intensity leading term of a more general and rigorously constructed wavefunction for an electron in the presence of the two fields of the present problem; (e) according to the previous point, the emerging physical picture of the ionization is that of a one-step process, in which the electrons leave the hydrogen atom ground state to populate simultaneously different continua. Again, this one-step process is expected to be the dominant one in the limit of very intense fields, as compared with multistep processes; (f) the laser field is taken as a chaotic field with zero bandwidth. Real very intense laser fields are poorly described by the laser ideal model; moreover, in many of the multiphoton ionization experiments the laser systems are operated in a multimode configuration.[3] Finally, due to focussing and pulse duration, in any real experiment the field exhibits quite distributed parameters. In our present case, we average the ideal laser model results over the distribution of the intensity realizations. This averaging is expected to represent to some extent also other averaging mechanisms not taken into account here. (g) In our 'one-step process' approximation, the atomic structure enters the process only via the final state coulomb interaction. In any case, the use of the chaotic model for the field, implying an averaging over the fluctuation of the shifts of the ionization level, is expected to smear out and to reduce the role of the atomic structure (provided there are not intermediate resonances).

In this note we address only two aspects of the problem, namely: (a) the energy distribution of the ejected electrons for circular and linear polarization of the ionizing field; and (b) the angular distributions for linear polarization.

The transition probability per unit time, in the S-matrix formalism, is given by

$$w = \int \frac{d^3p}{(2\pi\hbar)^3} \lim_{T \to \infty} \left\{ |S_{if}|^2 \Big/ \int_{-T}^{T} dt \right\} \tag{1}$$

where

$$S_{if} = (i\hbar)^{-1} \int_{-T}^{T} dt \ \langle \Psi_f(\underline{r},t) \ |e\underline{E}(t)\cdot\underline{r}| \ \psi_i(\underline{r},t) \rangle \tag{2}$$

The hydrogen ground state is the intitial state while the final one is taken as:

$$\Psi_f(\underline{r},\underline{t}) = \exp\left\{\frac{-1}{2\pi\hbar} \int^t \left[\hbar\underline{k} + e\underline{A}(\tau)/c\right]^2 dt + \frac{ie}{\hbar c} \underline{A}(t)\cdot\underline{r}\right\} \psi_{\underline{k}}^-(r) \tag{3}$$

$$\psi_{\underline{k}}^-(\underline{r}) = \exp(\pi\nu/2) \ \Gamma(1 + i\nu) \ \exp(i\underline{k}\cdot\underline{r}) \ F(-i\nu,1,-i(kr + \underline{k}\cdot\underline{r})) \tag{4}$$

where the standard notation has been adopted.

Assuming, first, a purely coherent field of arbitrary polarization, following usual procedures, we arrive at the doubly differential ionization rate to the (s + 1)-th continuum as

$$[d^2w/d\Omega d\varepsilon]_s =$$

$$\frac{e^2}{\hbar a_o} \frac{2^4}{\pi^3} \frac{I}{I_a} \frac{2\pi\nu k a_o}{1-\exp(-2\pi\nu)} \ |T_{n_o+s}(\underline{k},\underline{E}_o)|^2 \delta(\varepsilon + \Delta - \varepsilon_s) \tag{5}$$

s denoting the number of photons absorbed beyond the minimum required for ionization.

In Eq. (5), I is the field intensity (at this stage, well fixed), I_a is equal to $3.51\ 10^{16}\text{W/cm}^2$, T_n is the pertinent n-photon transition amplitude and

$$\Delta = e^2 E_o^2/(4m\omega^2) \qquad \varepsilon_s = (n_o + s)\hbar\omega - I_o \qquad I_o \simeq 13.6\text{eV} \qquad (6)$$

Δ can be seen as a shift of the ionization threshold.

In order to account for the field fluctuations within the assumption of a vanishing field spectrum bandwidth, we have to now average the doubly differential ionization rate over the chaotic field intensity distribution or over the threshold distribution:

$$P(\Delta) = \exp[-\Delta/\langle\Delta\rangle]/\langle\Delta\rangle \qquad\qquad \langle\Delta\rangle = (2\pi e^2/mc\omega^2)\ \langle I\rangle \qquad (7)$$

Averaging of Eq. (5) then gives the working formula, not reported here for the sake of brevity.[4] Selected results, using the outlined treatment, are reported in Figs. 1 and 2.

In Fig. 1 we show the effect of the field polarization on the photo-electron energy spectra; they are normalized to the first peak and are computed in the forward direction (parallel to the field polarization) for the linear case and on the polarization plane for the circular case (the angular distribution is uniform on this plane). The calculations have been performed at a wavelength of 1065 nm; the polarizations and the mean field intensities (in W/cm^2) are: a) linear, $7.5\ 10^{11}$; b) circular, $7.5\ 10^{11}$; c) linear, $5.\ 10^{12}$; d) circular, $5.\ 10^{12}$. The results clearly show that a circular field is much more effective in suppressing the low energy peaks than a linear one and that the number of peaks significantly present at a given intensity is much larger for circular polarization. This is in good agreement with the experimental information.[1]

In Fig. 2, we show angular distributions at 532 nm; the intensity is chosen in such a way that the ratio of the forward ionization rate to the first continuum state is ten times larger than the same for the second continuum; this value of intensity turns out to be $7.5\ 10^{11}\ \text{W/cm}^2$. The numbers on the curves denote s, the photons absorbed above the threshold. The s = 0 distribution is similar to the experimental one, though slightly shifted in the forward direction; the other distributions are generally strongly peaked at 0°, with a rich structure arising after 50°. Generally, by increasing the number of photons involved, the number of minima and maxima increases, while their ratio to the forward maximum and their distance decrease, making probably more difficult an experimental observation.

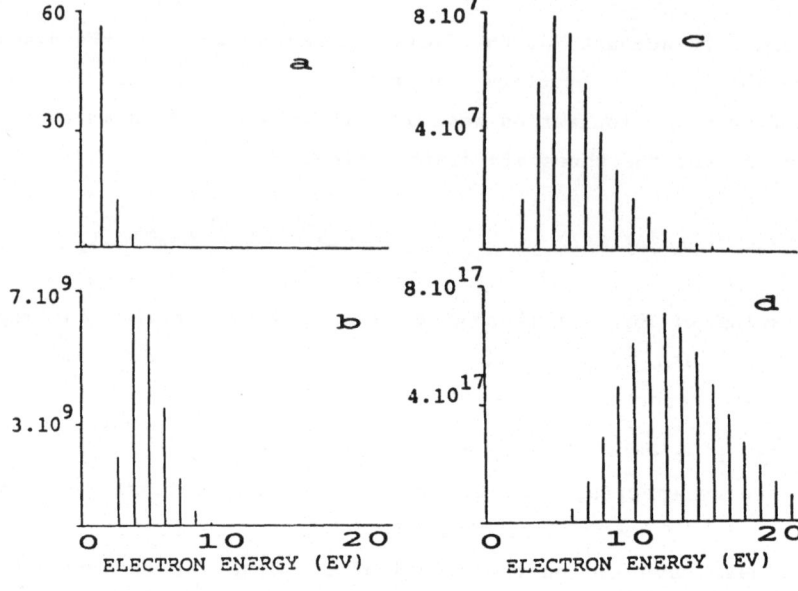

FIGURE 1

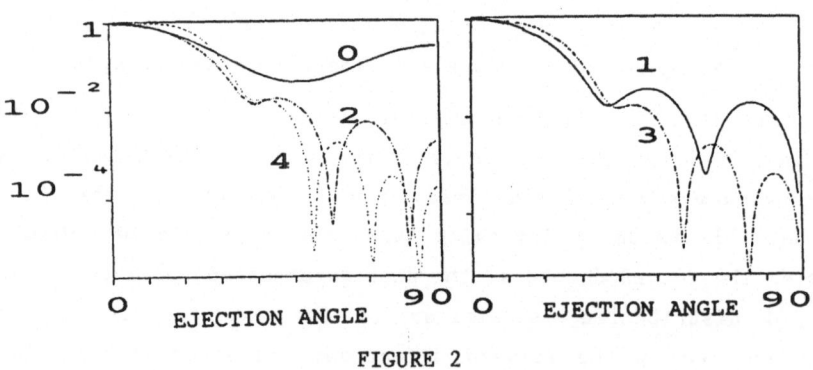

FIGURE 2

References

1. See, for instance, T. J. McIlrath, P. H. Bucksbaum, R. R. Freeman and M. Bashkansky, Phys. Rev. A35, 4611 (1987), and references therein.

2. C. Leone, R. Burlon, F. Trombetta, S. Basile and G. Ferrante, "Strong Field Multiphoton Ionization of Hydrogen. The S-Matrix Treatment of the Elementary Process," Il Nuovo Cimento D (1987) (in press).

3. P. Kruit, J. Kimman, H. G. Muller and M. J. van der Wiel, Phys. Rev. A28, 248 (1983); D. E. Kelleher, M. Ligare and L. R. Brewer, Phys. Rev. A31, 2747 (1985); D. Feldmann, B. Wolff, M. Wemhoner and K. H. Welge, Z. Phys. (submitted).

4. Details of the calculations will be reported in a paper now in preparation.

A. Alijah
Fakultät für Chemie
Universität Bielefeld
D-4800 Bielefeld   FRG

M. Al-Laithy
Department of Mathematics
Royal Holloway and Bedford New College
Egham, Surrey TW20 0EX   England

J. Bannerji
Department of Computer Science
Royal Holloway and Bedford New College
Egham, Surrey TW20 0EX   England

J. E. Bayfield
Department of Physics and Astronomy
100 Allen Hall
University of Pittsburgh
Pittsburgh, PA 15260   USA

N. Bhattacharya
Department of Physics
Banaras Hindu University
Varanasi 221 005   India

S. Bivona
Dipartimento di Energetica ed
        Applicazioni de Fisica
Viale delle Scienze
90128 Palermo   Italy

C. Bottcher
Physics Division
Oak Ridge National Laboratory
P O Box X
Oak Ridge, TN 37830   USA

H. C. Bryant
Department of Physics and Astronomy
University of New Mexico
Albuquerque NM 87131   USA

P. H. Bucksbaum
AT&T Bell Laboratories
600 Mountain Avenue
Murray Hill, NJ 07974   ·USA

L. Bureyeva
Scientific Council on Spectroscopy
USSR Academy of Sciences
Pr. Sapunova, 13-15
Moscow 103012   USSR

R. Burlon
Dipartimento di Energetica ed
        Applicazioni de Fisica
Viale delle Scienze
90128 Palermo   Italy

M. Cavagnero
Department of Physics and Astronomy
The University of Nebraska
Lincoln, NE 68588-0111   USA

B. N. Chichkov
P. N. Lebedev Physical Institute
Leninsky pr. 53
Moscow 117924   USSR

C. W. Clark
Radiation Physics Division
National Bureau of Standards
Gaithersburg, MD 20899   USA

J.-P. Connerade
Blackett Laboratory
Imperial College
Prince Consort Road
London SW7 2AZ   England

J. Coutts
Joint Institute for Laboratory Astrophysics
University of Colorado
Boulder, CO 80309   USA

D. S. F. Crothers
Department of Applied Mathematics and
        Theoretical Physics
The Queen's University
Belfast BT7 1NN   Northern Ireland

R Damburg
Institute of Physics
Latvian SSR Academy of Sciences
229021 Riga, Salaspils   USSR

A. Dave
Cray Research (UK) Ltd.
London Road
Bracknell, Berks.   England

D. Delande
Laboratoire de Spectroscopie
          Hertzienne de l'E. N. S.
Tour 12-1er Etage
4 Place Jussieu
75230 Paris   France

J. Delos
Joint Institute for Laboratory Astrophysics
University of Colorado
Boulder, CO 80309   USA

W. Dussa
Physikalisches Institut
Synchrotron Radiation Group
Nussallee 12
5300 Bonn 1   FRG

A.R. Edmonds
Blackett Laboratory
Imperial College
Prince Consort Road
London SW7 2AZ   England

F.H.M. Faisal
Fakultät für Physik
Universität Bielefeld
Postfach 8640
D-4800 Bielefeld   FRG

U. Fano
James Franck Institute
University of Chicago
5640 Ellis Avenue
Chicago, IL 60637   USA

G. Ferrante
Instituto di Fisica
Via Archirafi 36
90123 Palermo   Italy

H. Friedrich
Lehrstuhl für Theoretische Astrophysik
Auf der Morgenstelle 12C
D-7400 Tubingen   FRG

W. R. S. Garton
Blackett Laboratory
Imperial College
Prince Consort Road
London SW7 2AZ   England

C. H. Greene
Department of Physics and Astronomy
Louisiana State University
Baton Rouge, LA 70803   USA

U. Griesmann
Physikalisches Institut
Synchrotron Radiation Group
Nussallee 12
5300 Bonn 1   FRG

D. A. Harmin
Department of Physics and Astronomy
University of Kentucky
Lexington, KY 40506-0055   USA

H. Hasegawa
Department of Physics
Kyoto University
Kyoto 606   Japan

A. Holle
Fakultät für Physik
Universität Bielefeld
Postfach 8640
D-4800 Bielefeld   FRG

D. C. Humm
Department of Physics
University of Illinois at Urbana-Champaign
Urbana, IL 61801   USA

E. Karule
Institute of Physics
Latvian SSR Academy of Sciences
229021 Riga
Salaspils   USSR

D. H. Klaus
Freie Universität Berlin
Fachbereich Physik
Arnimallee 74
D-1000 Berlin 33   FRG

M. Le Dourneuf
Observatoire de Paris-Meudon
92190 Meudon   France

N. J. Mason
Department of Physics and Astronomy
University College London
Gower Street
London WC1E 6BT   England

H. Ma
Spectoscopy Group
Physics Department
Imperial College
London SW7 2BZ   England

J. Main
Fakultät für Physik
Universität Bielefeld
Postfach 8640
D-4800 Bielefeld   FRG

T. Matsuzawa
Department of Engineering Physics
The University of Electro-Communications
1-5-1 Chofugaoka Chofu-shi
Tokyo 182   Japan

Z. A. Melhem
Optics Section
Blackett Laboratory
Imperial College
London SW7 2AZ   England

M. H. Mittleman
Physics Department
The City College
New York, NY 10031   USA

L. Moorman
Department of Physics
State University of New York
Stony Brook, NY   USA

L. A. Morgan
Computer Centre
Royal Holloway and Bedford New College
Egham, Surrey TW20 0EX   England

F. da Mota Furtado
Department of Mathematics
Royal Holloway and Bedford New College
Egham, Surrey TW20 0EX   England

M. H. Nayfeh
Department of Physics
University of Illinois at Urbana-Champaign
Urbana, IL 61801   USA

R. K. Nesbet
IBM Almaden Research Centre
San Jose, CA 95120-6099   USA

K. L. Ng
Department of Physics
University of Illinois at Urbana-Champaign
Urbana, IL 61801   USA

D. W. Norcross
Joint Institute for Laboratory Astrophysics
University of Colorado
Boulder, CO 80309   USA

S. Nuzzo
Instituto di Fisica
Via Archirafi 36
90123 Palermo   Italy

P. F. O'Mahony
Mathematics Department
Royal Holloway and Bedford New College
Egham, Surrey TW20 0EX   England

A. E. Orel
The Aerospace Corporation
P.O. Box 92957
Los Angeles, CA 90009   USA

G. Petite
DPhG/SPAS
CEN Saclay
91191 Gif sur Yvette   France

J. Pinard
Laboratoire Aimé Cotton
Campus d'Orsay
Batiment 505
91405 Orsay   France

B. Piraux
Optics Section
Blackett Laboratory
Imperial College
London SW7 2AZ   England

H. Rinneberg
Fachbereich Physik
Institut für Atom-und Festkörperphysik
Freie Universität Berlin
Arnimallee 14
D-1000 Berlin 33   FRG

M. Robnik
Max-Planck-Institut für Kernphysik
D-69 Heidelberg   FRG

K. Sakimoto
Institute of Space and Astronautical Science
Komaba, Meguro-ku
Tokyo 153   Japan

W. Sandner
Fakultät für Physik
Universität Freiburg
Hermann-Herder-Straße 3
D7800 Freiburg   FRG

W. Schweizer
Department of Mathematics
Royal Holloway and Bedford New College
Egham, Surrey TW20 0EX   England

R. Shakeshaft
Department of Physics
University of Southern California
Los Angeles, CA 90089   USA

R. M. Sinclair
Physics Division
National Science Foundation
Washington, D.C. 20550   USA

H. A. Slim
School of Mathematics and Physical Sciences
Murdoch University
Murdoch, WA 6150   Australia

M. Smith
Mathematics Department
Royal Holloway and Bedford New College
Egham, Surrey TW20 0EX   England

W. W. Smith
Department of Physics
University of Connecticut
Storrs, CT 06268   USA

A. F. Starace
Department of Physics and Astronomy
The University of Nebraska
Lincoln, NE 68588-0111   USA

K. T. Taylor
Department of Mathematics
Royal Holloway and Bedford New College
Egham, Surrey TW20 0EX   England

F. Trombetta
Istituto di Fisica Dell'Universita
via Archirafi 36
90123 Palermo   Italy

S. Ward
Physics Department
York University
4700 Keele Street
Downsview
Toronto, Ontario   Canada

S. Watanabe
Observatoire de Paris-Meudon
92190 Meudon   France

K.H. Welge
Fakultät für Physik
Universität Bielefeld
Postfach 8640
D-4800 Bielefeld   FRG

G. Wiebusch
Fakultät für Physik
Universität Bielefeld
Postfach 8640
D-4800 Bielefeld   FRG

M. Wilson
Physics Department
Royal Holloway and Bedford New College
Egham, Surrey TW20 0EX   England

D. Wintgen
Fakultät für Physik
University Freiburg
Hermann-Herder-Straße 3
D-7800 Freiburg   FRG

G. Wunner
Lehrstuhl für Theoretische Astrophysik
Auf der Morgenstelle 12C
D-7400 Tubingen   FRG

M. Zarcone
Instituto di Fisica
Via Archirafi 36
90123 Palermo   Italy

G. Zeller
Lehrstuhl für Theoretische Astrophysik
Auf der Morgenstelle 12C
D-7400 Tubingen   FRG

R. Ch. Ziegelbecker
Institut für Experimentalphysik
Technische Universität Graz
Petersgasse 76
A-8010 Graz   Austria

INDEX